# TechOne:
# Automotive Engine Performance

# TechOne: Automotive Engine Performance

### Russell Carrigan

Instructor – Automotive Technology
Texas State Technical College, West Texas
Sweetwater, Texas

### Richard R. Kent, B.S.E.E.

Associate Professor – Automotive Technology
Onondaga Community College
Syracuse, New York

### Jack Erjavec, Series Editor

Professor Emeritus – Columbus State Community College
Columbus, Ohio

Australia  Canada  Mexico  Singapore  Spain  United Kingdom  United States

DELMAR
CENGAGE Learning™

**TechOne: Automotive Engine Performance**

**Russell Carrigan and Richard R. Kent**

Vice President, Technology
and Trades SBU:

**Alar Elken**

Editorial Director:

**Sandy Clark**

Senior Acquisitions Editor:

**David Boelio**

Developmental Editor:

**Matthew Thouin**

Marketing Director:

**Dave Garza**

Channel Manager:

**Bill Lawrensen**

Marketing Coordinator:

**Mark Pierro**

Production Director:

**Mary Ellen Black**

Production Editor:

**Barbara L. Diaz**

Art/Design Specialist:

**Cheri Plasse**

Technology Project Manager:

**Kevin Smith**

Technology Project Specialist:

**Linda Verde**

Editorial Assistant:

**Kevin Rivenburg**

For product information and technology assistance, contact us at
**Cengage Learning Customer & Sales Support, 1-800-354-9706**
For permission to use material from this text or product,
submit all requests online at **cengage.com/permissions**
Further permissions questions can be emailed to
**permissionrequest@cengage.com**

ExamView® and ExamView Pro® are registered trademarks of FSCreations, Inc. Windows is a registered trademark of the Microsoft Corporation used herein under license. Macintosh and Power Macintosh are registered trademarks of Apple Computer, Inc. Used herein under license.

© 2007 Cengage Learning. All Rights Reserved. Cengage Learning WebTutor™ is a trademark of Cengage Learning.

ISBN-13: 978-1-4018-3401-2

ISBN-10: 1-4018-3401-9

**Delmar Cengage Learning**
5 Maxwell Drive
Clifton Park, NY 12065-2919
USA

Cengage Learning products are represented in Canada by Nelson Education, Ltd.

For your lifelong learning solutions, visit **delmar.cengage.com**

Visit our corporate website at **www.cengage.com**

**Notice to the Reader**

Publisher does not warrant or guarantee any of the products described herein or perform any independent analysis in connection with any of the product information contained herein. Publisher does not assume, and expressly disclaims, any obligation to obtain and include information other than that provided to it by the manufacturer. The reader is expressly warned to consider and adopt all safety precautions that might be indicated by the activities described herein and to avoid all potential hazards. By following the instructions contained herein, the reader willingly assumes all risks in connection with such instructions. The publisher makes no representations or warranties of any kind, including but not limited to, the warranties of fitness for particular purpose or merchantability, nor are any such representations implied with respect to the material set forth herein, and the publisher takes no responsibility with respect to such material. The publisher shall not be liable for any special, consequential, or exemplary damages resulting, in whole or part, from the readers' use of, or reliance upon, this material.

Printed in the United States of America
3 4 5 6 7 11 10 09 08

# Contents

# vi • Contents

## viii • Contents

# Preface

## THE SERIES

Welcome to Thomson Delmar Learning's *TechOne*, a state-of-the-art series designed to respond to today's automotive instructor and student needs. *TechOne* offers current, concise information on ASE and other specific subject areas, combining classroom theory, diagnosis, and repair into one easy-to-use volume.

You'll notice several differences from a traditional textbook. First, a large number of short chapters divide complex material into chunks. Instructors can give tight, detailed reading assignments that students will find easier to digest. These shorter chapters can be taught in almost any order, allowing instructors to pick and choose the material that best reflects the depth, direction, and pace of their individual classes.

*TechOne* also features an art-intensive approach to suit today's visual learners: images drive the chapters. From drawings to photos, you will find more art to better understand the systems, parts, and procedures under discussion. Look also for helpful graphics that draw attention to key points in features like You Should Know and Interesting Fact.

Just as importantly, each *TechOne* starts off with a section on safety and communication, which stresses safe work practices, tool competence, and familiarity with workplace "soft skills," such as customer communication and the roles necessary to succeed as an automotive technician. From there, learners are ready to tackle the technical material in successive sections, ultimately leading them to the real test—an ASE practice exam in the Appendix.

## THE SUPPLEMENTS

*TechOne* comes with an **Instructor's Manual** that includes answers to all chapter-end review questions and a complete correlation of the text to NATEF standards. A **CD-ROM**, included with each **Instructor's Manual**, contains **PowerPoint Slides** for classroom presentations, a **Computerized Testbank** with hundreds of questions to aid in creating tests and quizzes, and an electronic version of the **Instructor's Manual**. Chapter-end review questions from the text have also been redesigned into adaptable **Electronic Worksheets**, so instructors can modify questions if desired to create in-class assignments or homework.

Flexibility is the key to *TechOne*. For those who would like to purchase jobsheets, Thomson Delmar Learning's NATEF Standards Job Sheets are a good match. Topics cover the eight ASE subject areas and include:

- Engine Repair
- Automatic Transmissions and Transaxles
- Manual Drive Trains and Axles
- Suspension and Steering
- Brakes
- Electrical and Electronic Systems
- Heating and Air Conditioning
- Engine Performance

Visit **http://www.autoed.com** for a complete catalog.

## OTHER TITLES IN THIS SERIES

*TechOne* is Thomson Delmar Learning's latest automotive series. We are excited to announce these future titles:

- Advanced Automotive Electronic Systems
- Advanced Engine Performance
- Automatic Transmissions
- Engine Repair
- Fuels and Emissions
- Heating and Air Conditioning
- Suspension and Steering

Check with your sales representative for availability.

## A NOTE TO THE STUDENT

There are now more computers on a car than aboard the first spacecraft, and even gifted backyard mechanics long ago turned their cars over to automotive professionals for diagnosis and repair. That's a statement about the nation's need for the knowledge and skills you'll develop as you continue your studies. Whether you eventually choose a career as a certified or licensed technician, service writer or manager, automotive engineer, or even decide to open your own shop, hard work will give you the opportunity to become one of the 840,000 automotive professionals providing and maintaining safe and efficient automobiles on our roads. As a member of a technically proficient, cutting-edge workforce, you'll fill a need, and, even better, you'll have a career to feel proud of.

Best of luck in your studies,
The Editors of Thomson Delmar Learning

# About the Authors

Russell Carrigan is an automotive technology instructor at Texas State Technical College, located in Sweetwater, Texas. He is an ASE Master Technician with L1 certification and has been a part of the automotive industry since 1988. He has advanced to the position of automotive instructor, having started his career as an apprentice technician. During his service career, he achieved master technician status and was employed as a service advisor before entering the field of education as an instructor.

Richard R. Kent received his baccalaureate degree in electrical engineering from Syracuse University in Syracuse, New York. He has been involved in radio frequency and power control circuit design for more than twenty-five years. He holds both ASE Master and L1 Technician certificates.

Professor Kent is chairman of the automotive technology program at Onondaga Community College, in Syracuse, where he has taught automotive technology since developing the program in 1990. Previously, he taught electrical circuit design and radio frequency communications systems theory at Onondaga, where he has been on the faculty since 1978.

Outside of the classroom, his interests include amateur radio, digital video, and development of specialized training programs.

# Dedication

This, my first attempt at authoring a textbook, has been an interesting journey. I have seen the project transform from an idea and a vision into a book. I dedicate this book to my wife, Tonya, and sons Ryan and Cody. Without their unwavering commitment and support, this project would not have been possible. I would also like to thank my long-time mentor Dale Alcorn for his technical expertise and for many years of guidance.

—Russell Carrigan

For many years, I have been searching for a textbook that fits my needs when teaching engine performance courses at Onondaga Community College. I believe this textbook fills that void. Special attention has been paid to presenting material in a straightforward, logical manner. Russell and I have spent countless hours developing this text together—via both email and telephone conversations—and individually. We hope our approach to presenting difficult subject matter in a manner that breaks down complex theory and diagnostics will give automotive technology students an edge when making the transition from classroom to workplace.

I would like to thank my wife, Janie, for her untiring assistance in preparing this manuscript. Without her help, my part in this project would not have been possible. In addition, I would also like to extend my thanks to the staff at Thomson Delmar Learning. Their commitment to education is second to none in an extremely competitive field, where economics can easily overshadow product development and quality. I have enjoyed working with everyone behind the scenes.

—Richard R. Kent

# Acknowledgments

Russell Carrigan and Richard R. Kent would like to extend special thanks to the following reviewers for their contributions to the success of this textbook:

Dale Alcorn
Lawrence Hall Chevrolet
Abilene, Texas

Stephen Belitsos
Vermont Technical Center
Randolph Center, Vermont

Gerard DiCola
School of Cooperative Technical Education
New York, New York

Rick Francis
Owens Community College
Toledo, Ohio

Andrew O'Neal
University of Northwestern Ohio
Lima, Ohio

Dan Perrin
Trident Technical College
Charleston, North Carolina

# Features of the Text

*TechOne* includes a variety of learning aids designed to encourage student comprehension of complex automotive concepts, diagnostics, and repair. Look for these helpful features:

**Section Openers** provide students with a **Section Table of Contents** and **Objectives** to focus learners on the section's goals.

**Interesting Facts** spark student attention with industry trivia or history. Interesting facts appear on the section openers and are then scattered throughout the chapters to maintain reader interest.

## Section 2

### General Engine Diagnosis

| | |
|---|---|
| Chapter 4 | Engine Overview |
| Chapter 5 | Fundamental Engine Operating Principles |
| Chapter 6 | Engine Construction |
| Chapter 7 | Support Systems |
| Chapter 8 | Engine Operation |
| Chapter 9 | Support System Diagnosis |
| Chapter 10 | Basic Engine Testing |
| Chapter 11 | Engine Diagnosis and Repair |

**Interesting Fact** *Engine performance technicians must be able to diagnose mechanical engine problems as well as electrical and electronic problems.*

### SECTION OBJECTIVES

After you have read, studied, and practiced the contents of this section, you should be able to:

- Identify the engine by observing the VIN.
- Describe the different methods used to classify vehicle engines.
- Explain the basic scientific principles and theories that pertain to engine operation.
- Describe how atmospheric pressure affects engine operation.
- Describe energy conversion and how it is applied in the automobile engine.
- Describe the difference between torque and horsepower.
- Describe the function and operation of the components of an internal combustion engine.
- Describe the function of the support systems that are required to make the engine operate.
- Describe in detail the operation of the four-stroke cycle.
- Perform a cooling system pressure test and interpret results.
- Verify engine operating temperature and determine necessary action.
- Identify and interpret engine performance concerns as they relate to mechanical failures and determine necessary action.
- Inspect the engine assembly for oil and coolant leaks and determine necessary action.
- Diagnose abnormal exhaust color and determine necessary action.
- Perform engine manifold vacuum tests and determine necessary action.
- Perform a compression test and interpret results.
- Perform a cylinder leakage test and interpret results.
- Diagnose engine mechanical concerns using engine diagnostic equipment and determine necessary action.

An **Introduction** orients readers at the beginning of each new chapter. **Technical Terms** are bolded in the text upon first reference and are defined.

# Chapter 21

# Ignition System Theory and Operation

## Introduction

Internal combustion engines can be classified as using either **spark ignition (SI)** or **compression ignition (CI)**. The latter is used in diesel engines as a means to ignite the air/fuel charge by raising the temperature of the fuel to the flash point. However, in the case of gasoline-powered engines, the ignition system must supply a high-voltage pulse to the **spark plug**, causing an electrical arc to form on the **electrodes**. The spark plug is an ignition device that contains an air gap. The air gap allows current to flow when the voltage across the gap is of sufficient magnitude. The spark plug is located within the combustion chamber. The heat created by the spark causes the air/fuel charge within the combustion chamber to ignite **(Figure 1)**.

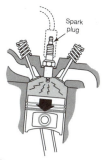

Spark plug

**Figure 1.** Heat created by the ignition spark causes the air/fuel charge within the combustion chamber to ignite.

**Interesting Fact**
A "hot" spark means there is an adequate flow of electrons traveling across the spark plug electrodes. Sharp electrodes help the electrons jump the air gap, minimizing ignition misfires. Engine misfire wastes fuel, robs the engine of power, and increases tailpipe emissions.

The ignition process converts the potential energy stored in the fuel into useful kinetic energy that develops horsepower. Sufficient spark energy must be available to prevent cylinder misfire, which results in decreased power and increased tailpipe emissions.

The ignition system must increase the low voltage provided by the storage battery and alternator to a value high enough to fire the spark plugs at the precise time. This high-voltage signal is typically around 10,000V but can range from approximately 6kV to over 40kV.

---

**You Should Know** Extreme caution should be exercised when working around the belt system and the fans. The engine should be turned off when performing any service work to the engine or accessories. The moving fan or belts can cause severe personal injury if your clothing or your body gets trapped in the fan or belts.

vehicles. This assembly uses an electric solenoid in place of the bimetallic spring. In this configuration, the computer monitors fan speed to determine the fan's efficiency.

Vehicles that do not use belt-driven fans use one or more electric motor–driven fans. The electric cooling fan is driven by an electric motor and is controlled by the computer using a relay. Fans are located directly at the front of the radiator or on the back side of the radiator. Those fans that are on the front of the radiator push air through the radiator and are called pusher fans. Fans that are located on the back side of the radiator pull air through the radiator and are called puller fans **(Figure 18)**.

The electric cooling fans are computer-controlled through the use of a relay. The computer monitors the engine temperature as well as other parameters such as air conditioner request and vehicle speed to determine

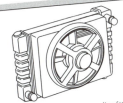

**Figure 18.** A puller-type electric cooling fan.

proper cooling fan operation. When engine temperature reaches approximately 220–230°F (105–111°C), the PCM will supply a ground to the cooling fan relay to command the cooling fans to turn on. When the temperature drops back down to approximately 210°F (100°C), the fan will turn back off.

**You Should Know** The electric fans on some vehicles can turn on when the engine and ignition switch are turned off.

**You Should Know** features inform the reader whenever special safety cautions, warnings, or other important points deserve emphasis.

A **Summary** concludes each chapter in short, bulleted sentences. **Review Questions** are structured in a variety of formats, including ASE style, challenging students to prove they've mastered the material.

## Summary

- The battery is critical to the operation of the starter system and other electrical systems. The starter system uses a DC motor to turn the engine to start the vehicle.
- In addition to providing battery current, the starter solenoid will also provide mechanical action to engage the starter drive. The drive gear clutch prevents starter damage by overrunning when the engine starts.
- The lubrication system cleans and distributes lubricant throughout the engine. The oil pump may be located on the front of the engine or inside the oil pan.
- Engine coolant is a combination of water and anti-freeze. The water pump circulates coolant throughout

the engine and the radiator. The radiator transfers heat from the coolant to the atmosphere.
- The thermostat operates as a temperature-sensitive control valve. Thermostats are rated by the temperature at which they begin to open.
- The valves within the radiator cap serve to keep the system full of coolant and keep air bubbles out. The overflow and reserve system serve to keep the radiator full.
- Cooling fans are used to move air across the radiator. Cooling fans can be mechanically or electrically driven. The engine management computer controls electric cooling fan operation.

## Review Questions

1. As long as the cables will connect, any battery may be installed in any vehicle.
   A. True
   B. False
2. Explain the operation and differences of a starter solenoid and starter relay.

3. All of the statements about the lubrication system are true *except:*
   A. The oil pump may be located on the front of the engine.
   B. The lubrication system cleans the engine oil.
   C. The lubricant is used to remove heat from the engine.
   D. The pickup screen removes small debris from the oil before it enters the pump outlet.

An **ASE Practice Exam** is found in the **Appendix** of every *TechOne* book, followed by a **Bilingual Glossary**, which offers Spanish translations of technical terms alongside their English counterparts.

# Appendix A

## ASE PRACTICE EXAM

1. Two technicians are discussing the diagnosis of a customer's vehicle concern. Technician A says that to properly diagnose the concern, you must first duplicate the concern. Technician B says that before you attempt to duplicate a vehicle concern you should gather as much data as possible from the customer. Who is correct?
   A. Technician A
   B. Technician B
   C. Both A and B
   D. Neither A nor B

2. Two technicians are discussing the availability of service information. Technician A says that technical service bulletins apply to specific vehicles and specific concerns. Technician B says that technical service bulletins can be found in the service manual. Who is correct?
   A. Technician A
   B. Technician B
   C. Both A and B
   D. Neither A nor B

3. All of the following information can be cross-referenced from the VIN except:
   A. The year model of the vehicle.
   B. The engine size.
   C. The factory-installed tire size.
   D. Which plant the vehicle was built in.

4. Two technicians are discussing the buildup of a milky substance found on the backside of the oil fill cap in a vehicle with a V6 engine. Technician A says that the substance indicates an internal coolant leak. Technician B says that a leaking intake gasket can cause the buildup of the substance on the oil fill cap. Who is correct?
   A. Technician A
   B. Technician B
   C. Both A and B
   D. Neither A nor B

5. Two technicians are discussing an engine vibration that increases with intensity as the engine warms up. Technician A says that a broken fan blade can cause a noticeable engine vibration. Technician B says that a broken fan blade can accelerate the wear of the bearings within the water pump. Who is correct?
   A. Technician A
   B. Technician B
   C. Both A and B
   D. Neither A nor B

6. Two technicians are discussing the emission of an abnormal amount of smoke from the exhaust system of a customer vehicle. Technician A says that a blown head gasket will result in the emission of black smoke from the exhaust system. Technician B says that a leaking fuel injector can cause the exhaust to emit black smoke. Who is correct?
   A. Technician A
   B. Technician B
   C. Both A and B
   D. Neither A nor B

A comprehensive **Index** helps instructors and students pinpoint information in the text.

# Bilingual Glossary

**Absorption** To take in by capillary action, as a sponge.
***Absorción*** *Incorporar mediante acción capilar, como una esponja.*

**Acceleration Simulation Mode (ASM)** A vehicle emissions test using a chassis dynamometer that follows a specific pattern of acceleration and deceleration.
***Modo de simulación de aceleración (ASM)*** *Prueba de emisiones de un vehículo mediante un dinamómetro de chasis que sigue un patrón específico de aceleración y desaceleración.*

**Accelerator pedal position (APP) Sensor** A sensor that sends signals to the throttle actuator control (TAC) indicating when the throttle is being opened or closed and how much and at what rate the position is being changed.
***Sensor de posición del pedal del acelerador (APP)*** *Sensor que envía señales al control del actuador del acelerador (TAC) que indica cuando el acelerador se abre o se cierra, así como la cantidad y proporción en que cambia la posición.*

**Actuator** An electromechanical device that changes electrical signals into mechanical actions.
***Actuador*** *Dispositivo electromecánico que convierte las señales eléctricas en acciones mecánicas.*

**Adaptive strategy** A strategy that allows the computer to learn and remember certain aspects of an operational condition.
***Estrategia adaptable*** *Estrategia que permite que la computadora aprenda y recuerde ciertos aspectos de una condición de funcionamiento.*

**Adsorption** The use of solids to remove substances from either gaseous or liquid solutions.
***Adsorción*** *Uso de sólidos para retirar sustancias de soluciones gaseosas o líquidas.*

**Air/fuel ratio (A/F ratio)** Numerical comparison of the amount of air to the amount of fuel, both measured by weight.
***Proporción de aire y combustible (proporción A/C)*** *Comparación numérica de la cantidad de aire en relación con la cantidad de combustible, medida por peso.*

**Air injection reaction (AIR) system** The system used to inject ambient air into the exhaust stream or catalytic converter.
***Sistema de reacción de inyección de aire (AIR)*** *Sistema que se usa para inyectar aire del medio ambiente en el caudal de escape o convertidor catalítico.*

**Air pollution** The introduction of impurities and contaminants into the atmosphere, many of which are caused by industrial activities.
***Contaminación del aire*** *Introducción de impurezas y contaminantes en la atmósfera, muchos de los cuales se deben a actividades industriales.*

**Alternating current (AC)** A current in which electrons can flow in either a positive or negative direction.
***Corriente alterna (AC)*** *Corriente en la cual los electrones pueden fluir en dirección positiva o negativa.*

**Amperage** A unit of measure expressing the cumulative flow of electrons within a circuit.
***Amperaje*** *Unidad de medida que expresa el flujo acumulativo de electrones dentro de un circuito.*

**Analog** A signal in which the voltage can be any value within a range.
***Análoga*** *Señal en la cual el voltaje puede ser cualquier valor dentro de un rango.*

**Antilock brake system (ABS)** A system installed on the vehicle to prevent the wheel brakes from locking up in a panic or inclement weather condition.
***Sistema de frenos antibloqueo (ABS)*** *Sistema instalado en el vehículo para evitar que los frenos de las ruedas se bloqueen en un momento de pánico o bajo condiciones de clima inclemente.*

**Atmosphere** The air enveloping the earth.
***Atmósfera*** *Aire que rodea a la tierra.*

**Atmospheric pressure** The amount of pressure that the atmosphere exerts on the surface of the earth.
***Presión atmosférica*** *Cantidad de presión que la atmósfera ejerce sobre la superficie de la tierra.*

# Index

# Section 1

## Safety and Communication

## SECTION OBJECTIVES

After you have read, studied, and practiced the contents of this section, you should be able to:

- Understand your role in providing a safe work environment.
- Work safely in the automotive shop environment.
- Understand what is expected from you as an engine performance technician.
- Understand the role of the technician in an automotive shop.
- Recognize the diagnostic resources available to you.
- Use service manuals.
- Apply service bulletins to a diagnostic procedure.
- Apply the diagnostic process.
- Define common engine performance symptoms.
- Understand the challenges associated with intermittent concerns.

**Interesting Fact** *Good technicians possess good mechanical aptitude, strong reading and other academic skills, and a strong desire to tackle challenging problems, and they consistently demonstrate good, safe work practices.*

1

# Chapter 1

# Safe Work Practices

## Introduction

Safety is an important topic, important to you and those around you. Mishaps in the automotive shop will happen from to time to time; however, most major injuries from these mishaps are preventable. Safe work practices are based on common sense and knowledge of safety procedures. Here are some guidelines to better prepare you in the case of an accident:

- Keep an up-to-date list of emergency phone numbers near the phone, including those of a doctor, a hospital, and the fire and police departments, and know the proper procedure for calling local Emergency Services (9-1-1 or otherwise).
- Make sure you know the location and contents of the shop's first-aid kit.
- Always summon the Emergency Medical Service (EMS) immediately after stabilizing a victim in the event of a severe emergency.
- Be familiar with the location and operation of all fire extinguishers. Know the fire evacuation routes for the entire building.
- Be familiar with the location and operation of the eyewash station and chemical shower, **Figure 1**.
- If someone is overcome by carbon monoxide, get him or her fresh air immediately.
- Cool burns immediately by rinsing them with water.
- Whenever there is severe bleeding from a wound, try to stop the bleeding by applying pressure with clean gauze on or around the wound and get medical help.

- Never move someone who might have broken bones unless the person's life is otherwise endangered. Moving that person can cause additional injury. Call for medical assistance.
- Immediately inform your supervisor of all accidents that occur in the shop.

**Figure 1.** You should know the exact location of the eyewash station and chemical shower, as well as the procedures required to use them.

**Figure 2.** Safety goggles, face shield, and safety glasses.

## EYE PROTECTION

Eye protection should be worn whenever you are working in the shop. Many types of eye protection are available, as shown in **Figure 2**. To provide adequate eye protection, safety glasses have lenses made of polycarbonate plastic. Safety glasses also offer side protection. For nearly all services performed on the vehicle, eye protection should be worn. This is especially true while you are working under the vehicle.

Some procedures might require that you wear other eye protection in addition to safety glasses. A face shield not only gives added protection to your eyes but also protects the rest of your face. A face shield should be used when grinding or when using pressurized cleaning equipment, or in any other situation when serious damage could occur to your face.

If chemicals such as battery acid, fuel, or solvents get into your eyes, flush them continuously with clean water. Have someone call a doctor and get medical help immediately. Many shops have eyewash stations or safety showers that should be used whenever you or someone else has been sprayed or splashed with a chemical. You must be familiar with the location and operation of the eyewash stations and safety showers.

## EAR PROTECTION

Ear protection should be worn any time that you are subjected to loud or high-pitched noises. Drying components off with compressed air is one of the most common practices performed in the automotive shop but one of the most damaging to your ears. Among other tasks that can damage your ears is the use of air chisels and air-powered

A

B

**Figure 3.**  (A) Earmuffs and (B) earplugs.

drills. Many comfortable types of protection are available. Two common methods of ear protection are illustrated in **Figure 3**.

## CLOTHING AND GLOVES

Your clothing should be well fitted and comfortable but made with strong material. If you have long hair, tie it back or tuck it under a cap. Never wear rings, watches, bracelets, and neck chains. These can easily get caught in moving parts and cause serious injury.

Automotive work involves the handling of many heavy objects, which can be accidentally dropped on your feet or toes. Always wear shoes or boots that are constructed of leather or similar material and that are equipped with no-slip soles. Steel-tipped safety shoes give added protection to your feet.

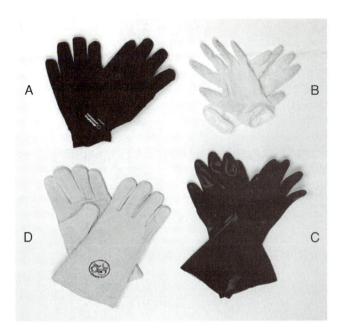

**Figure 4.** (A) Mechanic's gloves, (B) latex gloves, (C) chemical-resistant gloves, and (D) welding gloves.

Good hand protection is often overlooked. A scrape, cut, or burn can limit your effectiveness at work for many days. It is difficult to perform many automotive tasks with gloves on due to the lost sense of touch. However, in the last several years many different styles of gloves made of such materials as leather and Kevlar have become available. These materials offer good protection but are thin enough to offer some sense of touch. Because chemical agents can be absorbed through the skin, thin latex gloves have become popular among technicians. Latex gloves allow the technician an incredible sense of touch and provide protection from most chemicals, but they are ineffective against scrapes or cuts. The gloves come in a variety of thicknesses, **Figure 4**.

## SAFE WORK AREAS

Your work area is your responsibility and should be kept clean and safe. Here are some simple steps to establishing and maintaining a safe work area:

1. Keep the floor and bench tops clean, dry, and orderly. Any oil, coolant, or grease on the floor can make it slippery and should be cleaned up immediately. Slips can result in serious injuries.
2. Keep aisles and walkways clean and wide enough to easily move through.
3. Make sure the work areas around machines are large enough to safely operate the machine.

4. Make sure all drain covers are snugly in place. Open drains or covers that are not flush to the floor can cause toe, ankle, and leg injuries.
5. If you notice a dangerous situation or a piece of defective equipment, it is your responsibility to correct the situation or notify someone who can correct it. In the meantime, the electrical power to the equipment should be disabled and an out-of-order sign should be placed in a visible location.

## HAZARDOUS MATERIALS

As an automotive technician, you will come in contact with hazardous materials on a daily basis. These can include cleaners, solvents, fuels, used rags, and sealers. In dealing with these materials, it is important that you know how these materials should be stored, handled, used, and disposed of.

Most solvents and other chemicals used in an automotive shop have warning and caution labels that should be read and understood by everyone that uses them. Right-to-Know Laws concerning hazardous materials and wastes protect every employee in a shop. The general intent of these laws is for employers to provide a safe working place as it relates to hazardous materials. All employees must be trained about their rights under the legislation, including the nature of the hazardous chemicals in their workplace, the labeling of chemicals, and the information about each chemical listed and described on Material Safety Data Sheets (MSDS). These sheets are available from the manufacturers and suppliers of the chemicals. The MSDS detail the chemical composition and precautionary information for all products that can present health or safety hazards. The Canadian version of the MSDS is the Workplace Hazardous Materials Information Systems (WHMIS).

> **You Should Know** *When handling any hazardous material, always wear the appropriate safety protection. Always follow the correct procedures while using the material and be familiar with the information given on the MSDS for that material.*

A list of all hazardous materials used in the shop should be posted for the employees to see. Shops must maintain documentation on the hazardous chemicals in the workplace, proof of training programs, records of accidents or spill incidents, and satisfaction of employee requests for specific chemical information via the MSDS,

**Figure 5.** MSDS information should be located in an accessible place in a Right-to-Know station like this one.

as well as provide a general Right-to-Know compliance procedure manual to be used within the shop, **Figure 5**.

> **You Should Know** *Whenever a hazardous material is moved from its original container into a different container, the new container must be marked with the applicable information concerning health, fire, and reactivity hazards that the material poses.*

## FIRE HAZARDS AND PREVENTION

Many items around a typical shop are a potential fire hazard. These include gasoline, diesel fuel, cleaning solvents,

and dirty rags. Each of these should be treated as a potential firebomb and handled and stored properly **(Figure 6)**.

It is imperative that you know the location of the fire extinguishers and fire alarms in the shop and the procedures required to use them before a fire occurs. You should also be aware of the different types of fires and the fire extinguishers used to put out each type of fire. There are basically four types of fires. **Figure 7** lists the symbol found on the extinguisher, how each class is best extinguished, the fuels involved with each class, and the type of extinguisher material needed. Typically, fire extinguishers found in automotive repair shops are charged with chemicals to fight more than one class of fire.

## USING A FIRE EXTINGUISHER

Remember to never open doors or windows during a fire unless it is absolutely necessary; the extra draft will only make the fire worse. Make sure that the fire department is contacted before or during your attempt to extinguish a fire. To extinguish a fire, stand six to ten feet from the fire. Hold the extinguisher firmly in an upright position. Aim the nozzle at the base of the fire and use a side-to-side motion, sweeping the entire width of the fire. Stay low to avoid inhaling the smoke. If the area gets too hot or too smoky, get out of the building. Remember to never go back into a burning building for anything. To help remember how to use an extinguisher, memorize the word "PASS."

**P**ull the pin from the handle of the extinguisher.
**A**im the extinguisher's nozzle at the base of the fire.
**S**queeze the handle.
**S**weep the entire width of the fire with the contents of the extinguisher.

If there is not a fire extinguisher handy, a blanket or fender cover can be used to smother the flames. You must be careful when doing this because the heat of the fire can burn you and the blanket. If a fire is contained in a drain pan or container, smother the fire with the extinguisher's chemical or foam. If the fire is too great to smother, move

**Figure 6.** Hazardous materials must be stored in proper containers and in proper locations.

| | Class of Fire | Typical Fuel Involved | Type of Extinguisher |
|---|---|---|---|
| Class **A** Fires (green) | **For Ordinary Combustibles** Put out a class A fire by lowering its temperature or by coating the burning combustibles. | Wood Paper Cloth Rubber Plastics Rubbish Upholstery | Water*[1] Foam* Multipurpose dry chemical[2] |
| Class **B** Fires (red) | **For Flammable Liquids** Put out a class B fire by smothering it. Use an extinguisher that gives a blanketing, flame-interrupting effect; cover the whole flaming liquid surface. | Gasoline Oil Grease Paint Lighter fluid | Foam* Carbon dioxide[3] Halogenated agent[4] Standard dry chemical[5] Purple K dry chemical[6] Multipurpose dry chemical[2] |
| Class **C** Fires (blue) | **For Electrical Equipment** Put out a class C fire by shutting off power as quickly as possible and by always using a nonconducting extinguishing agent to prevent electric shock. | Motors Appliances Wiring Fuse boxes Switchboards | Carbon dioxide[3] Halogenated agent[4] Standard dry chemical[5] Purple K dry chemical[6] Multipurpose dry chemical[2] |
| Class **D** Fires (yellow) | **For Combustible Metals** Put out a class D fire of metal chips, turnings, or shavings by smothering or coating with a specially designed extinguishing agent. | Aluminum Magnesium Potassium Sodium Titanium Zirconium | Dry powder extinguishers and agents only |

*Cartridge-operated water, foam, and soda-acid types of extinguishers are no longer manufactured. These extinguishers should be removed from service when they become due for their next hydrostatic pressure test.
Notes:
(1) Freezes in low temperatures unless treated with antifreeze solution, usually weighs more than 20 pounds, and is heavier than any other extinguisher mentioned.
(2) The only extinguishers that fight A, B, and C classes of fires. However, they should not be used on fires of liquefied fat or oil of appreciable depth. Be sure to clean residue immediately after using the extinguisher so that sprayed surfaces will not be damaged (ammonium phosphates).
(3) Use with caution in unventilated, confined spaces.
(4) Can cause injury to the operator if the extinguishing agent (a gas) or the gases produced when the agent is applied to a fire are inhaled.
(5) Also called ordinary or regular dry chemical (sodium bicarbonate).
(6) Has the greatest initial fire-stopping power of the extinguishers mentioned for class B fires. Be sure to clean residue immediately after using the extinguisher so sprayed surfaces will not be damaged (potassium bicarbonate).

**Figure 7.** A guide to fire extinguisher selection.

everyone away from the fire and call the local fire department. A simple under-the-hood fire can cause the total destruction of the vehicle and the building and can take lives. You must be able to respond quickly and precisely to avoid a disaster.

## LIFT SAFETY

Always be careful when raising a vehicle on a lift or a hoist. Adapters and hoist plates must be positioned correctly on twin-post- and rail-type lifts to prevent damage to the underbody of the vehicle. All vehicles have specific lift points. Lifting from these specific locations allows the weight of the vehicle to be evenly supported by the adapters or hoist plates. The correct lift points can be found in the vehicle's service manual. Always follow the manufacturer's instructions. Before operating any lift or hoist, carefully read the operating manual and follow the operating instructions.

Once you feel that the lift supports are properly positioned under the vehicle, raise the lift until the supports contact the vehicle. Then, check the supports to make sure that they are in full contact with the vehicle. Once you are satisfied that they are properly positioned, raise the vehicle six to ten inches from the floor, shake the vehicle to make sure that it is securely balanced on the lift, and then raise the lift to the desired working height. Once you have achieved a comfortable working height, lower the lift until the mechanical locks are engaged.

## JACK AND JACK STAND SAFETY

A vehicle can also be raised off the ground with a hydraulic jack. The lifting pad of the jack must be positioned under an area of the vehicle's frame or at one of the manufacturer's recommended lift points. Never place the lifting pad under the floor pan or under steering and suspension components because these are easily damaged by the weight of the vehicle. Always position the jack so that the wheels of the vehicle can roll as the vehicle is being raised.

> **You Should Know** *Safety stands should be placed under the vehicle immediately after the vehicle has been raised to the desired height.*

Safety (jack) stands are supports of different heights that sit on the floor. They are placed under a sturdy chassis member, such as the frame or axle housing, to support the vehicle. Once the safety stands are in position, the hydraulic pressure in the jack should be slowly released until the weight of the vehicle is on the stands. Like jacks, jack stands also have a capacity rating. Always use a jack stand of the correct rating.

Never move under a vehicle when it is supported only by a hydraulic jack; rest the vehicle on the safety stands before moving under the vehicle. The jack should be removed after the jack stands are set in place. This eliminates a hazard such as a jack handle obstructing a walkway. A jack handle that is bumped or kicked can cause a tripping accident or cause the vehicle to fall.

## BATTERIES

When possible, you should disconnect the battery of a vehicle before you disconnect any electrical wire or component. This prevents the possibility of a fire or electrical shock. To properly disconnect the battery, disconnect the negative or ground cable first, then disconnect the positive cable. When reconnecting the battery, connect the positive cable first, then the negative.

> **You Should Know** *Hydrogen gas is produced any time that a battery is charging. Any spark created in the area of a charging battery could cause the battery to explode, exposing you and your co-workers to personal injury from debris and electrolyte, **Figure 8**.*

**Figure 8.**   Sparks can cause a charging battery to explode.

> **You Should Know** *The active chemical in a battery, the electrolyte, is basically sulfuric acid. Sulfuric acid can cause severe skin burns and permanent eye damage, including blindness, if it gets in your eyes. If battery acid gets on your skin, wash it off immediately and flush your skin with water for at least five minutes. If the electrolyte gets into your eyes, immediately flush them with water and then immediately see a doctor. Never rub your eyes; flush them well and go to a doctor. It is common sense to wear safety glasses or goggles while working with and around batteries.*

## AIR BAG SAFETY AND SERVICE WARNINGS

The dash and steering wheel contain the circuits that operate the air bag system. Whenever working on or around air bag systems, it is important to be familiar with specific manufacturer system operation and service procedures. Failure to comply with these service precautions can result in unintentional deployment of the air bag(s).

> **You Should Know** *Disconnecting the battery does not protect you from accidental air bag deployment. Air bag systems retain the ability to deploy for several minutes after the battery has been disconnected.*

# Summary

- Safe work practices are based on common sense and knowledge of safety procedures. Most major injuries and mishaps are preventable. Be aware of what is happening around you so that you will be alert to potential hazards.
- Eye protection should be worn at all times when working in the automotive shop. A face shield provides protection for your entire face.
- Many common shop procedures can cause loss of hearing, so it is important to wear ear protection.
- Your personal protection includes the clothing that you wear and the way in which it is worn.
- Your work area is your responsibility and should be kept clean and safe.
- You must know how to handle the chemicals that you come into contact with on a daily basis. Handling includes storage, usage, and disposal. The MSDS will provide you with information about chemical composition and precautionary information. All chemicals should be properly labeled.
- It is imperative that you know the location of all fire extinguishers and the procedures required to use them.
- Lift supports should be properly positioned before raising a vehicle on a lift. Shaking the vehicle will help verify that the vehicle is securely balanced. When using a floor jack, safety stands should be placed under the vehicle immediately after the desired height has been reached.
- When servicing air bags, all manufacturer precautions should be strictly adhered to.

# Review Questions

1. Explain when eye protection should be worn.
2. Your clothing is a significant part of your personal protection. Describe the safety aspects that are related to how you dress.
3. What information can be found on an MSDS?
4. The general intent of _____-____-_____ laws is for employers to provide a safe working place as it relates to an awareness of hazardous materials.
5. Explain the correct procedure for putting out a fire with an extinguisher.
6. All of the following statements about lift operation are correct *except:*
   A. Every vehicle has designated lift points.
   B. The mechanical locks should be engaged once the vehicle has reached the desired working height.
   C. After the vehicle has been lifted approximately six feet from the ground, shake it to ensure that it is properly positioned on the lift.
   D. Shaking the vehicle will help to ensure that the vehicle is properly positioned.
7. While discussing vehicle air bags, Technician A says that all manufacturer service procedures and precautions for a vehicle must be strictly adhered to. Technician B says that the air bag system may be capable of deploying even after the battery has been disconnected. Who is correct?
   A. Technician A
   B. Technician B
   C. Both A and B
   D. Neither A nor B

# Chapter 2

# Working as an Engine Performance Technician

## Introduction

An automotive technician is a person who uses a variety of tools and equipment to diagnose and repair various automobile problems. As an automotive technician, you have entered an incredibly demanding, yet satisfying, field.

You are entering the vehicle repair industry at a significant time of positive growth and change, making you a part of a new breed of technician. More will be expected of you and your counterparts than of any generation of technicians before you.

As an automotive technician you must first understand and accept that the modern automobile is not merely a mechanical machine but rather a high-tech instrument with highly integrated electronic systems and computers. Technicians must not only have mechanical aptitude but also have strong analytical skills. You must have the desire and the ability to constantly adapt to ever-changing technologies. Additionally, employers are looking for employees with excellent communication skills and the ability to understand what it takes to make the company successful **(Figure 1)**.

### CERTIFICATION

In many industries, service technicians, such as plumbers, electricians, and exterminators, are required by law to be licensed or certified. However, certification for the automotive technician is completely voluntary. Today, many employers prefer that you do have certification through the National Institute for Automotive Service Excellence (ASE).

The ASE exams are divided into several specialty areas such as engine performance and advanced engine performance. Each specialty area includes a comprehensive written exam that is used to test your knowledge of that system as well test your diagnostic reasoning skills. In addition to passing the exam, you must have completed two years of practical hands-on work experience in the

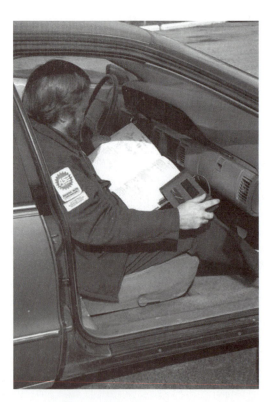

**Figure 1.** Technicians must have strong analytical reading, writing, and communication skills, in addition to strong mechanical aptitude.

**Figure 2.** Certified technicians receive shoulder patches recognizing their achievement.

transportation industry. Once technicians pass at least one exam and meet the experience requirement, they are recognized as ASE certified in the specialty area in which they tested. A person who has successfully completed the specified series of exams in a specific vehicle type will earn the distinction of master technician, for example, master automotive technician or master truck technician **(Figure 2)**.

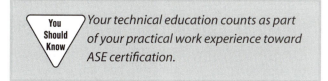

You Should Know *Your technical education counts as part of your practical work experience toward ASE certification.*

## EDUCATION

To be a successful, well-rounded automotive technician, you need a strong academic background, especially in the areas of reading, physical science, language, and communication skills. Reading aptitude is especially important because, although for many years service information had been written on an eighth- to ninth-grade level, almost overnight that level has jumped to a college level. If your reading skills are inadequate, you need to work on improving those skills. Understanding the technical information provided is essential to your success.

The time you are currently spending on your education is only the beginning. Each year, technology leaps forward; you must leap forward as well. It is not unusual for technicians to take part in some form of training on a weekly basis either at home or in the workplace. Educational opportunities are available from many outlets, including vehicle manufacturers, community colleges, parts manufacturers, and specialized providers. Many training programs are available in CD-ROM, video, and on-line formats.

## JOB EXPECTATIONS

As an employee of a company, certain things will be expected of you. You should always be on time or early to work. Tardiness is one of the largest problems faced by employers. If you are absent or late, work scheduled for you must be distributed among other technicians. If you discover that you will be late for work or unable to attend for any reason, you will be expected to notify your employer before you are due to begin your shift.

*Remember that the automotive service facility is a production-based profit center. Each day, your service supervisor will schedule work in anticipation of the number of technicians scheduled for work. The absence of just one technician will disrupt the profitability of the entire service facility.*

While you are at work, you will be expected to maintain a clean and professional image. This includes your appearance as well as your language **(Figure 3)**. Almost all employers

**Figure 3.** Your personal image is a reflection on your employer.

discourage foul language. Because you will be required to talk with customers on occasion, you must understand that the impression you leave with them will have an effect on their attitude toward you and your company. Remember that as an employee of a company you represent that company at all times even when you are not at work.

## CUSTOMER CONTACT

In the automotive industry, you will come into contact with customers regularly. Although each shop has its own policies for customer contact with technicians, as an engine performance technician, it is likely that you will have frequent contact with customers. You must always treat customers with respect, speaking to them in a courteous manner and listening carefully to their questions and concerns.

## SPECIALIZATION

As in many other professions, auto technicians are now often specialists. In years gone by, it was typical for a technician to be comfortable working on any part of the vehicle. Nowadays, however, because of the complexity of the modern vehicle, it is commonplace for a technician to specialize in one or two system areas. That is not to say that technicians cannot competently repair systems that they are not specialized in; it means that they might not be as familiar with the intricacies of the other systems. This text focuses on the skills required by the engine performance specialist, sometimes known as the driveability technician. The engine performance specialist repairs what directly relates to the performance of the engine and related powertrain systems. Other specialty areas include engine mechanical, transmission, brake, air conditioning, electrical, and chassis systems.

## COMEBACKS

**Comebacks** are probably the biggest source for customer dissatisfaction in the auto repair industry. A comeback is a situation in which a customer has to make a return trip to the service facility either because the concern was not corrected or because another problem might have been created during correction of the original concern. Comebacks are something that every technician will experience from time to time. However, every good technician should be conscientious enough to ensure that comebacks are limited. In many facilities comebacks are monitored, and in extreme cases a technician with an excessive amount of comebacks might be terminated.

## REPAIR ORDER

The repair order (RO) is a tool by which you communicate with the customer. In most cases, the manager or service advisor will retrieve information from the customer. This information will be written on the repair order and will include:

- Customer name, address, and telephone information.
- Vehicle make, model, type, and vehicle identification number (VIN).
- A thorough description of the customer concern.

The repair order is your tool in which to communicate with the customer. There are three "Cs" that must exist on a repair order, concern, cause and correction. Because this will be your direct line of communication with the customer, a repair order must be written in as much detail as possible.

- *Concern:* the reason that the vehicle was brought in for service. The concern should be written with as much relevant information as possible, including when the concern might occur, how often the problem might occur, and any other specific conditions under which it might occur.
- *Cause:* the second part of the work order that directly concerns the technician. This is where the technician fills in the **root cause** for the customer concern. The root cause is the primary factor that is causing the customer's concern. When a problem is diagnosed, you will be required to fill in the cause of the customer's concern. The cause should be written in as descriptive a manner as possible.
- *Correction:* the area in which you write in as much detail as possible regarding what you did to repair the vehicle. This should include any diagnostic steps that were performed and a brief description of the steps required to make the repair. A good description here allows the customer to get an idea of the amount of work that was required to repair the vehicle **(Figure 4)**.

Interesting Fact

*In facilities that use computerized or paperless RO systems, the remarks that the technician makes in the repair order screens are the exact remarks that the customers will see on their copy of the bill.*

## ETHICS

Ethics is the set of moral rules or values that you apply when making decisions. Ethics in the automotive industry is extremely important because customers trust you with their safety and that of their passengers. Whether you realize it or not, your ethics are applied to every decision that you make. This includes the quality and attention that you exercise when repairing a customer's vehicle and the

| RHO 1020304<br><br>GRW MOTORS<br>100 N. MAIN ST<br>SPOON RIVER, IL | Name: _Tom Tompson_<br>Address: _1234 West Elm St._<br>City: _Spoon River_   State: _Il_<br>Phone: Home: _603-5327_ Business: _603-4327_ | Year: _1996_ Model: _Olds/Riv_<br>VIN: _1G4ED22K7T1000001_<br>Color: _Silver_<br>Lic. #: _TT-500_ |

| Parts | | Tech # | Customer comments | Lb. code |
|---|---|---|---|---|
| | | #8 | 1. Check engine light "ON"<br>   Check for codes with scan tool<br>      Call for authorization<br><br>2. Rattle in right rear of vehicle<br>   Check for loose shock mounting or tailpipe clearance | |
| | | #12 | Change oil and filter | |

**Figure 4.** The repair order should contain information about the vehicle and the customer, as well as the customer's concern, the cause of the concern, and the correction.

recommendations that you make to that customer. Furthermore, your ethics are extended not only to your customers but to your employer as well. You should treat everyone the same way that you would want to be treated.

## PRIDE IN YOUR WORK

By now, you should be thinking of yourself as a professional. Like many other professionals, such as physicians, you are obtaining the education that is necessary to successfully diagnose and repair complex systems. Contrary to what you might have been led to believe, at this point you are entering a highly technical and professional field. The amount of pride that you take in your career will directly affect the amount of success that you obtain.

Strive to take an exceptional amount of pride in every job that you undertake. Whether it is a simple oil change or a major overhaul, you should treat every job with an equal amount of care and do the best job that is possible. There are no meaningless repairs made to an automobile. A repair to any vehicle system has potential safety implications and should not be taken lightly.

## *Summary*

- Technicians must adapt to ever-changing technologies.
- Employers are looking for employees with strong communication skills and who will make a significant contribution to the success of the company.

- Certification is not required to be an automotive technician but it is increasingly expected. Certifications can be obtained through ASE by passing a comprehensive written exam and obtaining two years of verifiable work experience.

- An automotive technician needs a strong academic background to be successful. Your training will continue throughout your career.
- You will be expected to be punctual for work, notify your supervisor if you will be late, and present a clean and professional-looking appearance.
- Because of the increasing complexity of modern vehicles, technicians often specialize in one or two areas.
- Comebacks are a large reason for customer dissatisfaction. All technicians will have comebacks, but an excessive amount can lead to termination of employment.
- The repair order contains pertinent information about the customer, the vehicle, and the concerns that the customer has. Three elements that you will use to communicate with the customer through the repair order are the concern, the cause, and the correction.
- Apply ethics to every diagnosis and repair that you make. Your ethics will directly affect your customers, your employer, and yourself.
- You should take pride in your work no matter how large or how small the job is. Every repair made to a vehicle is important.

# Review Questions

1. Explain what is required of the modern automotive technician.
2. All of the following statements about ASE certification are true *except:*
   A. Federal law requires that all technicians be certified.
   B. Two years of hands-on experience are required to be fully certified in your specialty area.
   C. Your education will count toward your experience requirement.
   D. Exams are given in several specialty areas.
3. Explain why technicians must be better educated now than in the past.
4. Explain how you as an individual reflect on the image of your employer.
5. Technician A says that specialization has become commonplace due to the lack of formal education among technicians. Technician B says that specialized technicians are competent only in their specialty area. Who is correct?
   A. Technician A
   B. Technician B
   C. Both A and B
   D. Neither A nor B
6. Which of the following statements about comebacks is true?
   A. Comebacks are an opportunity to increase shop revenue.
   B. Comebacks are always preventable.
   C. Comebacks increase customer confidence.
   D. Comebacks are a primary source for customer dissatisfaction.
7. Explain what information each of the three "Cs" represents.
8. Ethics in the automotive industry is important because customers trust you with their safety.
   A. True
   B. False
9. Which of the following statements about taking pride in your work is true?
   A. All repairs deserve an equal amount of attention to detail.
   B. Your success is relative to the amount of pride that you put into your work.
   C. Automotive technicians are professionals.
   D. All of the above.

# Chapter **3**

# Diagnostic Resources

## *Introduction*

As technicians, we have one fundamental task and that is to satisfy our customers. Your customers depend on you to accurately diagnose and repair their vehicles in a timely manner. To do this, you must be competent in your skills, which means continuing your training, and you must have a positive attitude and a desire to fix their vehicles. If any one of these is missing, the customer will not be satisfied. Many technicians are of the opinion that if a problem is not bothersome to them, then it is really not a problem. Remember that customers know their vehicles best, and they know when something has changed. No matter how insignificant a problem might appear to you, the customer expects you to fix it.

## INFORMATION RESOURCES

Repairing a vehicle today starts with the information that you gather because information is the most important element in accurate diagnosis of the modern vehicle. This information can come in the form of a description from a customer, from a service manual, data that are downloaded from a vehicle, or a suggestion from a help line. Technicians that know what their resources are and how to use them effectively will be highly regarded and generously compensated. As a technician, you have many resources available.

## Service Manuals

The service manual is the most important resource that a technician has. Service manuals are generally written for a specific vehicle model. The manual will generally contain a section devoted to every system in that specific vehicle, with a description of system operation, information on the diagnosis of the system, and specific repair information. Service manuals will also contain wiring diagrams and schematics. Almost any piece of information that you will seek as a technician will be available in the service manual.

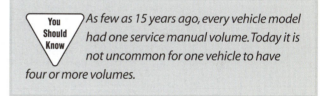

**You Should Know** *As few as 15 years ago, every vehicle model had one service manual volume. Today it is not uncommon for one vehicle to have four or more volumes.*

There are two types of service manuals, factory and aftermarket. Factory service manuals are the most specific; they contain information about every system in that particular vehicle **(Figure 1)**. Aftermarket service manuals are usually based on the information found in a factory service manual; however, they are usually not as specific. Aftermarket manuals can have information on several makes and models in one large volume and will usually contain much of the pertinent

**Figure 1.** A manufacturer's service manual.

**Figure 2.**   An aftermarket service manual.

information for the systems that are most commonly serviced **(Figure 2)**. Aftermarket manuals are also available for specific vehicles in small single-volume books.

## Service Bulletins

Service bulletins are an extremely helpful resource. Bulletins generally come from three sources: vehicle manufacturers, parts manufacturers, and repair associations such as the Automatic Transmission Rebuilders Association (ATRA) or the Automotive Engine Rebuilders Association (AERA) **(Figure 3)**.

Service bulletins are written to assist the technician in the field to make cost-effective repairs. Bulletins are usually written to address a common problem with a specific vehicle or specific component. A bulletin might also be issued when a change in a particular service procedure is

```
ARTICLE BEGINNING

TECHNICAL SERVICE INFORMATION

FALSE DTC P0121 (REPROGRAM PCM)

Model(s):      1997 Buick Century, Skylark
               1997 Chevrolet Lumina Monte Carlo Malibu, Venture
               1997 Oldsmobile Achieva, Cutlass, Cutlass Supreme,
                    Silhouette
               1997 Pontiac Grand Am, Grand Prix, Trans Sport
                    with 3100/3400 V6 Engine (VlNs M, E
                    - RPOs L82, LA1)
Section:       6E - Engine Fuel & Emission
Bulletin No.:  77-65-14A
Date:          May, 1998

NOTE:  This bulletin is being revised to add additional models and
       calibration numbers. Please discard Corporate Bulletin
       77-65-14 (Section 6E - Engine Fuel & Emission).


CONDITION

   Some owners may experience a MIL (Malfunction Indicator Lamp) light
illuminated on the vehicle's instrument panel. Additionally, the
engine's normal controlled idle speed may be slightly elevated when
the MIL is illuminated.


CAUSE

   The current DTC (Diagnostic Trouble Code) P0121 is too sensitive.
The rational check that the diagnostic calibration performs has been
changed. Part of those changes involve eliminating the defaulted
higher idle.


CORRECTION

   Check the calibration identification number utilizing a scan tool
device. Re-flash with the updated calibration if the current
calibration is not one listed in this bulletin. If the vehicle already
has the most recent calibration, then refer to the appropriate service
repair manual to diagnose and repair for DTC P0121. Test drive the
vehicle after repair to ensure that the condition has been corrected.
The new calibrations are available from the GM Service Technology
Group starting with CD number 6 for 1998.

IMPORTANT:  Do not attempt to order the calibrations from GMSPO. The
            calibrations are programmed into the vehicles PCM via a
            Techline Tool device.
```

**Figure 3.**   A typical service bulletin.

*Service bulletins will always address a specific problem for a specific vehicle. However, bulletins are not a substitute for proper diagnosis, but an aid in speeding up your diagnosis.*

suggested. Before a bulletin is released, it is heavily researched and tested to verify that the information is correct. It is a good idea to search for bulletins fairly early in the diagnostic procedure, particularly when a problem appears to be out of the ordinary or extremely intermittent.

## Labor Time Guide

The labor time guide is a book that is used to assist repair facilities in determining how much to charge a customer for a specific repair. The guide lists the most common service procedures performed and lists a suggested completion time for each repair, as shown in **Figure 4**. This information gives the repair facility a fair estimate of how long the repairs will take. The customer can then be consulted as needed on specific repair costs. Each vehicle make will have its own section in the guide, and specific year models and vehicle models will be listed.

Many repair facilities base technician pay on the labor time guide. What this means to you is that the amount of money that you will be paid for making a specific repair is based on what is recommended in the labor time guide. This is referred to as the flat rate–based pay system. For example, if a repair is listed as taking 1 hour, the customer is charged for 1 hour and you get paid for 1 hour. If the repair takes 3 hours, the customer still pays for only 1 hour and you still get paid for only 1 hour. The opposite is true should the repair take less time than quoted in the book.

## Computer-Based Information

Up to this point, all of the material referred to has been in the form of books or paper materials. Paper service manuals are rapidly becoming a thing of the past. Today all of the major vehicle manufacturers support computer-based information systems as the primary means of information delivery. The information included in most computer-based systems is the same as the information previously found in a service manual plus service bulletins, labor time guides, and other service materials. Materials in this format are cheaper to produce and easier to update and can be updated through satellite or Internet links, giving the technician the most up-to-date material available.

Aftermarket manufacturers such as ALLDATA and Mitchell OnDemand provide high-quality information services, very similar to those provided by the manufacturers, to aftermarket repair centers. Aftermarket providers, however, provide service information for all makes rather than

**7-80**

**GENERAL MOTORS CORPORATION**
AURORA : RIVIERA (1995 -1998)

| | Labor time | Service time |
|---|---|---|
| **IGNITION** | | |
| **Diagnose Driveability (A)** | | |
| 1995-02 (.5) .......................7 | | .7 |
| **Camshaft Interrupter, Replace (B)** | | |
| 3.8L | | |
| VIN 1 (3.2) ................................ 4.2 | | 4.9 |
| VIN K (2.9) ............................... 3.9 | | 4.4 |
| Replace | | |
| camshaft sensor add ...........1 | | .1 |
| **Camshaft Position Sensor, Replace** | | |
| 1995 | | |
| 3.8L VIN 1 (1.7) ......................... 2.3 | | 2.6 |
| 3.8L VIN K (.8) ........................... 1.1 | | 1.3 |
| 4.0L (.7) ..................................... 1.0 | | 1.1 |
| 1996-02 | | |
| 3.5L (1.1) ................................... 1.5 | | 1.7 |
| 3.8L (.8) ..................................... 1.1 | | 1.3 |
| 4.0L (.7) ..................................... 1.0 | | 1.1 |
| **Ignition Coil, Replace** | | |
| 1995-97 | | |
| one (.2) ...................................... .3 | | .3 |
| each additional, add .................. .2 | | .2 |
| 1998-02 | | |
| 3.5L | | |
| front (.3) ................................. .5 | | .5 |
| rear (.4) ................................... .6 | | .6 |
| both (.5) ................................... .7 | | .7 |
| 3.8L, 4.0L | | |
| one (.3) .................................... .5 | | .5 |
| each additional, add ............... .2 | | .2 |
| **Ignition Switch, Replace (B)** | | |
| 1995-02 (1.0) ................................ 1.4 | | 1.4 |
| **Spark Plug Cables, Replace (B)** | | |
| R3.8L (.8) ..................................... 1.1 | | 1.1 |
| 4.0L (1.0) ..................................... 1.4 | | 1.4 |

**Figure 4.**  A labor time guide indicates how long a specific repair should take.

one specific manufacturer **(Figure 5)**. Computer-based information systems can be delivered on CD-ROM or DVD or through a high-speed Internet connection. Many of these systems are additionally coupled with shop and parts management software.

## Technical Assistance

Technical assistance centers are set up to provide technical advice to technicians to help them solve difficult problems. The assistance center personnel receive their

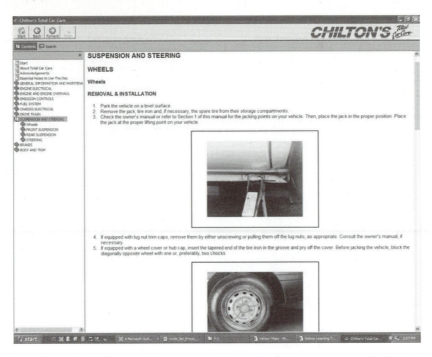

**Figure 5.** Computer-based service information systems give technicians access to the latest service information.

information from large databases comprised of repair information and engineering information. These services are available to dealership technicians through the manufacturers. Technicians not associated with a manufacturer have access to these services through respective parts manufacturers and automotive service associations. The latest trend in obtaining service information is the availability of Internet clearinghouses such as the International Automotive Technicians Network (IATN). Here technicians can search databases or ask for help by submitting a request to the network. The network will send out your inquiry to thousands of technicians around the world skilled in the area of your question. These technicians will then send you ideas or suggestions. This service is free to technicians and depends on technicians helping one another. This is a useful source for information because it gives you the ability to compare notes on one problem with thousands of other technicians. Additionally, hundreds of thousands of Internet websites and pages are devoted to automotive service and information. This is an increasingly valuable resource.

## Associations

Associations have been mentioned several times throughout this chapter. There are many different automotive service associations, and at least one major association for each major automotive specialty area. These associations are extremely valuable to automotive technicians, especially those who are not associated with a manufacturer. They provide valuable support to their members, many times including training, service bulletins, service information, publications, conventions, technical networks, and hotlines.

## You as a Resource

What is your most valuable resource? More than all of the tools and reference material available, your most valuable resource is you. You must equip yourself to move through and analyze the amount of information presented to you. Many people mistakenly think that connecting a vehicle to a computer will reveal what is wrong. The truth of the matter is there are no magic wands. Your tools provide mechanical assistance or provide information but it is still up to you to decide which one is best to use and to decipher what your tools are trying to tell you. Because you are the most important resource, you owe it to yourself to take advantage of any training available. The more knowledge you obtain, the more vehicles you will be able to repair and ultimately the more money you can make and the more satisfied you will be with your career of choice.

> **You Should Know** *Make it a point early in your career to explore all of your diagnostic resources and determine what is available. This may include reading your service manuals, exploring your computer-based information system, or surfing the Internet. By learning what is available and how to access it when it is needed, you will greatly increase your productivity.*

## DIAGNOSTIC PROCESS

A diagnosis is the process that is used by a technician to locate the root cause of a customer's concern. To diagnose a vehicle, the technician must gather information from the customer, the vehicle, and various service information sources in order to begin the process of finding the root cause of a customer concern. To make the most accurate and efficient diagnosis possible, the skilled technician uses a diagnostic process that begins with the most simple ideas and solutions and incrementally increases in complexity. This is to help ensure that no simple problems are left unnoticed. A basic framework for a diagnostic process is provided below. The process can be customized to fit your own personal needs and diagnostic style.

1. Verify the customer's concern. All diagnostic procedures should begin by verifying what the customer has noticed. This should be accomplished using the information obtained from the customer and should include a thorough description of the problem, the conditions in which the problem occurs, and how often it occurs. Most engine performance concerns will require you to test drive the vehicle to verify the concern. In some cases, it may be necessary to test drive with the customer. When you are able to experience the problem for yourself, you should be able to identify two or three vehicle systems that are possible causes for the concern.

2. Having verified the customer's concern, you need to perform a visual inspection of the possible systems. In this step, you are specifically looking for obvious problems such as loose electrical connectors and vacuum leaks **(Figure 6)**. For engine performance concerns, a quick scan of vehicle data and computer-generated trouble codes can also be useful, especially if a "service vehicle" or malfunction indicator lamp is illuminated.

3. Once you have made sure that there are no obvious problems, you can begin to dig a little deeper into the diagnosis with the specific intent of isolating the concern to one system. In this step, you might want to refer to service bulletins and diagnostic symptoms guides found in the service manual. These will give you specific complaints and list the most common systems associated with them.

4. After you have isolated the concern, you will use specific diagnostic equipment and service information to locate the failure within the system. This process is often accomplished through the process of elimination by testing many components within the system. When this becomes necessary, start with the most likely component in the system that might cause the concern.

5. Once you have made a diagnosis, replace or repair the failed component.

6. When repairs are completed, the last step is to verify that the vehicle is fixed. This is accomplished by operating the vehicle under the same conditions during which the problem occurred originally.

## SYMPTOMS

Common engine performance symptom terminology is often used within the repair industry. Here are some common definitions of engine performance symptoms.

- *Engine runs rough* is used to describe any problem that causes the engine to have an obvious irregular vibration. Potential causes of this symptom are ignition system failures, fuel system failures, compression problems, or mechanical imbalance problems.

- *Bucking* is used to describe a condition in which a rapid change in rpm takes place, causing the vehicle to jerk when driven under specific conditions. This symptom can often be associated with ignition system problems or a malfunctioning transmission.

- *Surge* is a symptom that is similar in nature to bucking, in which drastic rpm changes are common. A surge, however, will usually be more linear in nature and the vehicle might act as if the accelerator is being depressed and released. Common failures associated with this symptom are vacuum leaks, lean fuel conditions, clogged exhaust, or malfunctions in the fuel delivery system.

- *Chuggle or fish bite* is a slight bump or jerking sensation that can often be felt in the steering wheel or seat. It is associated with the sensation felt while holding a fishing pole when a fish is nibbling on the fishhook. This symptom is often associated with an ignition system failure, malfunction in the **exhaust gas recirculation (EGR)** system, or a transmission concern.

- *Stalling* is any condition in which the engine dies. Stalling can occur at any time and can occur with or without warning. When stalling is preceded by some other symptom such as surging, the problem is likely in the fuel system. When stalling occurs without warning,

**Figure 6.**   The intent of the visual inspection is to identify obvious problems such as loose connections.

the problem is likely within the ignition or other computer or electronic system.

- *No-start* is a symptom in which the vehicle fails to start. Two descriptions fit the no-start category: the engine turns over but will not start, and the engine will not turn over. If the engine turns over but will not start, an ignition system, fuel delivery system, or engine compression problem exists. If the engine will not turn over, a problem in the primary starter or starter control circuit might exist or the engine or a driven accessory may be locked up.

## INTERMITTENT DIAGNOSIS

An **intermittent problem** is one that happens for a short period of time and then stops. From a diagnostic standpoint, there is no tougher problem to solve than an intermittent one; even the most experienced technicians

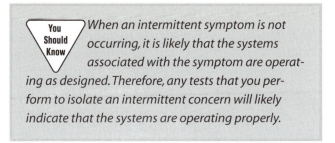

*When an intermittent symptom is not occurring, it is likely that the systems associated with the symptom are operating as designed. Therefore, any tests that you perform to isolate an intermittent concern will likely indicate that the systems are operating properly.*

have a difficult time finding the root cause of an intermittent problem. Intermittent problems do not follow distinct patterns, making them very difficult to verify. This makes the intermittent problem one of the greatest sources of frustration and dissatisfaction for technicians and customers alike. The use of a diagnostic worksheet similar to the one shown in **Figure 7** will get specific

---

Date ____/____/____

**DRIVEABILITY WORKSHEET**
(*information required for technical assistance*)

*VIN _____    *Mileage_____    R.O.# _____

*Model year _____  *Vehicle Model_____  *Engine ____

* Trans. Model _____        *Trans. Serial _____

CUSTOMER'S CONCERN

_____
_____
_____

**Check the symptoms that apply to your vehicle:**

Vehicle's "CHECK ENGINE" light
- ☐ Glows steady        ☐ Never comes on
- ☐ Glows intermittently

While operating the starter, my vehicle
- ☐ Will not crank      ☐ Cranks normally
- ☐ Cranks slowly

When starting my vehicle
- ☐ Will not start      ☐ Starts and dies
- ☐ Starts normally     ☐ Is difficult to start
                          - ☐ Hot
                          - ☐ Cold

While idling my vehicle
- ☐ Will not idle       ☐ Surges (up and down)
- ☐ Idles rough         ☐ Idles normally
- ☐ Backfires

While driving my vehicle
- ☐ Pings (spark knock)  ☐ Hesitates
- ☐ Backfires            ☐ Stalls
- ☐ Runs too hot         ☐ Stumbles
- ☐ Runs too cold        ☐ Surges
- ☐ Has a fuel odor      ☐ Vibrates
- ☐ Smokes excessively   ☐ Black ☐ Blue ☐ White
- ☐ Lacks power          ☐ Misses  ☐ Cuts out

**Other symptoms that apply to your vehicle:**

Transmission shifts  ☐ Too soon  ☐ Too late
- ☐ Emission test failed
- ☐ Poor fuel mileage

**Conditions of occurrence**

| Time: | Speed: | Distance: |
|-------|--------|-----------|
| ☐ Morning | ☐ Idle | ☐ Less than 2 miles |
| ☐ Midday | ☐ Low speed | ☐ From 2 to 10 miles |
| ☐ Evening | ☐ High speed | ☐ More than 10 miles |
| ☐ Night | ☐ Stop and go | Conditions: |
|  | ☐ Acceleration |  |
|  | ☐ Deceleration | ☐ Uphill |
|  | ☐ Highway (steady) | ☐ Downhill |

Frequency of occurrence
- ☐ Always          ☐ Since new
- ☐ Intermittently  ☐ After _____ miles

Engine conditions
- ☐ Cold engine       ☐ While braking
- ☐ Hot engine        ☐ While turning
- ☐ All temperatures  ☐ With A/C on
- ☐ When shifting     ☐ With headlights on

Fuel information
Type of fuel used _____
Octane rating ☐ 87 ☐ 89 ☐ 91 ☐ Greater than 91
Brand of fuel _____
Last fill-up date _____  Miles _____

---

**Figure 7.** A driveability worksheet can be a valuable tool when gathering specific information from the customer.

information about when a problem occurs. This can be invaluable when diagnosing an intermittent concern. As a driveability technician, you will likely encounter more intermittent problems than any other specialty technician. Any of the symptoms defined above can appear as intermittent concerns.

## Summary

- Information is the most important resource in the repair of the modern vehicle. Successful technicians will be fully aware of all of their resources and will know how to use them.
- Service manuals are written focused on one vehicle and will contain information on system operation, system diagnosis, and repair instructions.
- Service bulletins are written to address specific problems within a vehicle system or component. Vehicle manufacturers, parts manufacturers, and repair associations issue service bulletins.
- The labor time guide is used to help determine the cost for a specific repair. Common service materials are available in a computer-based format and can be supported by the vehicle manufacturer or an aftermarket provider.
- A technical assistance hotline might be able to provide technical advice to assist in the repair and diagnosis of difficult problems. Service associations provide a wealth of resources to their members.
- A well thought-out diagnostic process starts with relatively simple tasks and will progress in complexity until the problem is solved. An engine performance technician must be familiar with symptom definitions.
- Intermittent problems occur for brief periods of time and then disappear. This usually happens without the benefit of a distinct pattern.

## Review Questions

1. Describe the type of information found in each of the following resources: service manual, service bulletins, and labor time guide.
2. Explain why the computer is the preferred method of delivery for service information.
3. Describe each of the six steps in the diagnostic process.
4. Describe each of the six engine performance symptoms.

5. Which of the following statements about intermittent problems in general is true?
   A. Intermittent problems occur at regular intervals.
   B. The most difficult part of repairing an intermittent concern is being able to verify the concern.
   C. Most experienced technicians are able to locate intermittent problems very quickly.
   D. Very few symptoms can appear as intermittent concerns.

# Section 2

## General Engine Diagnosis

**Interesting Fact**

*Engine performance technicians must be able to diagnose mechanical engine problems as well as electrical and electronic problems.*

## SECTION OBJECTIVES

After you have read, studied, and practiced the contents of this section, you should be able to:

- Identify the engine by observing the VIN.
- Describe the different methods used to classify vehicle engines.
- Explain the basic scientific principles and theories that pertain to engine operation.
- Describe how atmospheric pressure affects engine operation.
- Describe energy conversion and how it is applied in the automobile engine.
- Describe the difference between torque and horsepower.
- Describe the function and operation of the components of an internal combustion engine.
- Describe the function of the support systems that are required to make the engine operate.
- Describe in detail the operation of the four-stroke cycle.
- Perform a cooling system pressure test and interpret results.
- Verify engine operating temperature and determine necessary action.
- Identify and interpret engine performance concerns as they relate to mechanical failures and determine necessary action.
- Inspect the engine assembly for oil and coolant leaks and determine necessary action.
- Diagnose abnormal exhaust color and determine necessary action.
- Perform engine manifold vacuum tests and determine necessary action.
- Perform a compression test and interpret results.
- Perform a cylinder leakage test and interpret results.
- Diagnose engine mechanical concerns using engine diagnostic equipment and determine necessary action.

# Chapter 4

# Engine Overview

## Introduction

The internal combustion engine is a complex machine that converts chemical energy into mechanical energy. This is accomplished through a controlled combustion process. The combustion process begins by supplying a precise mixture of fuel and air into the cylinders of the engine. Within the cylinders the mixture is compressed. When the mixture reaches a critical point of compression, the mixture is ignited by a source of ignition. The resulting explosion is ultimately transferred to the vehicle's transmission as a rotating mechanical force. This is accomplished through a series of pistons and levers.

As we first examine a vehicle's engine, it seems that the only requirement of the engine is to make the vehicle move when the accelerator is depressed; however, the requirements for a modern internal combustion engine go far beyond that. Modern engines must meet many different requirements. They must meet the performance demands of the customer; they must be reliable, operate with a minimal amount of noise, and be fuel-efficient. In addition to meeting the customer's expectations, modern engines must also conform to strict federal emissions standards and fuel economy guidelines.

As the demands placed upon the engine have increased, the engine and its support systems have become much more precise and complex. To diagnose driveability concerns that are often encountered in modern vehicles, you must first understand the interaction of the engine, its support systems, and the other powertrain systems. We will begin our study of engine performance by learning the basics of engine operation.

## ENGINE IDENTIFICATION

Knowing how to identify the particular engine you are working on is important because you must be able to positively identify your specific engine to obtain proper specifications and procedures for diagnosis, making repairs, obtaining replacement parts and locating labor times. The most accurate method of identifying the specific engine installed in a vehicle is by observing the VIN, a 17-digit number assigned to all vehicles. The VIN will contain specific information about that particular vehicle and can be found in the lower left-hand corner of the windshield **(Figure 1)**. Most vehicle manufacturers specify which engine is used in a particular vehicle by the eighth digit in the VIN. Each specific engine or engine variation is represented by a different number or letter in the eighth position. For example, a 1995 Buick Park Avenue might have a 3800-cc displacement engine. However, more than one 3800-cc engine was offered for that vehicle; one had a supercharger and one did not. Although both engine configurations displace 3800 cc, each one will have a different number or letter designation in the eighth position of the VIN. This information can be extremely critical when ordering parts for a vehicle. Although there are other means of finding out engine size, observing the eighth digit of the VIN will prove to be the most accurate.

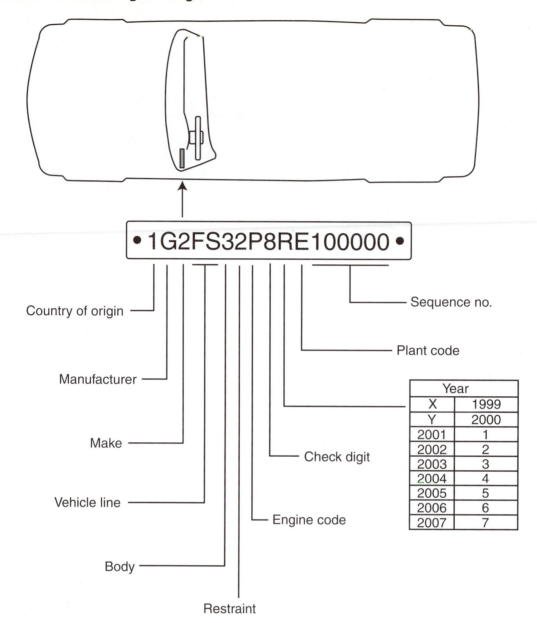

| Year | |
|---|---|
| X | 1999 |
| Y | 2000 |
| 2001 | 1 |
| 2002 | 2 |
| 2003 | 3 |
| 2004 | 4 |
| 2005 | 5 |
| 2006 | 6 |
| 2007 | 7 |

**Figure 1.**   Typical VIN sequence. The current identification system started in 1980 with the letter A and continued through 2000, ending with Y. Letters I, O, Q, U, and Z were not used. Beginning in 2001, numbers 1–9 were introduced. In the year 2010, the alphabet will be used again, starting with A.

## FUEL AND IGNITION TYPE

All engines share basic design and operational characteristics. The major difference in the engines currently being used is the method used to ignite the air and fuel mixture. The methods are either spark ignition or compression ignition.

Spark ignition engines are the most popular engines used in car and light truck applications. These engines require an external spark to start the combustion process. Gasoline is the primary fuel used in the spark engine. However, with some modification to the fuel supply system, the spark ignition engine can operate on liquefied petroleum gas (LPG), compressed natural gas (CNG), or some form of alcohol.

The compression ignition engine, also known as the diesel, has been an option in some automobile lines for many years and is a popular option in many light-duty trucks. Compression engines use the pressure and heat created when the air and fuel are compressed to start the combustion process. Compression ignition engines are the primary type of engine used in medium-duty and heavy-duty trucks as well as in off-road farm and construction equipment. The primary source of fuel used in the compression ignition engine is diesel. Compression engines can be modified to use LPG and CNG as well. This textbook focuses on the operation and diagnosis of the spark ignition engine.

## DISPLACEMENT AND CYLINDER ARRANGEMENT

A major factor in the performance of a vehicle is the size of the engine. This is called the **displacement**. Engine displacement is the sum volume of all of the engine's individual cylinders. A single cylinder's displacement is the volume of air that an individual cylinder can hold between bottom dead center (BDC) and top dead center (TDC) of its piston travel **(Figure 2)**. The total engine displacement is

the sum of all cylinder displacement and is expressed as cubic inches, cubic centimeters, or liters. Engine displacement is often noted on a label located in the engine compartment.

*Many vehicles currently on the road are capable of using more than one type of fuel. These vehicles, called variable fuel vehicles (VFVs) or flexible fuel vehicles (FFVs), may be equipped with this capability at the time of production or modified after delivery to use multiple fuels.*

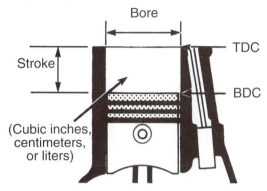

Bore² x Stroke x 0.7854 x Number of cylinders

**Figure 2.** Displacement is the volume of a cylinder between TDC and BDC.

The number of cylinders typically used in a vehicle engine ranges from three to twelve. However, most modern vehicles use four-, six-, eight-, or ten-cylinder engines. Although the number of cylinders in an engine is proportionate to the displacement, one must understand that the number of cylinders is not a pure representation of an engine's size. For example, an eight-cylinder engine can be nearly the same displacement as that of a ten-cylinder engine. In this case, the cylinders of the ten-cylinder engine are smaller than that of the eight.

The arrangement of the cylinders is the description of how the cylinders are arranged within the engine block **(Figure 3)**. Various cylinder arrangements have distinct

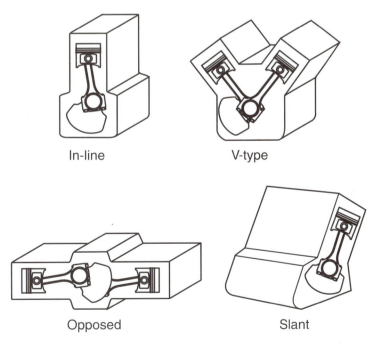

**Figure 3.** Various cylinder arrangements.

advantages and disadvantages. The various arrangements are:

- *In-line*. In-line engines are easily identified because each of the cylinders is placed upright in the cylinder block, and they are placed in a straight row. In-line engines typically house four or six cylinders, although the number of cylinders has ranged from three to eight. This design is common to all manufacturers and all vehicle types.
- *V-Type*. The cylinders in a V-type arrangement are placed in opposing rows and are placed at an angle of 60 or 90 degrees away from each other. Each row of cylinders is called a **bank**. V-type engines typically house six, eight, or ten cylinders. The number of cylinders has ranged from six to twelve and has been used by almost all manufacturers at some point in time.
- *Slant*. The slant engine is a combination of the in-line and the V-type engine. The slant design has all of the same characteristics of the in-line engine, with the exception that the cylinders are placed at a slant, similar to a single bank of a V-type engine. The slant engine was used extensively by Chrysler and Dodge in the 1970s and 1980s and can be found in both automobiles and trucks.
- *Horizontally opposed*. The horizontally opposed engine has two banks of cylinders, like that of the V-type engine; however, the banks are 180 degrees apart. This design makes the engine virtually flat, requiring very little engine compartment space. Four cylinders are commonplace for this design, but six-cylinder designs have been used. This type of engine will be familiar to most people as the design found in the Volkswagen beetle.

## CAMSHAFT AND VALVE LOCATIONS

Another critical difference in modern engine design is the placement of the valves and the camshafts. The following variations can be used with any cylinder arrangement.

- *Overhead valve (OHV)*. All engines used today use the OHV design. This simply means that the valves are located above the piston. However, this specific designation refers to the location of the camshaft. In the OHV design, the camshaft is located in the engine block, and the valves are actuated by a rocker arm and push rod **(Figure 4)**.
- *Single overhead cam (SOHC)*. The SOHC is an overhead valve design. In this engine, one camshaft per cylinder head is supported above the valves. The valves are actuated through the use of followers, tappets, or rocker arms **(Figure 5)**.
- *Dual overhead cam (DOHC)*. The DOHC is also an overhead valve design. This engine is similar to that of the SOHC except that each cylinder head has a separate intake and exhaust camshaft **(Figure 6)**.

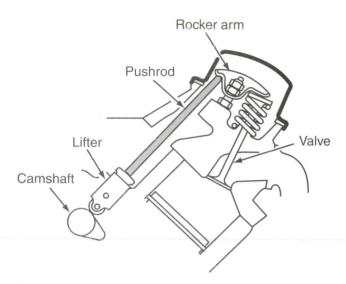

**Figure 4.**   A typical OHV configuration.

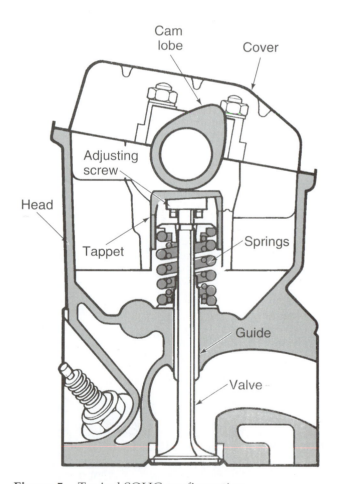

**Figure 5.**   Typical SOHC configuration.

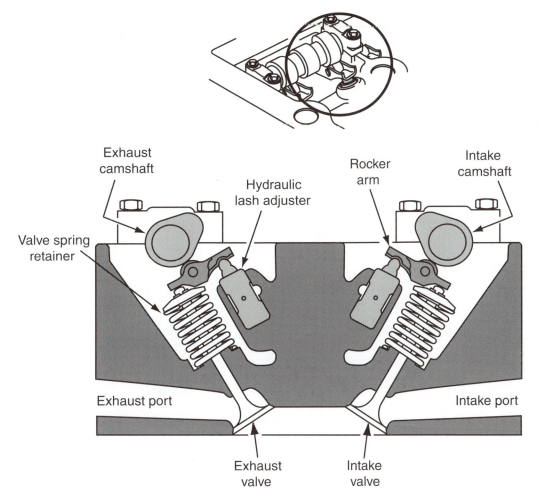

**Figure 6.** Typical DOHC configuration.

# ENGINE LOCATION AND POSITION

Automobile engines are found in a variety of locations and positions. Each engine location has specific advantages for the manufacturer, as does the position in which the engine is placed in a particular location **(Figure 7)**. The various locations for engine mounting are:

- *Front.* This design places the engine directly over the centerline of the front wheels.
- *Rear.* The rear-mounted engine is placed in the rear of the vehicle, placing the majority of the engine mass behind the centerline of the rear wheels.
- *Mid engine.* In this mounting position, the engine is mounted toward the rear of the vehicle and places the majority of the engine mass forward of the centerline of the rear wheels.

# MOUNTING POSITION

There are two positions in which to mount the engine in the vehicle. Both of these positions can be used in any of the three engine locations. The first position is longitudinal. This means that the engine is installed lengthwise in the chassis parallel to the frame rails. The longitudinal position is common to rear-wheel-drive vehicles but is also found in some older front-wheel-drive vehicles as well.

The second mounting position is known as transverse. In the transverse position, the engine is mounted in the chassis at a 90-degree angle to the vehicle centerline. Transverse-mounted engines are typically used in front-wheel-drive vehicles but have also been used in rear- and mid-mounting positions **(Figure 8)**. Transverse mounting requires the use of a special transmission called a **transaxle**. The transaxle combines the differential and transmission into one compact unit.

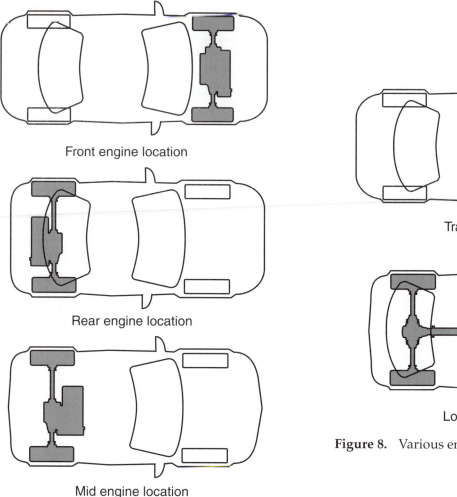

Front engine location

Rear engine location

Mid engine location

**Figure 7.**   Various engine locations.

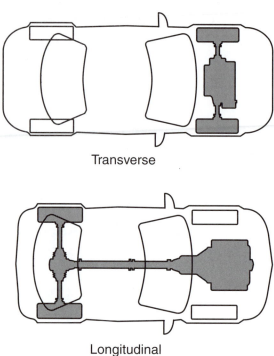

Transverse

Longitudinal

**Figure 8.**   Various engine installation positions.

## *Summary*

- The internal combustion engine converts chemical energy to mechanical energy through a controlled combustion process. Modern engines must meet expectations for performance, reliability, noise, and fuel efficiency.

- Positive engine identification is necessary to obtain correct specifications and procedures. The most accurate method of engine identification is obtained by referencing the eighth digit in the VIN.

- Internal combustion engines share basic designs and operational characteristics. Spark ignition and compression engines are common ignition methods.

- Displacement is a reference to an engine's size. Total displacement is the sum volume of all of the engine's cylinders. Displacement is expressed as cubic inches, cubic centimeters, or liters.

- Cylinders can be arranged in-line, in a V, at a slant, or horizontally opposed. Camshaft locations are OHV, SOHC, and DOHC.

- The engine can be located in the front of the vehicle, the rear of the vehicle, or toward the center. The engine can be installed longitudinally or transversely.

# *Review Questions*

1. What are the four requirements of a modern automobile engine?
2. Why does the VIN provide the most accurate form of engine identification?
3. The spark ignition engine does not need compression to operate.
   A. True
   B. False
4. All of the following statements about a V-6 engine may be true *except:*
   A. A V-6 engine may be an OHV design.
   B. A V-6 engine may be mounted transversely.
   C. A V-6 engine may be of a DOHC design.
   D. Cylinder banks are placed 180 degrees apart.

5. An engine is mounted toward the rear of a vehicle. Technician A says that if the mass of the engine is in front of the centerline of the rear wheels the engine is considered to be a mid-engine vehicle. Technician B says that if the mass of the engine is located behind the centerline of the rear wheels the engine is said to be a rear engine. Who is correct?
   A. Technician A
   B. Technician B
   C. Both A and B
   D. Neither A nor B

# Chapter 5

# Fundamental Engine Operating Principles

## Introduction

As we continue to examine engine operation, we must look at and understand several operating principles. You might have encountered many of these elements in a science or physics class without giving much thought to how these principles apply to an automobile engine. The principles that we examine in this chapter deal directly with the physics fundamental to the operation of an internal combustion engine. These fundamentals apply to every internal combustion engine, whether it is a two-cycle, a four-cycle, American, Asian, or European engine. Learning and understanding these principles is essential because these principles show us why an engine works. Once we understand why, we can then understand how it works.

## VOLUME

**Volume** is defined as the amount of space that is occupied by an object. In the case of an engine, we use a measurement of volume to define the amount of air and fuel that an engine's cylinders will hold. When discussing an engine, volume is represented by the engine displacement. Refer to Chapter 4 for more information.

## ATMOSPHERIC PRESSURE AND VACUUM

**Atmosphere** is defined as the air enveloping the earth. **Atmospheric pressure** is the weight of that air pressing upon the earth as well as everything on it **(Figure 1)**. At sea level, the atmospheric pressure exerts 14.7 pounds per square inch (psi) of force upon the earth.

Two things affect atmospheric pressure: altitude and temperature. As the landscape rises farther above sea level, the amount of atmosphere above the surface of the landscape decreases. Therefore, there is less air applying weight to the earth, resulting in lower atmospheric pressure.

*Interesting Fact*

*Many drag racers prefer to race at tracks with low elevations because the higher atmospheric pressure at lower elevations makes it easier for an engine to fill with air. The more air that can be put into the engine, the more fuel that can be burned, thus increasing horsepower and lowering elapsed times.*

Just as elevation affects pressure, so does temperature. As the air becomes warmer, it naturally expands. This reduces the number of air molecules in a given space, thus reducing the atmospheric pressure. As the air cools down, the air molecules contract. This increases the number of molecules in a given space, and the air becomes heavier, causing the atmospheric pressure to rise.

To understand the relationship that atmospheric pressure has with engine performance, an understanding of

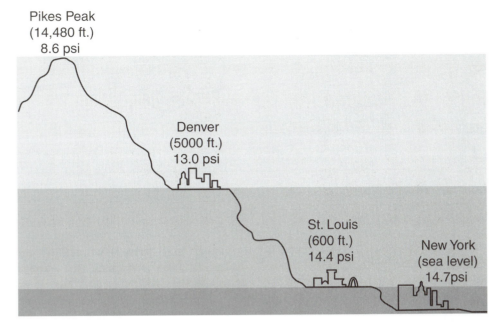

**Figure 1.** Atmospheric pressure changes as the altitude increases or decreases.

**vacuum** is necessary. For all practical purposes, vacuum is any pressure that is less than that of the atmosphere. As the pistons in the engine move down in their respective bores, a void is created above the piston. Within the void a low pressure area exists. Atmospheric pressure present outside of the engine pushes air in to fill the void left by the piston **(Figure 2)**. Because air is forced into the engine by atmospheric pressure, when pressure is high, the engine's cylinders fill more completely with air. This allows the engine to burn more fuel and produce more power. When pressure is low, the cylinders do not fill completely, the engine cannot burn as much fuel, and performance is reduced.

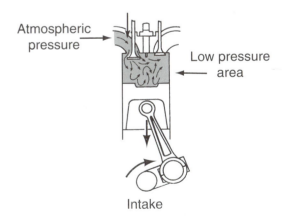

**Figure 2.** Atmospheric pressure rushes to fill the void that was left by the piston.

## LIQUIDS, GASES, AND COMBUSTION

We study the principles of liquids and gases to better understand how **combustion** makes an engine start and run. Gases and liquids follow the same basic principles; however, different materials will react differently.

- As the temperature of a gas or liquid rises, the gas or liquid will expand.
- As the temperature decreases, the gas or liquid will contract.
- Pressure and temperature are directly related. If pressure goes up, so does temperature; if pressure is reduced, so is the temperature.
- Gases can be compressed.
- Liquids cannot be compressed.

As gases and liquids are heated, their molecules will expand. If these materials expand in an open space, no pressure is formed. However, if we place these materials in a closed container, pressure will be formed inside the container.

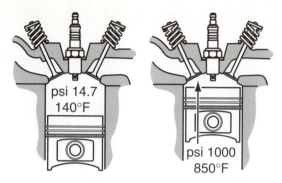

**Figure 3.** As the air/fuel mixture is compressed, the temperature increases.

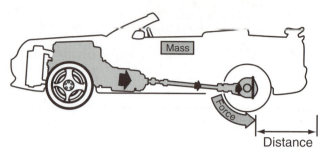

**Figure 5.** When force is applied to the wheels of a car, causing it to move some distance, work has been accomplished.

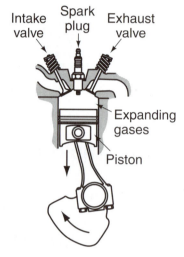

**Figure 4.** When the air/fuel mixture is ignited, the rapid expansion forces the piston down in its bore.

The expansion of gases is the basis for combustion in an engine. Inside an engine, we introduce fuel and air mixed together. We rely on the expansion of this air and fuel mixture to move the pistons and, in turn, spin the crankshaft. To further enhance this expansion, we do two things. The piston moves upward to tightly compress the air and fuel mixture. This forces the molecules closer together and raises the temperature of the mixture **(Figure 3)**. As the mixture is being compressed, we introduce spark. This ignites the air and fuel mixture and creates a massive explosion inside the cylinder. As the gases expand from the explosion, they exert a force on the piston, which forces the piston down in its bore **(Figure 4)**.

## FRICTION

**Friction** is the resistance to motion and occurs when two surfaces touch one another and move in different directions or at different speeds. Friction can be found in liquids, solids, and gases. As friction occurs, heat is produced, effectively changing mechanical energy into heat energy. For our discussions on the engine, our greatest concern is the friction between internal engine parts, such as pistons and cylinder walls, as well as crankshafts and bearings. Because we lose energy when friction is produced, an engine can be made to be more fuel-efficient by reducing internal friction.

## WORK AND FORCE

When we move an object, **work** has been done. Work is the result of applying **force** through a distance; force can be defined as the act of applying power to an object **(Figure 5)**. However, it is possible to apply force to an object without any work being done. For example, if we try to push a vehicle and it fails to move, we have applied a force but, because the vehicle did not move, no work was done. Inside an engine, we burn fuels to produce force. This force is then applied to the vehicle's transmission, and drive wheels to do work, which is accomplished when the vehicle moves. Work can be calculated using this formula:

$$\text{work} = \text{force} \times \text{distance}$$

## TORQUE

**Torque** is a force that is applied in a twisting motion. We measure torque in a unit called foot-pounds (ft.-lb) or Newton-meters (N•m). When we twist a bolt with a wrench, we apply force in a twisting motion, called torque. Torque can be calculated by the following formula:

$$\text{torque} = \text{force} \times \text{radius}$$

For example, if we apply 10 pounds of force to a wrench 1 foot in length, 10 ft.-lb of torque have been produced. If the same 10 pounds of force is applied to a wrench 2 feet in length, 20 ft.-lb of torque have been produced **(Figure 6)**. Torque is a common measurement of an engine's output.

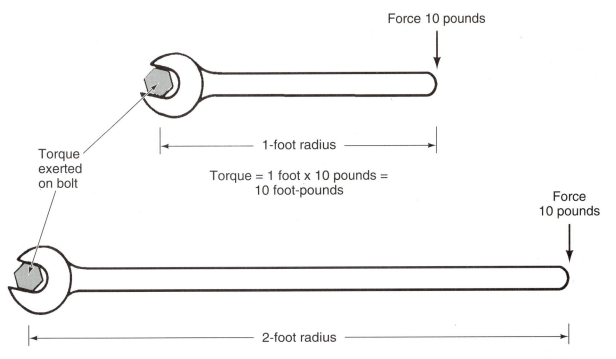

Force 10 pounds

Torque exerted on bolt

1-foot radius

Torque = 1 foot x 10 pounds =
10 foot-pounds

Force
10 pounds

2-foot radius

Torque = 2 feet x 10 pounds = 20 foot-pounds

**Figure 6.** Torque can be increased by increasing the length of the lever.

Engine torque is a measurement of how much force is available at the end of the crankshaft. We can measure engine torque with a dynamometer.

## POWER

**Power** is a calculation of the rate at which work is done. For the purpose of this text, we refer to horsepower as a measurement of engine output. Horsepower is a common unit of measure used to compare one engine with another. Horsepower is a calculation based on an engine's torque and the speed at which this torque is produced **(Figure 7)**. Once we know what the measured torque is and the engine speed at which it occurs, we can use the following formula to determine horsepower:

horsepower = (torque × engine speed)/5252

*Because the number 5252 is our mathematical constant, when we compare torque and horsepower we will find that the torque and horsepower curves will always cross at 5252 rpm.*

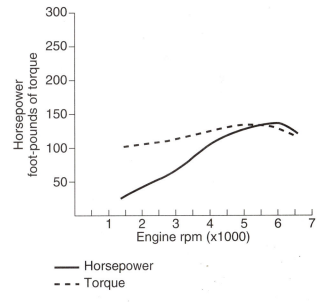

**Figure 7.** Horsepower and torque have a direct relationship.

## ENERGY

Energy is defined as the ability to do work and is available in six basic forms. Five of those are used within the automobile:

- **Chemical energy** is found in liquids or gases. Within a vehicle, chemical energy is fuel for the engine. Inside an engine, a chemical hydrocarbon fuel is combined with air and the mixture is compressed and ignited. By doing this, the intensity at which that energy is released is increased.

   Battery electrolyte is another chemical used in a vehicle. Electrolyte is a chemical used in the battery for production of electricity. The electrolyte reacts with other materials inside the battery to produce electricity.

- **Electrical energy** in a vehicle is produced either by a battery or by an alternator or something similar. Electrical energy is used to push electrons through an electrical circuit and is released into the vehicle's electrical circuits to power the many electrical devices found on the vehicle.

- **Mechanical energy** is defined as the ability to move objects. Mechanical energy is released any time an object is moved and is exemplified in many ways throughout the vehicle. Mechanical energy is applied to the drivetrain to cause vehicle movement and to drive the engine accessories.

- **Thermal energy** is produced by heat from the engine and is used to heat the interior of the vehicle.

- **Radiant energy** is produced by light. Although we do not harness the radiant energy produced by the vehicle, it does exist any time a light bulb is illuminated.

## ENERGY CONVERSION

**Energy conversion** occurs when one type of energy is changed to another type. In regard to vehicles, energy is seldom, if ever, used in its original state. The following are the most common forms of conversion found in a vehicle. Energy used within a vehicle can change from one form to another several times in a particular event **(Figure 8)**.

- *Electrical-to-mechanical* conversion occurs in the starter motor. Electrical energy is used to turn the electric motor, which then, in turn, spins the engine over to aid in starting. This type of conversion also occurs in window motors, windshield wiper motors, and various other electric motors that cause a mechanical movement.

- *Chemical-to-thermal* conversion occurs when an engine's fuel is ignited, as soon as ignition occurs and the fuel burns and changes to thermal energy.

- *Thermal-to-mechanical* conversion occurs as the burning fuel begins to expand. As it expands, it exerts a force on the pistons, which can then be used as a mechanical force to turn the crankshaft and eventually move the vehicle.

- *Mechanical-to-electrical* conversion takes place in the alternator. The mechanical movement of the engine's crankshaft spins the alternator and thus produces electricity.

- *Mechanical-to-thermal* energy occurs as a result of friction. This can best be illustrated by a brake system on a vehicle. The friction between the brake linings and drums or rotors is used to slow the vehicle. The by-product of this friction is heat. This form of conversion will be found anywhere friction is found.

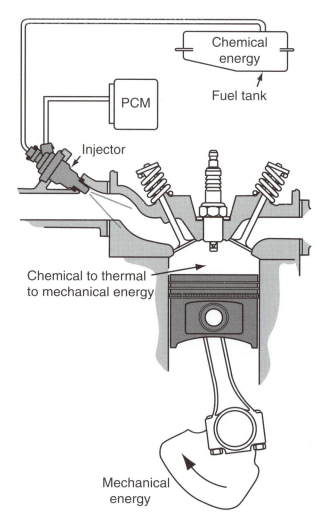

**Figure 8.**   Many forms of energy conversion must take place for an engine to operate.

# Summary

- Volume is the amount of space that is occupied by an object. Engine volume is expressed as displacement.
- Atmospheric pressure is the weight of the air pressing on the earth's surface. Atmospheric pressure changes in relation to altitude and temperature. Vacuum is the absence of atmospheric pressure.
- Gases and liquids follow the same basic principles. Pressure and volume are directly related to temperature.
- Force is the act of applying power. Work is accomplished whenever an object is moved as a result of applying force. Torque is a force that is applied in a twisting motion. Power is the rate at which work is done.
- Five forms of energy are used within a vehicle. Many forms of energy conversion take place within a vehicle.

# Review Questions

1. Only a small portion of an engine's displacement makes usable power at a given time.
   A. True
   B. False
2. Explain the effects that temperature and altitude have on atmospheric pressure.
3. All of the following statements concerning gases and liquids are correct *except:*
   A. Gases expand as temperature rises.
   B. Gases contract as temperature decreases.
   C. The pressure of a liquid in an open container increases as temperature is increased.
   D. The volume of a liquid in an open container increases as the temperature increases.
4. Explain how force, power, and work are used within an engine.
5. Which of the following energy types is used to make an engine start and run?
   A. Chemical
   B. Mechanical
   C. Electrical
   D. All of the above

# Chapter 6

# Engine Construction

## Introduction

Although an engine performance technician might not perform internal engine repairs, an understanding of the components is necessary to understand how everything works together to make the engine run. By learning how the engine is constructed, you will better understand how to apply what you have learned thus far. Furthermore, learning about engine construction will give you the needed insight on where all of the major engine components are located and how those components operate. By understanding the construction, you will be able to make more sense of many of the diagnostic procedures that you will use on a daily basis.

## CYLINDER BLOCK

The **cylinder block** is the main structure of the engine. The engine block can be compared to the foundation of a house. The inner cavities of the block provide the necessary means to support combustion. The outer surfaces of the block provide mounting locations for the accessory system components and transmission. Additionally, the block provides mounting provisions to attach the engine to the vehicle chassis **(Figure 1)**.

The engine block is a precision component that houses the engine's cylinders, passages to lubricate the internal moving components, and coolant passages to keep the entire engine within a specific operating temperature range. The cylinder is a hollow tube that is cast as part of the block and is referred to as the **cylinder bore**. The bore provides the space for the piston to be housed and to operate and for the combustion process to take place. The cylinder bores are precision machined and have a very

precise diameter. The top of the block, above the cylinders, is called the **deck**. The deck provides a perfectly flat surface onto which the cylinder heads will attach **(Figure 2)**.

## CRANKSHAFT

The crankshaft is centered lengthwise in the block, placed in a cradle called the **bearing saddles**, and retained by bearing caps **(Figure 3)**. The function of the crankshaft is to change the vertical motion of the pistons to rotational motion. The force that is exerted on the pistons is then changed to torque at the crankshaft **(Figure 4)**. This torque can then be directed into the vehicle's drivetrain and used to move the vehicle.

To better understand how the crankshaft converts this motion, we can think about a bicycle crankshaft. The crankshaft on a bicycle is centered below the seat in the frame. On each side of the shaft, offsets are attached 180 degrees apart. These offsets are called **throws**. In this example, the pedals are attached to the throws. In operation, each time one pedal is pushed down, the other pedal is forced in the opposite direction. This is the basis for the operation of the automotive crankshaft.

Each cylinder represented in the engine will have a throw. As the combustion forces applied to the tops of the pistons force the piston down in the cylinder, the connecting rod transfers this force to the crankshaft. As this happens, the crankshaft is forced to turn. The crankshaft throws in a crankshaft are separated at various angles from one another. Typical configurations are 90 degrees, 120 degrees, or 180 degrees, depending on the engine design **(Figure 5)**. In most engine configurations, the crankshaft throws will be arranged in such a manner so that two cylinders will be at top dead center at the same time, one will be at the end of the compression stroke, and the other will be at

1 PCV valve grommet
2 thermostat housing gasket
3 intake manifold gasket
4 intake manifold end seal
5 water pump gasket
6 front crankshaft seal
7 timing cover gasket
8 cylinder head gasket
9 oil pan gasket
10 valve cover gasket

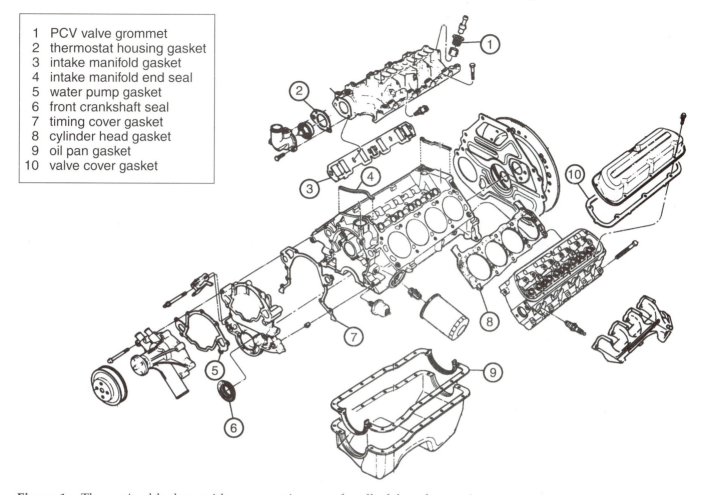

**Figure 1.** The engine block provides a mounting area for all of the other engine components.

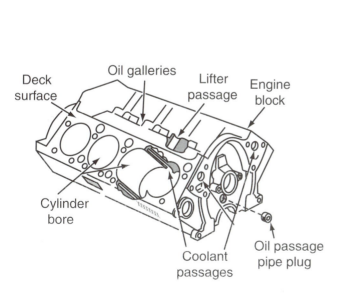

**Figure 2.** The engine block provides many precision machined passages, bores, and surfaces that ultimately support engine combustion.

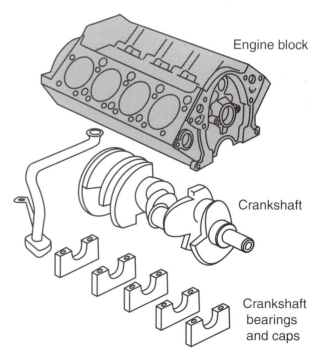

**Figure 3.** The crankshaft is located in the bottom of the engine block.

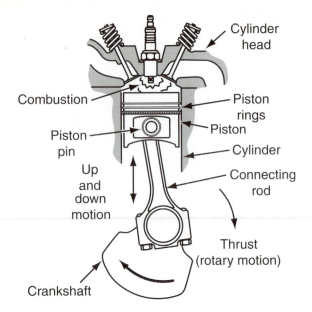

Cylinder head

Combustion

Piston rings

Piston pin

Piston

Cylinder

Up and down motion

Connecting rod

Crankshaft

Thrust (rotary motion)

**Figure 4.** As the piston is forced downward, the crankshaft is forced to turn.

the end of the exhaust stroke. This is done to evenly space the firing pulses of the engine for smooth operation. Crankshafts are made from cast iron or steel.

> **You Should Know** *In many V-type engines, it is common for two cylinders on opposite banks of the engine to share crankshaft throws.*

## Bearings

The crankshaft is supported and protected by insert-type bearings that separate the connecting rods from the crankshaft and the engine block from the crankshaft. Engine bearings are two-piece designs that look more like shims than bearings. A bearing set consists of a lower half that is fitted to the cap and an upper half that is fitted to the block or rod. The bearings receive pressurized oil through the oil galleries from the lubrication system.

> **Interesting Fact** *Casting is the most common type of piston construction; however, forged pistons are much stronger. Forged pistons should be used in performance applications where high engine speeds and/or high compression will be frequently experienced. Forged pistons should also be used anytime an external power enhancer, such as a supercharger, turbocharger, or large amounts of nitrous oxide, is used. Cast pistons can be used in these applications but can be subject to high failure rates due to the added stresses of increased cylinder pressures.*

## PISTONS AND CONNECTING RODS

**Pistons** are aluminum slugs that are fitted into each cylinder. The pistons move up and down in the cylinder. As the pistons move down, cylinder volume is increased and a vacuum is created within the cylinder. As the pistons move up, cylinder volume is decreased and pressure is built in the cylinder (**Figure 6**). In order for the cylinder to build vacuum and pressure, the piston must be tightly sealed within the cylinder bore; however, the pistons have to be free to move vertically within the cylinder. In order to have a tight

In-Line Four-Cylinder Crank Pin Arrangement

In-Line Six-Cylinder Crank Pin Arrangement

V-8 Crank Pin Arrangement

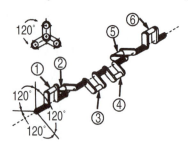

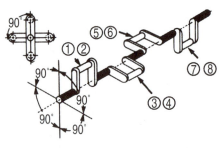

Rod journals are consecutively numbered from front to rear. Main journals are black.

A seven main bearing in-line six.

Five main bearing journals shown in black.

**Figure 5.** Crankshaft throws are offset 180 degrees in four-cylinder engines, 120 degrees in in-line six-cylinder engines, and 90 degrees in V-8 engines.

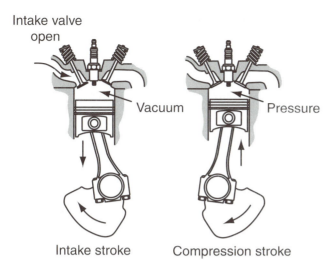

Intake valve open

Vacuum

Pressure

Intake stroke        Compression stroke

**Figure 6.** As the piston moves downward, volume is increased and a vacuum is created. As the piston moves upward, the cylinder volume decreases and pressure is created.

cylinder seal and maintain the necessary movement, several spring tension sealing rings located around the top of the piston are used. These rings allow the piston to move freely but they also exert enough force against the cylinder wall to tightly seal the piston to the bore **(Figure 7)**.

The pistons are connected to the crankshaft through the use of **connecting rods**. Connecting rods are metal beams that are connected to the piston by a press fit pin on one end and are bolted to the crankshaft on the other **(Figure 8)**. This effectively connects the pistons to the crankshaft. Thus, each time the crankshaft moves, the pistons will move accordingly. As combustion takes place above the piston, it forces the piston down in the bore, which in turn twists the crankshaft. As the crankshaft twists,

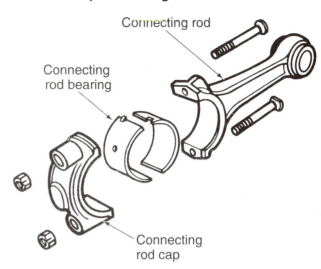

Connecting rod

Connecting rod bearing

Connecting rod cap

**Figure 8.** A typical connecting rod assembly.

the other pistons are forced to move as well. Some are forced up while others are pulled down.

## CYLINDER HEAD

The cylinder head is bolted to the cylinder block and covers the top of the cylinder bores. (Refer to Figure 1.) Directly above each cylinder, pockets are provided within the cylinder head for combustion to take place. These pockets are called **combustion chambers**. Poppet valves are located within each combustion chamber **(Figure 9)**. It is common for many modern engines to have two, three, four, and even five valves per cylinder. These additional valves allow the engine to breathe better.

Located behind the valves are the **ports**. The ports are the passages that connect the cylinders to the fuel and exhaust systems. The intake ports funnel the air and fuel

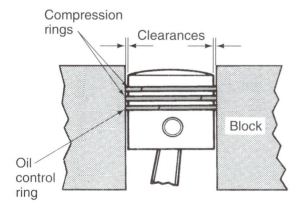

Compression rings

Clearances

Block

Oil control ring

**Figure 7.** Piston rings allow the piston to move within the cylinder bore and promote tight cylinder sealing.

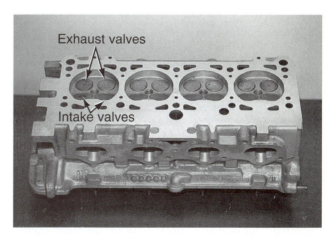

Exhaust valves

Intake valves

**Figure 9.** A typical multivalve cylinder head.

mixture from the intake system into the cylinders. The exhaust ports allow for passage of exhaust gases to be expelled into the exhaust system. The valves separate the ports from the combustion chamber and regulate the flow of gases entering and leaving the cylinders. The cylinder head also provides passages throughout the casting for coolant flow. In some overhead cam (OHC) heads, passages are also provided for the flow of pressurized oil for camshaft lubrication. Additional passages are present to allow exhaust gases to be recirculated into the engine **(Figure 10)**.

*Interesting Fact*

*It is commonplace in high-performance applications to open up the intake and exhaust ports to be able to move more gases. This process is called porting.*

## VALVE TRAIN

The **valve train** consists of all of the components used to open the valves. The camshaft actuates the valves. As discussed in Chapter 4, the camshaft can be located in different places within the engine: OHV, SOHC, and DOHC. Refer to Chapter 4 for more information on camshaft location.

Regardless of location, all camshafts must be timed to the crankshaft. This is called **cam timing** or **valve timing**. The cam is timed to the crankshaft so that the valve opening and closing events occur at the proper time in relation to piston position. OHV engines typically use a chain and a set of gears; however, gear drive mechanisms have been used in many OHV engines. SOHC and DOHC engines can use timing belts or chains to achieve proper cam timing **(Figure 11)**. Because the chains and belts must be typically very long in comparison to the OHV design, they are more prone to stretch than those in OHV designs.

To compensate for wear and stretching, a tensioner is used to keep the belt or chain tight. In addition to turning the camshaft, the timing belt or chain may also be used to

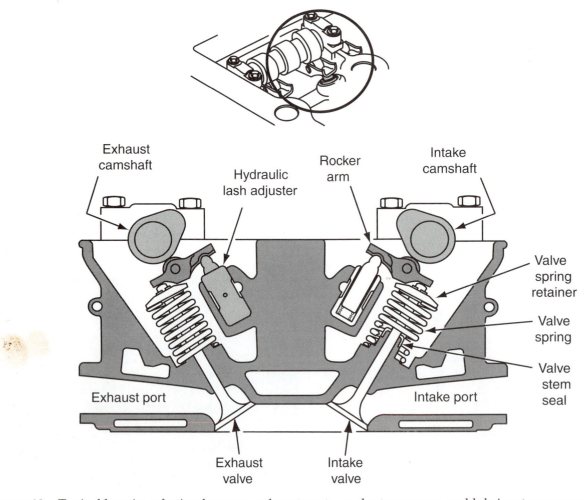

**Figure 10.**   Typical locations for intake ports, exhaust ports, coolant passages, and lubricant passages.

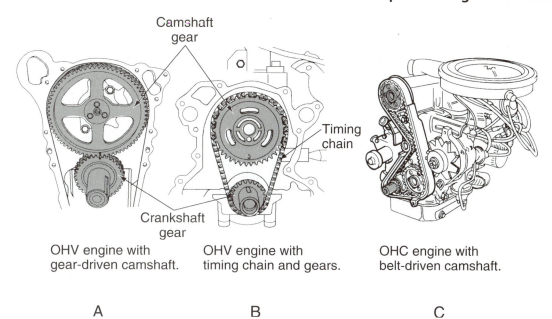

OHV engine with gear-driven camshaft.

OHV engine with timing chain and gears.

OHC engine with belt-driven camshaft.

A

B

C

**Figure 11.**   Common timing methods: (A) gear drive, (B) chain drive, and (C) belt drive.

**You Should Know**

*If the valves should ever become out of time with the crankshaft, the valves could make contact with a piston and bend (Figure 12). This would keep that particular cylinder from building compression. This can happen when a timing chain or belt breaks or the cogs are stripped from a belt or gear. However, many valves have been bent when a chain or belt was improperly reinstalled on an engine after service, so it is critical to double check timing marks when reinstalling timing components.*

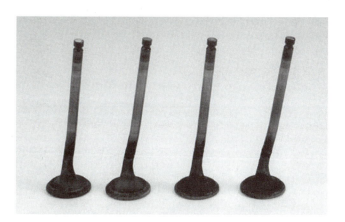

**Figure 12.**   Bent valves will often result when the camshaft and crankshaft are not properly timed.

operate the water pump in many designs. Timing arrangements in OHC engines are complicated due to the length of the belt or chain and the number of cams involved. Because of this, great precision is required when installing the timing components on one of these engines.

## BALANCE SHAFTS

Each time a cylinder fires, a vibration is created. Some engine designs inherently vibrate more than others. Some designs create enough vibration that it becomes an undesirable trait. To counter this vibration, some engines use a **balance shaft (Figure 13)**. Balance shafts are tuned to a specific engine model and create an equal and opposite vibration to that of the undesirable vibration. Balance shafts can be located in the oil pan or the engine block and are timed to both the camshaft and crankshaft so that the vibration that the balance shaft creates will occur at the right time.

## COVERS

All engines have covers of various designs and materials. These covers accomplish two major purposes: they keep coolant and lubricants in and dirt and water out. In applications in which the covers bear no stress or load, it will be stamped out of sheet metal or made from phonelic plastics. Some covers, however, add structural integrity to the engine assembly. In these situations, covers are made from heavy cast aluminum. These types of covers are used to provide strength and rigidity to the engine assembly.

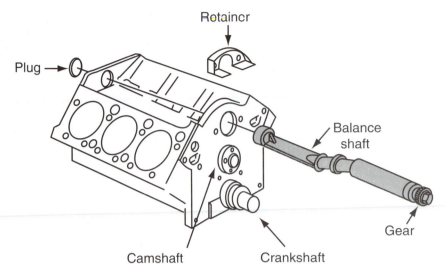

**Figure 13.** Typical in-block balance shaft installation.

## GASKETS

**Gaskets** are materials that are used when two components that hold a fluid or pressure are connected to one another. Gaskets can be made from many materials such as paper, rubber, metal, and composite fibers. Great care should be exercised when installing gaskets; leaks caused from improperly installed gaskets are a leading cause for repeat repairs and customer dissatisfaction.

## Summary

- The cylinder block is the main structure of the engine. The engine block provides necessary means to support combustion, lubrication, and cooling processes.
- The crankshaft converts vertical movement of the pistons to rotational movement. Forces created above the pistons are transferred to the crankshaft.
- Pistons move vertically within the cylinders. The pistons are connected to the crankshaft using connecting rods.
- Cylinder heads provide the means to bring air and fuel into the engine and expel exhaust gas out of the engine. Combustion chambers are located within the cylinder head. Valves are located within the combustion chambers and separate the ports from the cylinders.
- Camshafts have various locations within the engine. The camshaft(s) must be timed to the crankshaft.
- Balance shafts are installed to counteract normal engine vibrations. Balance shafts must be timed to the crankshaft and camshaft.

## Review Questions

1. Two technicians are discussing the cylinder block. Technician A says that the cylinders of the block are subject to a great deal of heat and pressure. Technician B says that excessive wear within the cylinder bore will cause a decrease in efficiency. Who is correct?
   A. Technician A
   B. Technician B
   C. Both A and B
   D. Neither A nor B

2. All of the following statements about a crankshaft are true *except*:
   A. The crankshaft changes vertical motion to a twisting motion.
   B. Automobile crankshafts and bicycle crankshafts share some of the same operating principles.
   C. Consecutive crankshaft throws are located at 360-degree angles from one another.
   D. Position of the crankshaft throws can have an effect on how smoothly an engine operates.

3. Pistons must seal tightly within the cylinders to create pressure and vacuum.
   A. True
   B. False
4. Two technicians are discussing cylinder heads. Technician A says that cylinder heads used in an OHV engine have passages built in the head for camshaft lubrication. Technician B says that if the intake valve is leaking, compression pressure could be forced into the intake port. Who is correct?
   A. Technician A
   B. Technician B
   C. Both A and B
   D. Neither A nor B

5. Explain why the camshaft, crankshafts, and balance shaft must be timed to one another.
6. Explain what the results might be if the valve timing on an engine were incorrect.

# Chapter 7

# Support Systems

## Introduction

For any engine to start and run, there are several systems that do not directly aid the combustion process but are essential to the operation of the engine. Although these systems might not appear to affect engine performance, we will find that each could cause engine driveability concerns, so being familiar with the operation of these systems is essential to the engine performance technician. These systems include the starting, lubrication, and cooling systems, and in this chapter, we examine what they are comprised of. In Chapter 9 we examine some common diagnostic procedures for testing these systems.

## BATTERY

The battery is the heart of the vehicle electrical system as well as the starting system. The battery supplies the necessary voltage and amperage to make the starter operate. Batteries are composed of groups of positive and negative plates. The plates are contained in a plastic case and the case is filled with electrolyte **(Figure 1)**. The electrolyte solution reacts with the plates to produce voltage. This voltage can then be used in the vehicle's electrical systems. A faulty battery can directly affect the operation of all electrical engine management functions. The following must be considered when choosing a replacement battery:

1. *Battery size and post location.* Battery size refers to the physical size of the battery. A replacement battery should have the same physical dimensions of the original equipment battery.

   Post location refers to the position of the posts in which the battery cables are connected. The post on

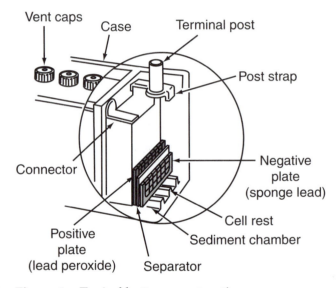

**Figure 1.** Typical battery construction.

a replacement battery should be in the same position and the same style as the original equipment battery.

2. *Cold cranking amps.* Cold cranking amps is a figure that tells you how many amps the battery can deliver at zero degrees for 30 seconds and still maintain 7.2V. A good rule to use is to install the battery with the most cranking amps that will properly fit in the space allotted.

3. *Reserve capacity.* Reserve capacity is how many minutes that a fully charged battery will supply 25A without dropping below 10.5V. This gives us an indication of battery life in the event of a charging system failure.

4. *Ampere hour rating.* This rating indicates the exact amount of current a particular battery will deliver for

20 hours without dropping below 10.5V. This rating is particularly useful for fleet maintenance personnel because many fleet vehicles have high-powered radios and other parasitic accessories that stay on continuously.

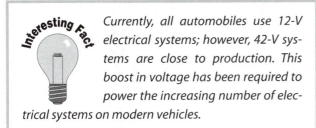

*Interesting Fact*

*Currently, all automobiles use 12-V electrical systems; however, 42-V systems are close to production. This boost in voltage has been required to power the increasing number of electrical systems on modern vehicles.*

## STARTING SYSTEM

The starter motor is a **direct current (DC)** motor that is fastened to the engine block or bell housing. The armature of the starter is equipped with a movable drive gear. The purpose of this drive gear is to engage the starter to the flywheel and spin the engine **(Figure 2)**. The starter is equipped with a starter solenoid or starter relay. Both the relay and the solenoid deliver high current from the battery to the starter. The solenoid differs from a relay in that not only does it deliver a high current connection from the battery to the motor but it also provides the mechanical action that moves the drive gear to engage the flywheel **(Figure 3)**. The starter solenoid is mounted to the starter, whereas a relay will usually be located close to the battery **(Figure 4)**. Starters that use a relay will have the mechanism for moving the drive within the starter motor assembly.

When the driver turns the key, the ignition switch will supply current to the solenoid or relay **(Figure 5)**. This current will energize an electromagnet within the solenoid or relay that, in turn, activates a set of high-amperage electrical

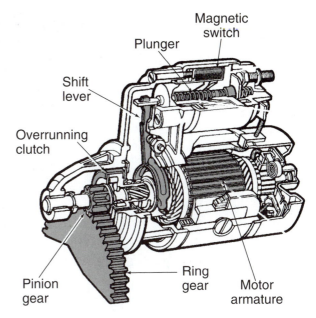

**Figure 3.** Starter motor construction.

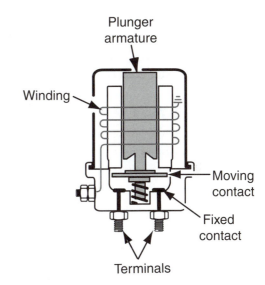

**Figure 4.** A remote mounted starter relay.

contacts. The contacts then deliver the necessary amperage to the starter. This may be done through a direct connection with a solenoid or a cable in models that use a relay. In solenoid-equipped models, a movable core located inside the center of the solenoid is connected to a lever that, in turn, thrusts the drive gear into the **flywheel** at the same time that the contacts are closed.

The flywheel is a large steel plate that is bolted to the rear of the crankshaft. Teeth are located around the perimeter of the flywheel. The starter drive gear engages these teeth to spin the engine. When the starter drive engages the flywheel, it spins the crankshaft. Once the engine starts and begins to spin faster than the starter motor, the drive gear

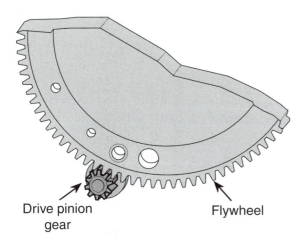

**Figure 2.** Starter drive gear–to–flywheel engagement.

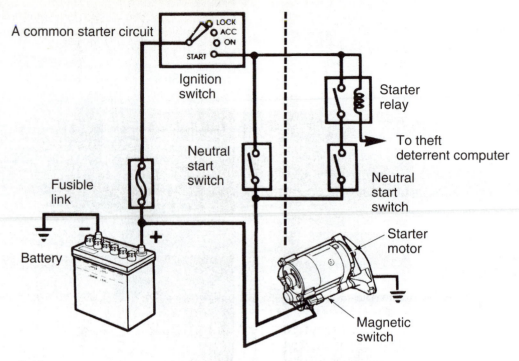

**Figure 5.**   A typical starter circuit.

clutch begins overrunning until the starter is disengaged. This prevents damage to the starter.

## LUBRICATION SYSTEM

The lubrication system is designed to provide a sufficient amount of clean pressurized oil to the moving components of the engine **(Figure 6)**. In addition to lubrication, the oil is used to remove heat as well as debris from the various engine components.

The oil pump is a gear-type pump that draws oil from the storage sump in the oil pan, pressurizes it, and sends it through various passages in the engine block and cylinder heads to lubricate the moving parts as well as remove heat

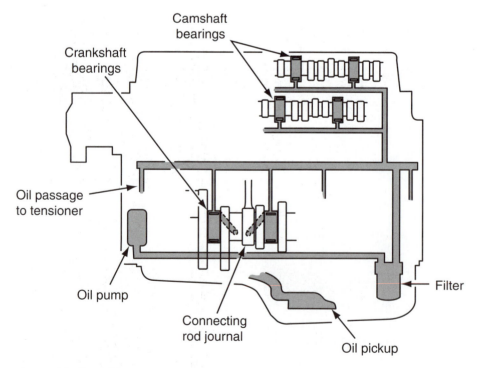

**Figure 6.**   Oil flow in a typical engine.

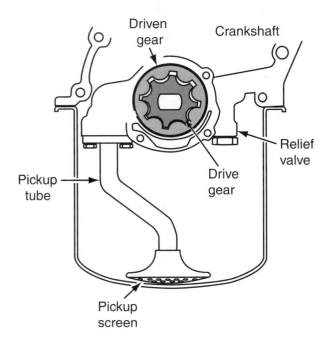

**Figure 7.** Inside a gear-driven oil pump.

and debris in the process. A gear-type pump is shown in **Figure 7**. The oil pump may be located either at the front of the engine behind the harmonic balancer or inside the oil pan. The oil pump uses a mesh screen, called a **pickup**

**screen**, attached to the inlet side of the pump. The pickup screen extends to the bottom of the oil pan and allows the pump to pull oil into the oil pump. The screen protects the pump from any large particles that might be in the oil. When the oil leaves the pump, it is passed through a filter designed to remove smaller particles from the engine oil. After the oil is cleaned, it is sent to the various passages within the engine that lubricate the crankshaft, pistons, camshaft, and other valvetrain components. The oil is then returned to the oil pan where the process starts over.

## COOLING SYSTEM

As the engine operates, the combustion process generates a large amount of heat. Combustion temperatures can reach as high as 4500°F (2468°C). We depend on the cooling system to remove most of this heat. There are two primary types of cooling systems: air cooled and liquid cooled. However, liquid-cooled engines are the most common systems.

The liquid-cooled engine uses a coolant solution that is circulated throughout coolant passages that are cast in the engine block and cylinder heads. As the coolant is circulated, heat is transferred from the engine components to the coolant and subsequently carried out of the engine to the radiator, where it is cooled and eventually recirculated back into the engine. Coolant flow through a typical engine is shown in **Figure 8**.

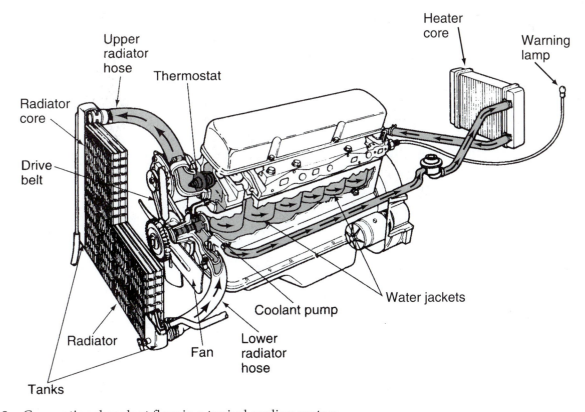

**Figure 8.** Conventional coolant flow in a typical cooling system.

# COOLANT

In a liquid-filled cooling system, we depend on a liquid solution to transfer heat away from the metal components within the engine. Water is a good chemical for transferring heat; however, when used alone, pure water promotes corrosion, quickly becomes acidic, and will freeze at 32°F (0°C). Because of these negative effects, water is mixed with antifreeze to form a solution that is 50 percent water and 50 percent antifreeze. Antifreeze is a chemical solution that has a very low freezing point and has special additives that protect the engine from corrosion and acidity. Additionally, antifreeze adds lubricants and conditioners to protect seals and gaskets.

> **You Should Know** *Some manufacturers are using long-life engine coolants designed to last for 5 years or 150,000 m. This coolant is orange in color and should not be mixed with regular coolant. If the two are mixed, the long-life coolant will lose much of the extended-life properties.*

# WATER PUMP

The water pump is responsible for moving coolant throughout the engine. The water pump uses a belt-driven impeller to circulate the engine coolant (**Figure 9**). The water pump can be driven by an accessory drive belt (**Figure 10**) or driven by the timing belt or chain (**Figure 11**). There are two methods in which coolant is circulated

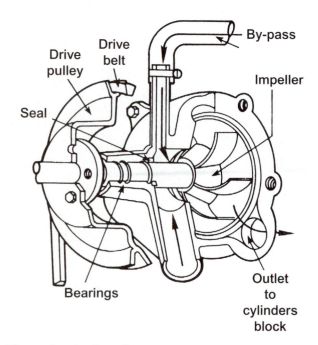

**Figure 9.**  An impeller-type water pump.

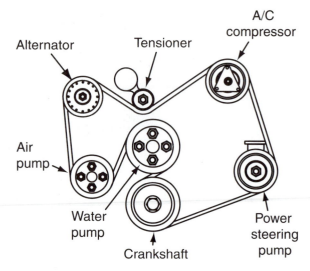

**Figure 10.**  The water pump is driven by an accessory drive belt.

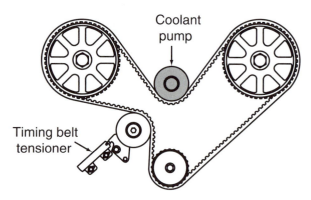

**Figure 11.**  The water pump can also be driven by the timing belt or chain.

through the engine: conventional flow and reverse flow. In the conventional method, coolant is circulated from the radiator and into the block, to the cylinder heads, and eventually to the inlet side of the radiator. (Refer to Figure 8.) Reverse-flow cooling systems draw coolant from the radiator outlet and push it through the cylinder heads first and down through the block and back to the radiator (**Figure 12**). Because most of the heat in the engine is generated in the cylinder heads and the top of the engine block, it is a definite advantage to cool these areas first.

>  **Interesting Fact** *Although used engine coolant might maintain proper freeze protection levels, its chemical additives that prevent corrosion and electrolysis might have dissipated to dangerously low levels.*

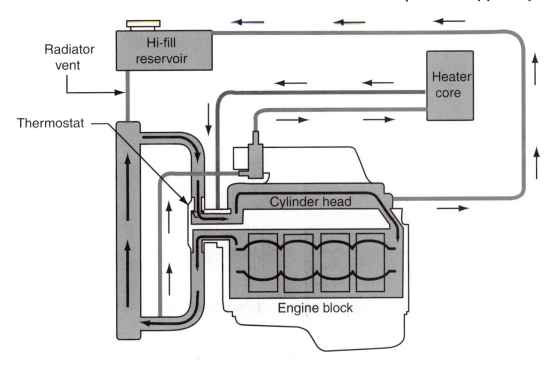

**Figure 12.** A reverse-flow cooling system with a surge tank.

## THERMOSTAT

The thermostat controls the flow of coolant within the engine. Thermostats are basically heat-sensitive flow control valves. Each thermostat is calibrated to open at a specific temperature. As the coolant temperature in the engine reaches the preset temperature, the thermostat begins to open and allow coolant to flow **(Figure 13)**. The temperature at which the thermostat begins to open is the method

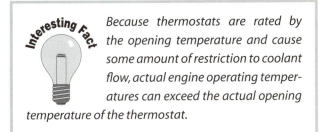

*Because thermostats are rated by the opening temperature and cause some amount of restriction to coolant flow, actual engine operating temperatures can exceed the actual opening temperature of the thermostat.*

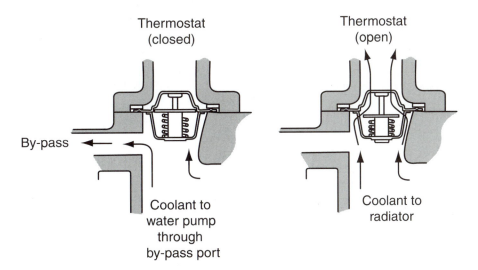

**Figure 13.** Coolant flow through the thermostat.

by which thermostats are rated. Many different temperature ranges are available, but only one meeting the manufacturer's recommended temperature should be installed. The thermostat temperature rating is stamped into the thermostat. Because all engine and powertrain decisions are made partially based on coolant temperature, the thermostat is the most critical cooling system component from an engine performance perspective.

> **You Should Know** *Aftermarket computer programs are often calibrated to use a lower temperature thermostat; this is the only time that installing a lower temperature thermostat should ever be considered. If a lower temperature thermostat is otherwise installed, the vehicle's emission compliance will be violated and the vehicle most likely will fail emission inspection.*

## RADIATOR

The radiator is responsible for removing heat from the coolant. A radiator, a heat exchanger that is mounted at the front of the vehicle, is comprised of many horizontal or vertical tubes placed about ½ inch apart. The tubes are joined at each end into common tanks or reservoirs. Placed between the external surfaces of the tubes are hundreds of very thin fins. The coolant pump circulates coolant through the radiator tanks and into the tubes; the heat is transferred from the coolant to the tubes and then to the fins. As air passes across the radiator tubes and fins, the heat is then transferred to the surrounding air, and the temperature of the coolant is lowered (**Figure 14**).

## OVERFLOW AND RESERVE SYSTEM

As engine coolant warms up, it expands. To keep cooling system pressure from rising above an acceptable level, there is an overflow system. The overflow system consists of the radiator pressure cap and an overflow bottle.

The radiator pressure cap fits on the radiator fill neck and performs three functions: it allows the cooling system to build pressure, allows the system to purge excess coolant from the radiator as it expands, and allows reserve coolant to be drawn back into the radiator, to maintain a consistent radiator fluid level. These functions are accomplished through the use of a calibrated pressure valve and vacuum valve located in the radiator cap (**Figure 15**). As the coolant expands, pressure builds up. When the pressure exceeds the calibrated pressure of the cap, usually 14 psi to 16 psi, coolant is exhausted through a small tube on the fill neck

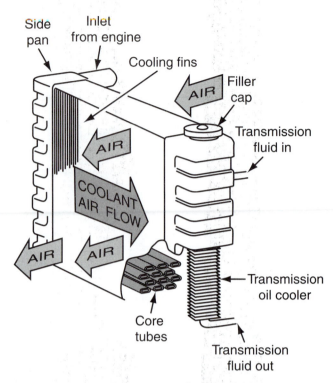

**Figure 14.**   Radiator coolant flow and heat transfer.

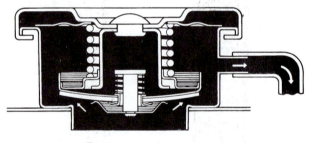

Pressure valve

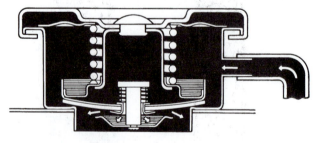

Vacuum valve

**Figure 15.**   Radiator cap pressure and vacuum valve operation.

into the overflow bottle (**Figure 16**). As the coolant cools down, it contracts and creates a vacuum within the radiator. This opens the vacuum valve and draws coolant in from the overflow bottle. The action of these valves also keeps the system from drawing in air.

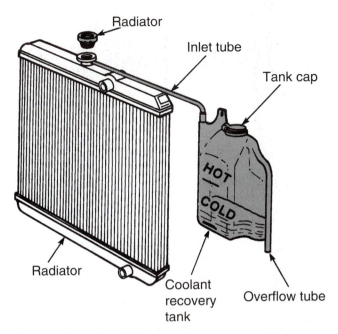

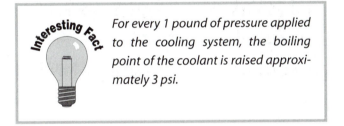

**Figure 16.** A typical coolant recovery system.

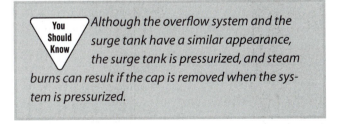

*For every 1 pound of pressure applied to the cooling system, the boiling point of the coolant is raised approximately 3 psi.*

**Interesting Fact**

Many newer vehicles are using a surge tank system. This system is similar to the overflow system. However, the tank is a pressurized part of the system and may have coolant flow through the tank. (Refer to Figure 12.) In this system, the radiator does not have a fill neck and the pressure cap is placed on the tank. The tank carries a relatively low level of coolant. As coolant expands, the cap will relieve excess pressure, but due to the low level in the tank, no coolant should escape. As the system cools, the vacuum valve in the cap will open to relieve the vacuum.

**You Should Know** *Although the overflow system and the surge tank have a similar appearance, the surge tank is pressurized, and steam burns can result if the cap is removed when the system is pressurized.*

## COOLING FANS

Cooling fans are required to pull air across the radiator when the vehicle is stopped or at low operating speeds.

There are two methods used to draw air through the radiator: a belt-driven fan and an electric cooling fan.

The belt-driven fan has been around since the automobile was invented. The fan is mounted to the water pump shaft and driven by the crankshaft and turns any time that the crankshaft turns. Fan blades come in a variety of different sizes and blade counts. The blade operates within an enclosure called a **fan shroud**, which increases blade efficiency by sealing the area around the fan. This makes sure that all of the air that is drawn in by the fan is pulled through the radiator. The belt-driven fan has seen few changes since the introduction of the viscous fan clutch and is still used on many sport utility vehicles and light trucks.

The clutch allows the fan to spin at a different speed from that of the engine. This is called slip. Changes in slippage are required based on the amount of air required for radiator cooling. The temperature of the air flowing across the clutch affects the amount of clutch slippage. Fan clutches commonly use a silicone fluid to control the amount of slippage.

The fan mounts to the clutch hub. Inside the clutch hub, a shaft that spins at engine speed is mounted with a bearing. The shaft is allowed to spin separately from the hub. The amount of slippage is controlled with the silicone fluid. As the fluid heats up, expansion occurs. As it expands, the fluid forces the hub to lock to the shaft, causing the fan to spin at an rpm close to that of the water pump **(Figure 17)**. The speed difference between the hub and the shaft is infinitely variable, depending on air temperature. A heat-sensitive coil can be used to help control the action of the silicone fluid.

Recently, the traditional viscous fan clutch has given way to a computer-controlled fan clutch on some new

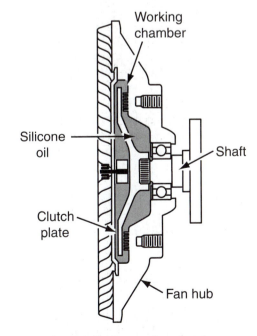

**Figure 17.** A typical viscous fan clutch.

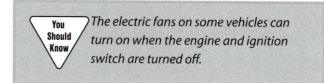

*Extreme caution should be exercised when working around the belt system and the fans. The engine should be turned off when performing any service work to the engine or accessories. The moving fan or belts can cause severe personal injury if your clothing or your body gets trapped in the fan or belts.*

vehicles. This assembly uses an electric solenoid in place of the bimetallic spring. In this configuration, the computer monitors fan speed to determine the fan's efficiency.

Vehicles that do not use belt-driven fans use one or more electric motor–driven fans. The electric cooling fan is driven by an electric motor and is controlled by the computer using a relay. Fans are located directly at the front of the radiator or on the back side of the radiator. Those fans that are on the front of the radiator push air through the radiator and are called pusher fans. Fans that are located on the back side of the radiator pull air through the radiator and are called puller fans **(Figure 18)**.

The electric cooling fans are computer-controlled through the use of a relay. The computer monitors the engine temperature as well as other parameters such as air conditioner request and vehicle speed to determine

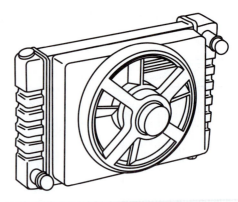

**Figure 18.**   A puller-type electric cooling fan.

proper cooling fan operation. When engine temperature reaches approximately 220–230°F (105–111°C), the PCM will supply a ground to the cooling fan relay to command the cooling fans to turn on. When the temperature drops back down to approximately 210°F (100°C), the fan will turn back off.

*The electric fans on some vehicles can turn on when the engine and ignition switch are turned off.*

# Summary

- The battery is critical to the operation of the starter system and other electrical systems. The starter system uses a DC motor to turn the engine to start the vehicle.
- In addition to providing battery current, the starter solenoid will also provide mechanical action to engage the starter drive. The drive gear clutch prevents starter damage by overrunning when the engine starts.
- The lubrication system cleans and distributes lubricant throughout the engine. The oil pump may be located on the front of the engine or inside the oil pan.
- Engine coolant is a combination of water and anti-freeze. The water pump circulates coolant throughout

the engine and the radiator. The radiator transfers heat from the coolant to the atmosphere.
- The thermostat operates as a temperature-sensitive control valve. Thermostats are rated by the temperature at which they begin to open.
- The valves within the radiator cap serve to keep the system full of coolant and keep air bubbles out. The overflow and reserve system serve to keep the radiator full.
- Cooling fans are used to move air across the radiator. Cooling fans can be mechanically or electrically driven. The engine management computer controls electric cooling fan operation.

# Review Questions

1. As long as the cables will connect, any battery may be installed in any vehicle.
   A. True
   B. False
2. Explain the operation and differences of a starter solenoid and starter relay.

3. All of the statements about the lubrication system are true *except:*
   A. The oil pump may be located on the front of the engine.
   B. The lubrication system cleans the engine oil.
   C. The lubricant is used to remove heat from the engine.
   D. The pickup screen removes small debris from the oil before it enters the pump outlet.

4. Which of the following statements about a cooling system is correct?
   A. The thermostat controls the flow of coolant.
   B. The operation of the thermostat can influence the amount of fuel that an engine uses.
   C. The radiator cap helps keep the radiator full of coolant.
   D. All of the above

5. Describe how a radiator removes heat from the engine coolant.

6. Which of the following statements concerning the cooling fan is true?
   A. The fan shroud increases the efficiency of the cooling fan.
   B. The clutch-type mechanical fan can operate at an infinite number of speeds.
   C. The electric fan can operate when both the engine and the ignition are turned off.
   D. All of the above

# Chapter 8

# Engine Operation

## Introduction

Now that we have established some basic knowledge about engine design, location, principles, and construction, we can now put all of these pieces in place to see how the engine functions. In order for the engine to work, we must combine a fuel source with air, place it inside a sealed chamber, tightly compress the mixture, and then at the correct time ignite that mixture. As the mixture burns, it expands, pushing down on the piston and turning the crankshaft.

### PISTON STROKE

We depend on the movement of the pistons and the valves to create an environment inside of the engine that will support combustion. Each time a piston moves from TDC to BDC or vice versa, we can say that the piston has completed a **stroke**, or **cycle (Figure 1)**. There are two designs of a piston engine: the four-stroke engine and the two-stroke engine. In the four-stroke cycle, it takes four complete strokes of the piston to produce one power stroke. The other three strokes do not make any power but are necessary to have an efficient power stroke. The two-stroke cycle engine requires only two strokes of the piston to produce one power stroke. The two-cycle engine operates by combining the primary functions of the intake, power, and exhaust strokes into a single stroke. The compression stroke not only compresses the air/fuel mixture, but also aids in the evacuation of exhaust gases from the cylinder and the introduction of the air/fuel mixture into the crankcase. Almost all automobile engines in use today are of the four-stroke design.

Piston at BDC
(bottom dead center)

Piston at TDC
(top dead center)

**Figure 1.** A piston's stroke can be measured as the distance the piston travels between BDC and TDC.

### CAM TIMING

Camshaft timing is the relationship of the valve opening and closing events in relation to piston position. The camshaft lobes are manufactured to give the camshaft very specific dimensions and characteristics **(Figure 2)**. These include:

- **Lift.** The distance that the valve is lifted off the seat. Cam lift is usually multiplied by the rocker arm ratio.
- **Duration.** The amount of time expressed in degrees of crankshaft rotation that the valves are held open.
- **Overlap.** The amount of time expressed in degrees of crankshaft rotation that the intake and exhaust valves are opened at the same time.

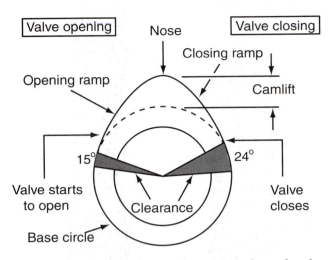

**Figure 2.**  The lobe design determines how far the valve will open and for how long.

The camshaft is timed to the crankshaft by lining up specific marks on the cam or its gears and the crankshaft. The two are connected by a belt, chain, or gears and rotate together. This ensures that they will stay in time. Some facts to remember as we begin to examine each stroke of the engine are:

- Every time the crankshaft completes one rotation, a given piston has completed two full strokes.
- Each time the crankshaft completes one rotation, the camshaft will turn only 180 degrees.

## FOUR-STROKE OPERATION

As we have just learned, one piston must move through four strokes to produce power for one stroke. The four required strokes are intake, compression, power, and exhaust. In the following example, we take a detailed look at what happens during each stroke. For this example, we observe the camshaft timing specifications as they are presented in **Figure 3**. This is a generic cam-timing example; each engine model will have its own set of cam-timing specifications.

## Intake

In this example, we begin with the piston at 21 degrees *before* the piston reaches top dead center (BTDC) on the end of the exhaust stroke. At this point, the exhaust valve has been open for quite some time. When the piston reaches 21 degrees BTDC, the intake valve is opened **(Figure 4)**. This is done because the exiting exhaust gases will create a siphoning effect and will help start the flow of the incoming air/fuel mixture. As the piston reaches TDC of the exhaust stroke, both the intake and exhaust valves remain open. When the piston starts back down, the intake stroke begins. The exhaust valve remains open until 15

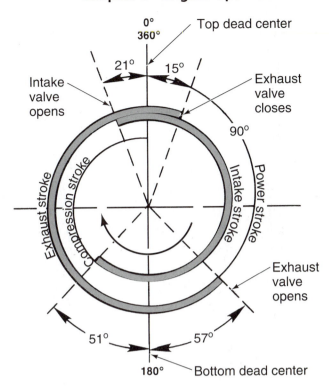

**Figure 3.**  This diagram indicates the valve timing in relation to the position of the piston.

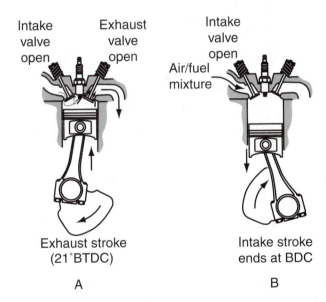

**Figure 4.**  (A) The intake valve opens before the exhaust stroke has ended. (B) The intake valve remains open for the duration of the intake stroke.

degrees *after* top dead center (ATDC) on the intake stroke. This allows a maximum amount of exhaust gases to exit the combustion cylinder. When the piston reaches BDC, the intake stroke ends (Figure 4).

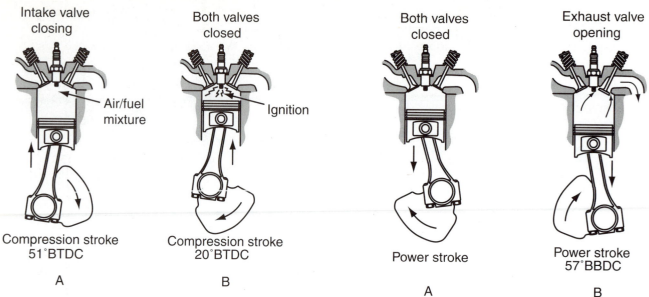

**Figure 5.** (A) The intake valve remains open for the first portion of the compression stroke. (B) Both valves are closed and ignition is introduced near the end of the compression stroke.

**Figure 6.** (A) Both valves are closed until the piston nears the end of the power stroke. (B) The exhaust valve opens just before the end of the power stroke.

## Compression

The compression stroke begins with the piston at bottom BDC. In this example, the intake valve is still open and will remain open until the piston reaches 51 degrees *after* bottom dead center (ABDC) **(Figure 5)**. Because the column of incoming intake air has a given amount of momentum, the cylinder will continue to fill even after the piston has started to move upward in the cylinder. After the intake valve is closed, both valves remain closed until the piston is well into the power stroke. At some point between 15 and 50 degrees BTDC, on the compression stroke, a spark is introduced into the combustion chamber to ignite the air/fuel mixture (Figure 5). It is necessary to ignite the mixture BTDC so the fuel will have sufficient time to burn. Once combustion has begun, the piston continues to move upward until it reaches TDC when the compression stroke ends.

## Power

The force of the combustion forces the piston back down to begin the power stroke **(Figure 6)**. The power stroke begins with both valves still closed and the piston at TDC. Both valves remain closed during the power stroke until the piston reaches 57 degrees *before* bottom dead center (BBDC) (Figure 6). By this time, most of the energy acting on the piston has been used and the exhaust valve is opened. Opening the valve at this point allows the pressure that remains in the cylinder to establish

exhaust flow from the cylinder to clear out all of the exhaust gases. The power stroke ends when the piston reaches BDC.

## Exhaust

The exhaust stroke begins with the piston at BDC after the power stroke. The exhaust valve is opened and the piston is forced upward by the crankshaft **(Figure 7)**. The piston moving upward in the cylinder forces the exhaust gases

**Figure 7.** The exhaust valve is open for the entire exhaust stroke and the process starts over.

out, past the open valve. The piston continues upward in the cylinder until it reaches 21 degrees BTDC, and the entire process is repeated. (Refer to Figure 4.)

## STRATIFIED CHARGE

These engines differ from the regular four-stroke engines by the method in which the combustion process starts. The stratified charge engine uses a spark plug inserted into a small precombustion chamber that is interconnected to the normal chamber. A small rich mixture is introduced into the precombustion chamber at the same time a lean mixture is introduced into the normal combustion chamber. The mixture in the precombustion chamber is ignited by the spark plug. The smaller explosion in the precombustion chamber then ignites the lean air/fuel mixture in the normal chamber **(Figure 8)**. Overall, this allows these engines to operate with leaner

*Camshaft upgrades are one of the most common high-performance modifications that performance enthusiasts undertake. Every engine built has a camshaft with very specific specifications to meet very specific needs for power, emissions, and fuel economy. It is critical that a camshaft be matched to the engine in which it will be installed and its intended usage considered. If you make an uneducated selection, replacing the camshaft can decrease engine performance.*

air/fuel ratios that burn faster and more completely. This results in better fuel economy and fewer emissions.

## MILLER CYCLE

The Miller cycle is another variation of the four-stroke cycle. Normally aspirated engines depend on atmospheric pressure to force the air/fuel mixture into the cylinders. The Miller cycle engine relies on **forced induction** to fill the engine's cylinders. Forced induction is when a mechanical pump such as a supercharger or turbocharger is used to force the air/fuel mixture into the engine.

In the Miller cycle engine, the intake valve is held open for a longer period of time during the compression stroke. Because the air/fuel mixture is forced into the engine by mechanical means rather than by relying on the pressure from the atmosphere, the cylinders continue to fill after the piston has moved a significant distance toward TDC. If the valves were held open in a normally aspirated engine, the pressure of the piston would quickly overcome the pressure of the atmosphere and would begin to push the air/fuel mixture back out of the cylinder.

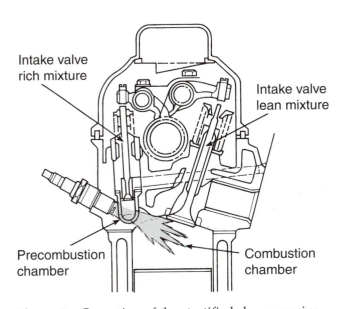

**Figure 8.** Operation of the stratified charge engine.

## *Summary*

- A stroke, or cycle, is completed when the piston moves from TDC to BDC or vice versa. In a four-cycle engine, one piston must make four complete cycles to produce one power stroke.
- Camshaft timing is the relationship of the valve opening and closing events to the position of the piston. The camshaft is timed to the crankshaft.
- Intake, compression, power, and exhaust are the four strokes required to make a four-cycle engine run. By

having both the intake and exhaust valves opened at specific times, the engine will operate more efficiently and create more power.
- The stratified charge engine uses a small rich combustion process to ignite a leaner air/fuel mixture in the cylinder. The Miller cycle engine depends on forced induction and longer intake valve duration for operation.

# Review Questions

1. Which of the statements about a four-stroke engine is true?
   A. A stroke is complete when a piston moves from TDC to BDC.
   B. Two engine events are combined into one stroke.
   C. A stroke is complete when a piston moves from BDC to TDC.
   D. Both A and C

2. Explain in detail the operation of the intake, compression, power, and exhaust strokes of a four-stroke engine.

3. When discussing the operation of a four-stroke engine, Technician A says that opening the intake valve before the exhaust stroke has ended will help draw in the incoming air/fuel mixture. Technician B says that opening the exhaust valve before the power stroke has completed will help to evacuate the cylinder. Who is correct?
   A. Technician A
   B. Technician B
   C. Both A and B
   D. Neither A nor B

4. The stratified charge engine has fewer emissions and better fuel economy because:
   A. The design of the engine allows for the use of a leaner air/fuel ratio.
   B. It is a two-stroke engine.
   C. The precombustion chamber uses a lean fuel mixture.
   D. Spark plugs are no longer required.

5. A naturally aspirated engine depends only on atmospheric pressure to fill the cylinders with air.
   A. True
   B. False

# Chapter 9

# Support System Diagnosis

## Introduction

This chapter focuses on the basic diagnostic procedures for the essential support systems. To receive certification as an engine performance specialist, you will be required to have a significant amount of knowledge in the diagnosis of these support systems.

### BATTERY DIAGNOSIS

The battery is the heart of all vehicle electrical systems. Any deficiency in the strength of the battery or integrity of its connections will eventually lead to an electrical problem in one or more systems. Because the engine management systems are heavily integrated by electronics, the condition of the battery is a concern for the engine performance technician.

A quick analysis of the battery's state of charge can be made while attempting to start the vehicle by observing the sound of the starter and the brightness of the interior lights. While the starter is engaged, the interior lights should initially dim slightly but remain relatively bright, and the starter should spin over smoothly and evenly. If the lights dim a great deal and the starter sounds as if it is struggling to turn the engine over, a problem with the battery or starter exists. Load test the battery if the battery state is ever suspected.

The condition of the battery cables and terminals is of particular concern to the engine performance technician. Loose or corroded battery terminals consume a significant amount of voltage. This is called a **voltage drop**. Excessive voltage drops in any electrical circuit can cause a variety of electrical and engine management system concerns. Many vehicle manufacturers attach major accessory battery supply cables and system grounds directly to the battery cables or terminals. Because of this, when cables become corroded, the likelihood of a driveability concern is greatly increased.

Top-post battery cables can be readily inspected by observing the amount of corrosion buildup on the terminal **(Figure 1)**. Side-post mountings are more difficult to inspect because the terminals are covered. Although these terminals are covered, they can still become severely corroded in the event that the battery terminals leak. If terminals are severely corroded, they should be cleaned or replaced as needed. Technicians should be aware that

**Figure 1.** Corroded battery cables can be a source of many electrical and electronic system problems.

replacement, bolt-on terminals are not sufficient for use in the modern vehicle. If battery terminals are damaged beyond service, the entire cable should be replaced with the proper replacement cable with proper ground and accessory power feed provisions.

## STARTER SYSTEM DIAGNOSIS

Many problems that will prohibit the engine from starting can be directly attributed to the starter system. These concerns can range from an engine that will not turn over to one that turns over slowly. These problems can also be caused by faults in the electrical system of the vehicle. The charts in **Figure 2** and **Figure 3** will provide you with some diagnostic procedures that will successfully pinpoint the cause of many starter system concerns.

## COOLING SYSTEM DIAGNOSIS

Diagnosis of any cooling system concern should begin with a thorough visual inspection. You should begin by checking the level of the coolant in both the radiator and the coolant reservoir. The coolant level should remain fairly stable. If the reservoir is empty or the level in the radiator is low, a leak might be indicated. If the fluid level is low, it should be filled to the proper level at this time.

Once fluid level is checked and corrected, a visual inspection of all hose connections and gasket mating areas should be performed. **Figure 4** illustrates many typical leak locations. When inspecting these areas, you should look for any noticeable trails that begin at a coolant hose or gasket area. These trails will appear as streaks and run toward the ground. Areas that have been leaking for a long time can also have mineral

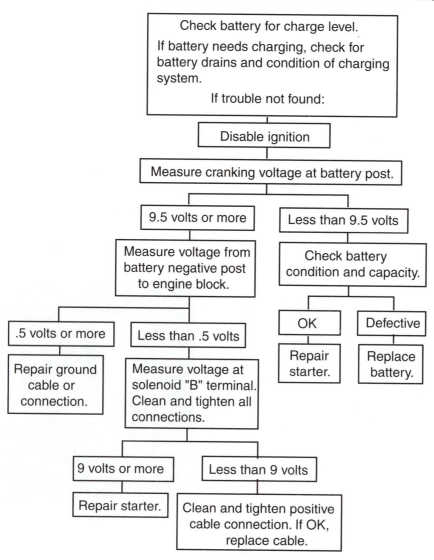

SLOW CRANKING, SOLENOID CLICKS OR CHATTERS

**Figure 2.**  A diagnosis chart for slow cranking, clicking, or chattering.

ENGINE WILL NOT TURN OVER

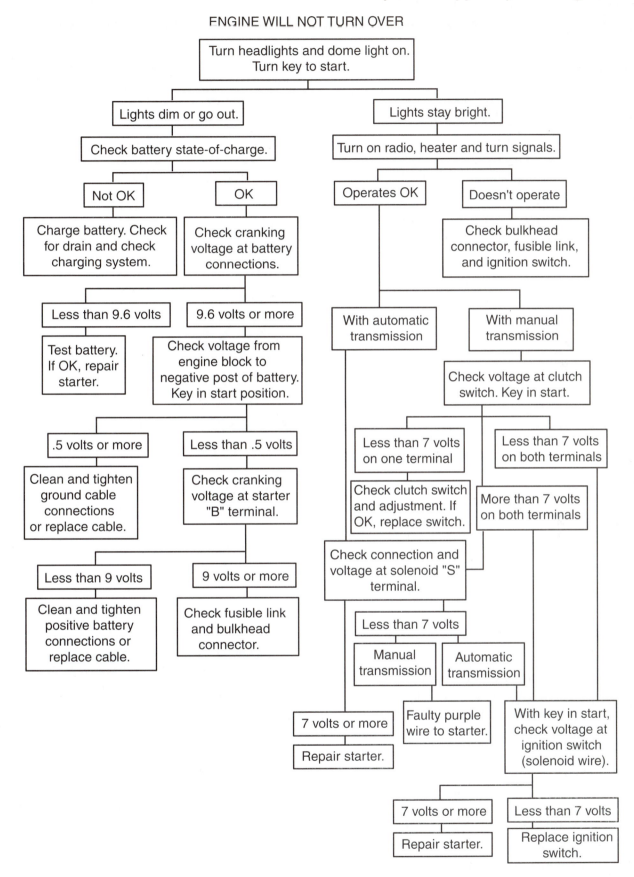

**Figure 3.** A diagnosis chart for an engine that will not turn over.

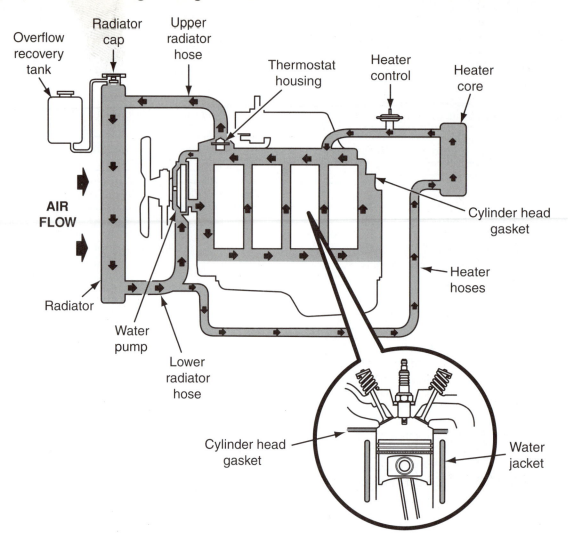

**Figure 4.**   Typical coolant leak locations.

buildup at or near the origin of the leak **(Figure 5)**. If a coolant trail is dry, it might indicate a previous leak that has been repaired or a leak that has temporarily sealed off. If a leak is readily evident, it should be repaired at this time.

> **You Should Know** *Removing the radiator cap from a pressurized cooling system will cause coolant to violently erupt from the radiator, exposing you and anyone in the area to the risk of severe burns. Extreme caution should be exercised when working with a pressurized cooling system. It is best to wait until the system has cooled off and pressure has been relieved before removing any cooling system component. Residual pressure can be determined by squeezing the upper radiator hose; if the hose is firm, pressure still exists.*

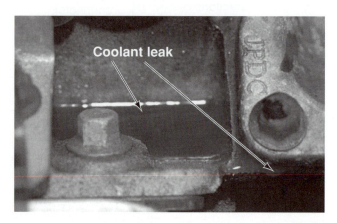

**Figure 5.**   Coolant leaks will often leave white mineral stains.

As you continue your visual inspection, check the fan and shroud for condition and placement. The fan shroud should be properly positioned and fastened securely. Any holes or gaps within the shroud or its mounting position will decrease the efficiency of the fan and can eventually lead to engine overheating. Observe the condition of the fan, making sure that all the blades are intact and secure. Make sure that the fan blade is free of any debris such as mud or grass. If the fan is belt-driven, grasp the fan and observe the amount of free play that is present between the fan clutch and drive hub. Any excessive movement will require clutch replacement.

*Most fan blades used today are made from plastic. If mud or other debris clings to the fan, one or more blades can break off. The missing blade will cause a severe imbalance and a subsequent vibration. Often the vibration is severe enough that it can be felt within the passenger compartment and can often be mistaken for an engine misfire condition. Continued use of a damaged fan will additionally lead to subsequent damage of the water pump or electric fan motor.*

Many vehicles today have an air dam or deflector placed directly below the radiator. As vehicle designs have become more aerodynamic, the grill opening on the vehicle has decreased, thus reducing the available airflow to the radiator. The deflectors are installed to direct air, which would normally pass beneath the vehicle, up into the radiator area. Often the deflectors hang low and are vulnerable to road damage. As part of a visual inspection, make sure that any air deflection devices are present, in good condition, and securely fastened.

## LEAK DIAGNOSIS

There are two types of leaks that you will encounter as a technician: external leaks and internal leaks. External leaks are those that leak outside of the engine. These leaks will lead to loss of coolant and eventually cause engine overheating. They are usually easily detected using simple methods.

Internal leaks are those in which the coolant leaks into the engine and often mixes with the engine lubricant. These types of leaks are often more difficult to locate. Depending on the nature of the internal leak, driveability symptoms such as engine misfire or white smoke from the exhaust system might appear. Internal leaks that are left unchecked will lead to dilution of engine lubricant, and severe engine damage will occur.

*Some seals that are exposed to engine coolant will expand with heat and contract as they cool off. This makes diagnosis of some leaks very difficult on a warm engine. When diagnosing a warm engine, where a leak is not readily evident, allow the system to cool off while pressurized. This will allow the seals to contract and expose the source of the leak.*

## External Leaks

The customer might recognize the external leak as a puddle in the garage or a loss of coolant in the reservoir. Many times a coolant leak can be obvious; such is the case when a radiator hose ruptures. Smoke, engine overheating, and large puddles of coolant usually accompany this type of leak. Less-evident cooling system leaks might require a more detailed set of diagnostic steps. Further leak diagnosis will often include manually pressurizing the cooling system or adding dye solution.

The pressure test is performed by applying pressure to the cooling system with a specially equipped hand pump **(Figure 6)**. The hand pump attaches in place of the radiator cap. Pressure is then applied to the system equal to that of the relief pressure of the radiator cap. Pressurizing the system will force coolant to leak from small pinholes or hose clamps that have loosened slightly. For those coolant leaks that are very small in nature, a dye solution can be added to the cooling system. The engine is run, the coolant is circulated, and then the system is examined with an ultraviolet light. If the dye has penetrated a leak, it will show under the light as a fluorescent yellow or green.

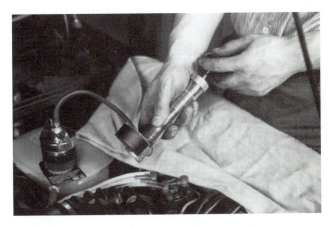

**Figure 6.**   Applying pressure to a cooling system using a hand pump. This tester can also be used to test the radiator cap.

## Internal Leaks

If coolant loss is persistent and no external leaks are evident, an internal leak might be present. Internal leaks are often more difficult to locate and can require some engine disassembly to pinpoint. Internal engine leaks can occur at a number of sources, such as the intake manifold gaskets on V-type engines, the intake plenum or gasket, head gaskets, or timing covers on some models. Internal leaks may be characterized by a milky substance on the back side of the oil filler cap or **positive crankcase ventilation (PCV)** valve and/or excessive amounts of moisture in the fresh-air inlet tubes.

A blown head gasket is one of the most common sources of an internal coolant leak. This is often accompanied by poor engine performance and plumes of white smoke from the exhaust. Other symptoms include a rapid buildup of pressure in the cooling system and excessive wisps of smoke from the radiator. Two practical tests used to locate a blown head gasket are a chemical combustion gas detector and a four- or five-gas analyzer.

**Figure 7.** Testing for a combustion leak with a chemical tester.

The chemical tester involves a tube with a suction bulb that holds a chemical **(Figure 7)**. The tube is placed over the radiator inlet with the engine running. A suction bulb is used to pull gases from the radiator up through the chemical. If the chemical changes color, combustion gas is present in the cooling system, indicating a head gasket leak or in some cases a cracked cylinder head.

The other testing method involves measuring the hydrocarbon content of the radiator gases with a gas analyzer. With the engine running, wave the probe of the analyzer across the radiator filler, allowing the probe to draw in the gases. If hydrocarbons are present in these gases, a blown head gasket or cracked head is indicated. This procedure is illustrated in **Figure 8**.

If these two tests prove inconclusive and an internal leak is still suspected, disassembly of the engine will be required. Other common points of entry for engine coolant are leaking intake manifold gaskets or a timing cover that has been compromised by a loose timing chain that is wearing a hole in the cover.

## ENGINE OVERHEATING

Engine overheating is often caused by low coolant level, and repair of the leakage will usually fix these concerns; however, other concerns can cause an overheating condition. These problems will be directly related to airflow across the radiator or coolant flow within the system. Typically, these problems will have unique signatures that can help point you in the right direction for an efficient diagnosis. Described below are the most common concerns and the specific symptoms of their occurrence.

If the vehicle overheats while it is stopped and idling but cools down when the vehicle begins to move, the concern is related to airflow across the radiator. Another clue to this concern is that the A/C does not cool well at idle but works normally when the vehicle is moving. These conditions usually stem from an inoperative cooling fan.

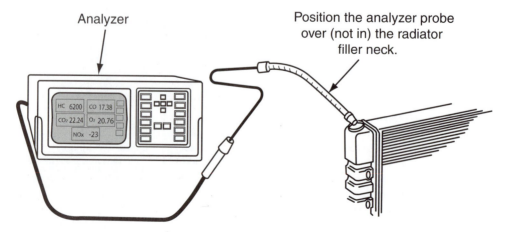

Analyzer

Position the analyzer probe over (not in) the radiator filler neck.

| HC | 6200 | CO | 17.38 |
| CO₂ | 22.24 | O₂ | 20.76 |
| NOx | -23 | | |

**Figure 8.** An exhaust gas analyzer can be used to check for combustion gases within the radiator.

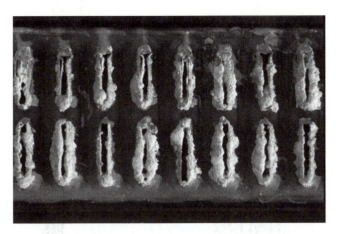

**Figure 9.** Corroded radiator tubes will restrict coolant flow.

If the engine overheats after driving for several miles, the problem is associated with a radiator that has a restriction to coolant flow **(Figure 9)**. What occurs in this situation is that minerals build up on the tubes within the radiator and restrict the flow of coolant. When coolant flow is restricted the coolant gathers more heat, while in the engine the radiator cannot remove the required amount of heat and the volume of cooled coolant supplied back to the engine is reduced. Mineral buildup may further reduce radiator efficiency by acting as an insulator that ultimately hinders the radiator's ability to transfer heat.

If the engine overheats under all operating conditions, the problem is related to the loss of coolant flow within the engine. This can be caused by a thermostat that never opens or by a water pump with a damaged impeller. If the engine thermostat is inoperative and will not open, the engine will overheat whether the engine is idling or being driven, although it will overheat faster when driven. Thermostat operation can be monitored by observing the coolant temperature using a diagnostic scan tool while observing the temperature of the outlet hose from the engine. The thermostat on most late-model vehicles will open at approximately 195°F (91°C). When the thermostat opens, the temperature of the outlet hose will also rise to the approximate temperature of the coolant. This temperature can be measured with an infrared thermometer.

On many of today's vehicles, it may be difficult to access the coolant at the radiator. In this case, the thermostat can be tested using the following method:

1. Following service manual procedures, remove the thermostat from the engine.
2. Place the thermostat in a pot of water.
3. Place a thermometer in a position to monitor water temperature.
4. Slowly raise the temperature of the water.
5. The thermostat should start to open just as the water reaches the temperature at which the thermostat is rated and continue to open until it is fully opened.
6. If the thermostat does not open or visibly sticks, it is defective and must be replaced.

> **You Should Know**
> *This six-step thermostat test is used only to verify your diagnosis. Should the thermostat test good, it is advised that a new thermostat be installed anyway. When comparing the relative cost of a thermostat to the difficulty of its removal, in many cases it is wise to replace the thermostat any time that it is removed.*

## Summary

- Any deficiency in the battery or its connections will lead to electrical problems. A quick assessment of battery state of charge can be made while starting the vehicle. Excessive voltage drops can cause various electrical and engine performance concerns.
- A visual inspection is the first step in identifying cooling system problems. Many cooling system concerns can be identified by a thorough visual inspection.
- The technician will encounter both internal and external cooling system leaks. External leaks are usually relatively easy to locate. Internal leaks are often more difficult to locate.

- External leaks can often be identified visually. Some external leaks can be identified by applying external pressure to the system.
- Internal coolant leaks can be characterized by a loss of coolant with no external evidence. Internal leaks might require partial engine disassembly to locate.
- Engine overheating concerns can be caused by a number of individual concerns, but all problems will relate directly to the loss of coolant flow or the loss of airflow. The conditions in which overheating occurs are excellent indicators of the part of the cooling system in which the problem lies.

# Review Questions

1. Technician A says that corroded battery cables can cause the vehicle to turn over very slowly. Technician B says that some cables might have to be removed to check for corrosion. Who is correct?
   A. Technician A
   B. Technician B
   C. Both A and B
   D. Neither A nor B

2. Describe the basic procedure that should be followed when visually inspecting a cooling system.

3. Which of the following methods can be used to diagnose an external engine coolant leak?
   A. Disassembling the engine.
   B. Applying pressure to the cooling system.
   C. Using an exhaust gas analyzer.
   D. Identifying a milky substance on the oil fill cap.

4. An internal coolant leak can be identified by all of the following methods *except:*
   A. An exhaust gas analyzer.
   B. A chemical combustion gas detector.
   C. Partial engine disassembly.
   D. A chemical dye solution.

5. A vehicle overheats only when the vehicle is stopped and allowed to idle for several minutes. Technician A says that a thermostat that is stuck in the closed position can be the cause. Technician B says that a poor coolant flow through the radiator is the cause. Who is correct?
   A. Technician A
   B. Technician B
   C. Both A and B
   D. Neither A nor B

# 10

# Basic Engine Testing

## Introduction

The engine performance technician is responsible for diagnosing a range of driveability problems. One of the areas that an engine performance technician must be proficient in is the diagnosis of engine mechanical concerns, specifically those problems that relate to the engine's ability to build and maintain cylinder pressure. This chapter shows what tools are necessary and how tests are performed using these tools. Chapter 11 will place emphasis on testing order and analyzing test results.

### COMPRESSION TESTING

A compression test is performed to measure the cylinder's ability to build pressure within the cylinder. For the cylinder to build pressure, the valves, piston rings, and the piston must create a tight seal. A compression test allows the technician to test the sealing ability of the valves and piston rings, as well as the integrity of the piston and the basic operation of the camshaft and valvetrain. If a cylinder is unable to build pressure, proper combustion will not occur and a misfire will be the result.

A compression gauge is used to test cylinder compression and consists of a gauge, a hose, and several adapters **(Figure 1)**. The adapters have different thread sizes and lengths to match that of the various spark plug styles. The adapters screw into the cylinder head in place of the spark plug. Once connected, the engine is then cranked and the pressure that is built within the cylinder is displayed on the gauge. By observing the gauge while the engine is cranking, the technician is able to observe the amount of compression and the rate at which the pressure was built. It

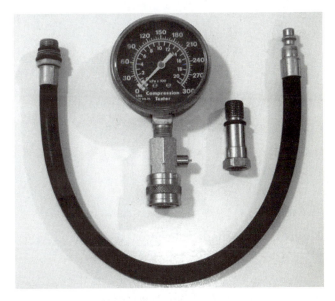

**Figure 1.** A typical compression gauge kit.

should be noted that a compression test alone would not pinpoint the actual cause of a compression problem; other tests must be performed in conjunction with the compression test to identify the exact cause of a compression concern.

Three types of compression tests that a technician may find useful are available: dry, wet, and dynamic. Each of these tests will give the technician specific information about engine operation. The dry compression test is always performed first in the series of specific compression tests. This gives you a baseline in which to judge other engine mechanical tests. The steps to performing a dry compression test are as follows:

1. Make sure that the ignition is in the off position.
2. Disable both the fuel and ignition systems. This will help avoid personal injury or a fire. The ignition can be disabled by disconnecting the ignition module; the fuel pump can be disabled by removing the fuse or relay. As always, consult your specific service manual.
3. Use compressed air to blow any debris from around the spark plug before removing.
4. Remove all of the spark plugs. This will allow the engine to spin at a smooth and consistent speed **(Figure 2)**.
5. Open the throttle to wide open position. This will allow the maximum amount of air into the engine.
6. Select the adapter that has the same size, thread pitch, and length as the threads of the spark plug. Install the adapter into the cylinder being tested and connect the gauge **(Figure 3)**.

7. Using the starter, allow the engine to turn over through four complete revolutions. A remote starter switch can be used for convenience **(Figure 4)**. It might also be desirable to connect a battery charger to make sure that the battery has sufficient voltage to spin the engine.
8. Although the needle on the gauge should bounce with every compression stroke, pay special attention to the first bounce and the last bounce. The compression should rise to at least 50 percent of total compression on the first bounce. Record the highest reading **(Figure 5)**.
9. Check each cylinder two times. By doing so, you can pinpoint some valvetrain problems that might otherwise be overlooked. If readings are different, repeat the test a third time and observe the readings that match.
10. All cylinders should be within the manufacturer's recommended range. All cylinders should be within 20 percent of one another from lowest to highest.

**Figure 2.** The spark plugs must be removed before performing a compression test.

**Figure 4.** A remote starter can be used to turn the engine over.

**Figure 3.** Install an adapter that is the same diameter, length, and thread pitch as the spark plugs.

**Figure 5.** Note the reading on the first compression stroke and record the highest reading.

*Using a test adapter that protrudes too far into the cylinder can cause damage to the piston, the adapter, and the spark plug threads.*

## Wet Compression Test

A wet compression test is used after one or more cylinders have tested low. The wet test will specifically test the condition of the piston rings and cylinder bore. The wet test is performed as directed above except that approximately 1 oz. of motor oil is added to the cylinder before installing the cylinder adapter. The heavy motor oil will temporarily seal the piston against the cylinder walls. If the compression in the cylinder rises after testing with oil, this indicates that the rings or cylinder walls are worn. If the compression remains the same, this indicates other problems and the cylinder should be tested further. Adding oil will not increase the compression of an engine with a burnt or bent valve. A wet test should be completed only on those cylinders that have low compression. When analyzing the results of a compression test, you should find the following:

- The dry test should produce compression pressures within vehicle specifications and within 20 percent from lowest to highest.
- The first pump of the cylinder should produce at least 50 percent of the total compression.
- Low compression could indicate worn cylinders, burnt or bent valves, a damaged piston, or a condition that does not allow the valve to open.
- Any cylinder with low compression should be leak checked to pinpoint the exact cause. (Refer to cylinder leakage testing later in this chapter.)

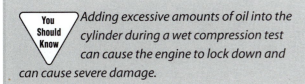

*Adding excessive amounts of oil into the cylinder during a wet compression test can cause the engine to lock down and can cause severe damage.*

## Dynamic Compression Test

The dynamic compression test is the third type of compression test. This test will differ from the others because we test one cylinder at a time with the engine running. The dynamic compression test will show us how efficiently a particular cylinder is able to draw air into the cylinder and subsequently move burned gases into the exhaust. It is possible for a cylinder with a restriction to show normal cranking compression but have poor operating efficiency. This is because the slow engine speeds encountered when the engine is cranking allow the cylinder plenty of time to move air in and out of the cylinder even If a restriction exists. At the higher engine speeds, when the engine is running, a restriction will not allow the air to move in and out of the cylinders efficiently.

*Because no cylinder can seal perfectly, adding oil to a cylinder with good compression can cause the readings to increase a small amount. As long as the cylinder's readings are within specifications, this should be considered normal.*

The dynamic compression test will usually be used only once a cranking compression test and a cylinder leakage test have shown good readings and you still suspect the problem is a mechanical failure. This test is particularly helpful in locating restrictions in the intake and exhaust tracts, such as a partially worn camshaft lobe or badly carboned intake valve. A dynamic compression test can also indicate other cylinder sealing problems such as worn valve seats or broken springs. The steps to performing this test are as follows:

1. Make sure that the ignition is turned off.
2. Remove the secondary plug wire from the cylinder you want to test. Use a jumper wire to ground the secondary wire. This will reduce the possibility of ignition system damage.
3. Remove the spark plug.
4. Install the compression tester using the proper adapter.
5. Start the engine and let it idle.
6. Depress the release valve every few revolutions to ensure consistent readings. Record the highest reading.
7. At the throttle body, quickly snap the throttle to its wide-open position and let it close. This should be done quickly enough so that the engine does not increase in rpm. Observe the reading while at wide-open throttle (WOT). Doing this allows the engine to draw in the maximum amount of air that it is capable of. You may want to repeat this portion of the test to ensure an accurate reading.

When examining your results, readings taken at idle should be approximately 50 percent of cranking compression. Readings from the WOT test should be approximately 80 percent of cranking compression test. If readings are lower than these, look for a restriction in the intake tract **(Figure 6)**. If the readings are higher, an exhaust restriction might be indicated. It might also be helpful to remember when diagnosing an elusive problem to check other cylinders as well and compare the readings. Even if the readings on the suspect cylinder are within the above specifications, large variations between other cylinders on the same engine can indicate a problem.

**Figure 6.** Excessive carbon buildup on the back side of an intake valve is a common source of an intake tract obstruction.

## CYLINDER LEAKAGE TESTING

A cylinder leakage test will test a cylinder's ability to hold pressure. The leakage tester will test the sealing ability of the valves, rings, and piston. In this test, the cylinder will be filled with compressed air and the leakage will be assessed.

The tester is equipped with two gauges, an air pressure regulator, and various spark plug hole adapters similar to those for a compression tester (**Figure 7**). The regulator is connected to a source of compressed air and is adjusted to provide a specific amount of pressure into the cylinder. One gauge will indicate the amount of air being supplied to the cylinder, and the other gauge will indicate the amount of air that the cylinder is holding. If an excessive amount of air is being lost, the technician can actually feel air escaping from the source of the leak. Typical leak locations are from the exhaust, the throttle body, radiator, or crankcase, depending on the source of the leak. A cylinder leakage test can be

used to check overall condition of an engine but is most often used after a compression test has verified that a particular cylinder has no compression. The general guidelines for performing a cylinder leakage test are as follows:

1. Make sure the ignition is in the off position.
2. Disable both the fuel and ignition systems. This will help avoid personal injury or a fire. The ignition can be disabled by disconnecting the ignition module; the fuel pump can be disabled by removing the fuse or relay. As always, consult your specific service manual.
3. Remove the spark plug from the cylinder being tested. Because the engine will try to spin when applying pressure to the test cylinder, it is desirable to leave the other spark plugs installed. The compression in the other cylinders will help hold the crankshaft steady.
4. Move the piston of the cylinder being tested to TDC on the compression stroke. A whistle or compression tester can be used to help locate the piston position (**Figure 8**).
5. Select the adapter that has the same size, thread pitch, and length as the threads of the spark plug. Install the adapter into the cylinder. Make sure that the regulator of the tester is turned to its lowest setting and connect

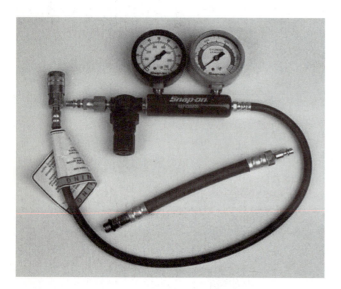

**Figure 7.** A typical cylinder leakage tester.

**Figure 8.** A compression whistle is used to help identify when a cylinder reaches TDC.

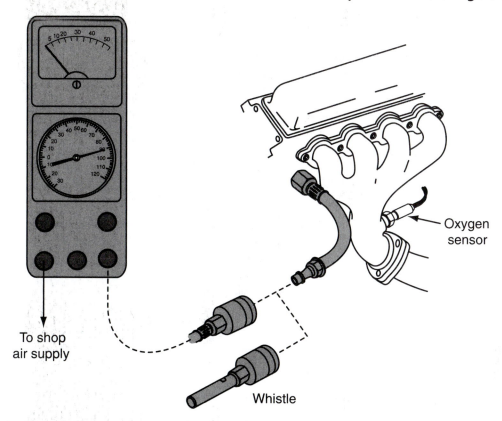

**To shop air supply**

**Whistle**

**Oxygen sensor**

**Figure 9.**   A cylinder leakage tester connected to the engine.

the tester to the vehicle adapter and the shop air supply **(Figure 9)**.

6. Calibrate and adjust the tester to the tool manufacturer's specifications.
7. Observe the reading on the gauge.
8. If cylinder leakage exceeds 15 percent, listen for air escaping from the throttle body (leaking intake valve), the tail pipe (leaking exhaust valve), or the oil fill cap (damaged piston or worn rings).

> **You Should Know** *Although the spark plug adapters for a compression tester and a leakage tester might appear to be identical, they are not. It is important to note that the adapter for a compression tester will be equipped with a Schrader valve in the end; the adapter for the leakage tester will not. Using the wrong adapter with the wrong tester will result in incorrect readings.*

## Compression Whistle

The compression whistle is a useful tool for determining the TDC of a cylinder. The tool consists of a whistle attached to a rubber hose. (Refer to Figure 8.) The hose will screw in in place of the spark plug of the cylinder you are

testing. As the piston moves up toward the top of the compression stroke, the air will push through the whistle and make a high-pitched sound.

## VACUUM GAUGE

The vacuum gauge is used to measure the entire engine's ability to breathe. Most vacuum gauges have the ability to read vacuum as well as pressures up to 10 psi. The vacuum gauge has one dual-scale gauge and a hose fitting **(Figure 10)**. One scale is usually calibrated from 0 to

**Figure 10.**   A typical vacuum gauge.

30 in. of vacuum; the other scale is calibrated from 0 to 10 psi. When vacuum is applied to the hose, the needle moves across the vacuum scale; when pressure is applied, the needle will move in the opposite direction across the pressure scale.

The technician attaches the vacuum gauge to a source of manifold vacuum. By reading the amount of vacuum on the gauge, many different things about the overall condition of an engine can be observed. **Figure 11** illustrates various engine conditions and their associated gauge readings.

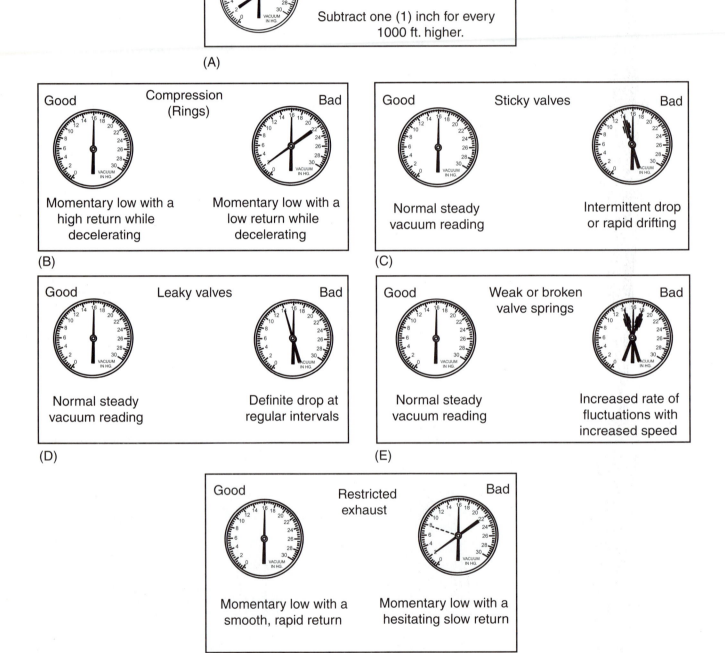

**Figure 11.** Various engine conditions and their associated vacuum readings. (A) Normal reading. (B) Worn piston rings. (C) Sticking valves. (D) Leaking valves. (E) Weak valve springs. (F) Restricted exhaust system.

# Summary

- A compression test is used to measure a cylinder's ability to build pressure. The amount of compression an engine produces can give the technician information about the condition of the valves, rings, cylinder condition and valvetrain operation. The compression gauge replaces the spark plug in the cylinder being tested. Variations of the compression test are the wet compression test and the dynamic compression test.
- A wet compression test is useful to help locate a low compression problem caused by worn piston rings or a worn cylinder. If wet test readings increase, a cylinder or ring problem is indicated. Wet test readings will not increase if the condition is caused by a valvetrain problem.

- A dynamic compression test measures a cylinder's ability to breathe. A dynamic compression test can pinpoint specific valvetrain problems. A technician should always compare readings from the engine's other cylinders when performing any compression test.
- The cylinder leakage test measures a cylinder's ability to hold pressure. Both valves must be closed when performing a cylinder leakage test.
- A vacuum gauge tests the entire engine's ability to breathe. Many engine problems can be detected by measuring engine vacuum.

# Review Questions

1. Two technicians are discussing engine compression testing. Technician A says that the piston must be at TDC to begin the compression test. Technician B says that the compression test can be used to diagnose suspected valve timing problems. Who is correct?
   A. Technician A
   B. Technician B
   C. Both A and B
   D. Neither A nor B

2. A wet compression test will isolate a damaged intake valve.
   A. True
   B. False

3. Two technicians are discussing the results of a dynamic compression test. The cylinder in question had a dynamic reading of 40 psi compared with a dry test reading of 100 psi. Technician A says that a partially worn intake lobe on the camshaft could be the cause. Technician B says that a restricted catalytic converter could be the cause. Who is correct?
   A. Technician A
   B. Technician B
   C. Both A and B
   D. Neither A nor B

4. All of the following statements about a cylinder leakage tester are true *except:*
   A. The cylinder leakage tester is connected to shop air.
   B. The cylinder leakage tester is installed in place of the spark plug.
   C. The cylinder leakage tester is often used after a compression test has indicated a compression problem.
   D. The cylinder leakage tester is used in place of a compression test.

5. Technician A says that approximately 1 oz. of oil should be added to the cylinder being tested before using a cylinder leakage tester. Technician B says that the technician can feel air escaping from the tail pipe if the intake valve is not sealing. Who is correct?
   A. Technician A
   B. Technician B
   C. Both A and B
   D. Neither A nor B

6. A vacuum gauge can be used to identify a restricted exhaust system.
   A. True
   B. False

# Chapter 11

# Engine Diagnosis and Repair

## Introduction

In this chapter, you are going to learn about basic engine diagnosis and maintenance procedures. In Chapter 10, you learned how to perform specific engine diagnostic tests. In this chapter, we will expand on that knowledge and learn how it can be applied in real situations. This chapter focuses on driveability concerns that can be associated with the engine, such as engine misfire and no-start diagnosis; however, the chapter also includes basic information on the diagnosis of oil leaks and sources of unusual smoke. In addition, some common maintenance procedures are covered.

### ENGINE LEAK DIAGNOSIS

Engine oil leaks are among the most common customer concerns. It is important that all technicians be adept at diagnosing oil leaks. Oil leakage from an engine can occur at any location where a gasket or seal is present. In addition to making a mess, leaking oil can create a fire hazard by leaking onto the exhaust and can cause damage to electrical insulation and connector seals. Using a few fundamental techniques, you will be able to diagnose most engine oil leaks.

The visual inspection is your most effective procedure for diagnosing oil leaks. When oil leaks from a gasket, it tends to be picked up by the wind as the vehicle is driven and can coat a large area. As dust and dirt are attracted to the oil, a thick layer of grime is formed, making accurate diagnosis seemingly impossible. However, as is often the case, the source of the leak can be identified as a wet but clean area above the bulk of the grime **(Figure 1)**.

**Figure 1.** Fresh oil indicates the proximity of an oil leak.

Once you have identified these clean areas as possible leak sources, the entire area should be cleaned with solvent and dried with compressed air. After the area is dry, first observe any leakage that might occur with the engine off. If nothing is noted at this point, start and run the engine. It might be necessary to test drive the vehicle to duplicate the leak. In those situations in which the leak is difficult to identify, two materials will help in positively locating the origin of the leak: foot powder and chemical dye.

Common aerosol foot powder can be applied directly to the suspected leak source. When oil starts to leak, it will cause the powder to seemingly disappear and produce a distinct contrast against the remaining powder. Adding a chemical dye to the engine oil can be used to help locate more challenging leaks. After the proper amount of dye has

**Figure 2.** An ultraviolet light used in conjunction with a chemical dye can be used to help identify engine oil leaks.

been added, the engine should be run until the leak again appears. Using an ultraviolet light, examine the suspected leak areas **(Figure 2)**. The dye will appear at the source of the leak and can be identified by its fluorescent green or orange color.

## SMOKE DIAGNOSIS

In the context of dealing with the engine, smoke is typically associated with the exhaust gases emitted from the tailpipe. Light wisps of exhaust gases and condensation are normal by-products of the combustion process and are hardly noticeable. However, when smoke from the exhaust becomes noticeable, a problem is indicated. The color of the smoke is a leading indicator of the source. Any condition that causes excessive amounts of smoke will at some point cause a driveability concern as well. The following guidelines can be used when diagnosing exhaust smoke conditions.

- White smoke indicates that engine coolant is being burned in the engine. Typical sources include a blown head gasket or cracked cylinder head.
- Black smoke indicates an excessively rich condition. Typical causes include leaking fuel injectors and excessive fuel pressure. Failure of some electronic engine management components can also cause this condition.
- Gray or blue smoke is an indication of oil that is being burned in the engine. If the smoke is constant, this typically indicates worn piston rings and cylinders. If the smoke appears only when the engine is first started or under high-vacuum conditions, worn valve guides and/or seals are indicated.

## MISFIRE DIAGNOSIS

In Chapter 10 you learned about a variety of different engine tests. Now we are going to put those tests to work and learn how to analyze the results. Driveability concerns that are caused by engine mechanical failures will usually manifest themselves as either a misfire condition or a no-start condition. In either case, the ignition or fuel systems can cause both of these concerns. Because of this, it is imperative that you confirm proper operation of these systems before suspecting an engine mechanical concern. Refer to Sections 3, 4, and 5 of this text for more information about the diagnosis of these systems.

## Testing for Vacuum Leaks

Vacuum leaks are a potential source of engine driveability symptoms such as engine misfire, surge, and stall. You should inspect for vacuum leaks early in your diagnostic process once the condition has been verified. A vacuum leak will cause an engine misfire by drawing unmetered air into the engine. If unmetered air enters the engine, the fuel system cannot compensate with additional fuel and an extremely lean condition will be present. Depending on the location of the leak, it can affect the operation of the entire engine or might affect only one cylinder if located on a particular intake manifold runner. There are two types of vacuum leaks: external and internal.

External vacuum leaks are those that are found outside of the engine such as in vacuum hoses, throttle body gaskets, and intake manifold gaskets. Many times an external vacuum leak can be heard. A vacuum leak will usually produce a loud hissing or whistling noise. However, some smaller vacuum leaks cannot be heard.

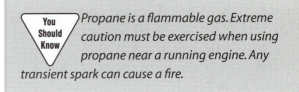

Vacuum leaks may be located by many different means. Two of the more practical methods are done using water or propane. Spraying water in the potential leak area can identify vacuum leaks. When a leak is located, the engine idle will smooth out and the rpm will change. Keep spraying an increasingly smaller area until the location of the leak can be positively identified. For the propane method, using a small propane bottle and a regulator with a flexible hose, propane can be directed into some very confined areas that other methods might be unable to locate **(Figure 3)**. When a leak is located, the engine idle will smooth out and a change in rpm should be noticeable.

**Figure 3.** Vacuum leaks can be detected using propane.

> △ **You Should Know** *Never spray a flammable substance such as carburetor or brake cleaner onto a running engine. A minor spark could ignite the chemical and cause a fire, resulting in serious personal injury and severe property damage.*

Internal leaks are only found on V-type engines with a sealed valley. The internal vacuum leak is much more difficult to locate. However, the most common source of the internal vacuum leak is the intake manifold gasket **(Figure 4)**. Use the following steps to locate an internal vacuum leak:

1. With the engine off, remove the PCV valve and fresh air tubes.
2. Plug all holes that are in the valve covers.
3. Remove the engine oil dipstick.
4. Attach a vacuum gauge to the dipstick tube.
5. Start the engine and allow it to idle.

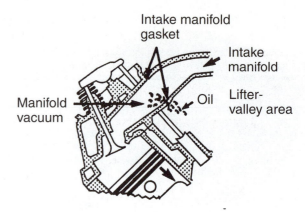

**Figure 4.** Internal vacuum leaks stem from leaking intake gaskets on V-type engines.

6. An engine with no leaks should build pressure within the crankcase. If a vacuum is present under these conditions, an internal vacuum leak is indicated.

## Compression Testing

Once you have identified the weak cylinder(s) through the use of a cylinder balance test and confirmed that both fuel and ignition systems are in working order, it is time to check the mechanical condition of the engine. The first in the series of tests is the compression test. Compression testing will determine the sealing ability of the pistons in the cylinders and if the valvetrain is operating. In the case of a single cylinder misfire condition, you might choose only to test the cylinder in question. This can be done as a quick verification of the condition of that cylinder but should not be considered to represent the entire engine.

Typical compression readings will be from 100 psi up to 150 psi. It is always recommended that you refer to the engine manufacturer's specifications. If compression readings are low, you should perform a wet compression test as outlined in Chapter 10. If the readings improve significantly, the cylinder probably has worn piston rings and/or a worn cylinder. The engine will need to be disassembled for further inspection.

> △ **You Should Know** *It is important to perform the compression test two times without removing the compression gauge; if the low-compression condition is caused by an intake valve not opening, the air allowed into the cylinder during gauge installation might show a significant amount of compression on the first test. Once pressure is bled off, only the air that is passed by the valve will be allowed into the cylinder.*

If the wet test readings remained low, a large leak in the cylinder is indicated and a cylinder leakage test should be performed. A large leak in the cylinder can be caused by a number of failures such as a burnt valve, burnt piston, severely worn cylinder, camshaft with a flattened lobe, or a loose or broken rocker arm.

## Cylinder Leakage Test

The cylinder leakage test is performed after the compression test to locate the cause of a low-compression concern. Perform the leakage test following the steps in Chapter 10. Once you have the tester installed, observe the amount of leakage. If air is leaking:

- From the tail pipe, the exhaust valve is burned.
- From the throttle body, the intake valve is damaged.
- From the oil fill cap, the cause is excessive wear in the piston rings, a worn cylinder, or a hole in the piston.
- From the radiator, any pressure that is built in the radiator is a result of a blown head gasket.
- From an adjacent cylinder, the head gasket is blown.

Many times in making a diagnosis you must use the results from more than one test to draw a conclusion. For example, if compression readings and cylinder leakage are low, this would indicate that the intake valve was not opening for some reason. If leakage was high, then you will certainly be able to detect where the leak is originating.

## Dynamic Compression Test

As we learned in Chapter 10, a dynamic compression test will measure the cylinder's ability to breathe. The dynamic compression test is not often used but can provide some valuable information. You will usually use this test after a compression test or cylinder leakage tests proved inconclusive and you still suspect that a mechanical condition is the root cause of the customer's concern. The dynamic compression test can also be used to test for a restricted exhaust system.

**Interesting Fact**  *Some vehicle antitheft systems have the ability to disable the starting system.*

## NO-START DIAGNOSIS

The next concern that can often be related to mechanical failure is a no-start condition. We will assume that the engine will turn over but will not start. You may be asked to repair a vehicle that will not turn over, but this concern is usually the result of a starting system failure. The exception

to that would be a seized engine or engine accessory. A no-start diagnosis will follow a similar progression to that of a misfire diagnosis. As with a misfire, many systems can keep the vehicle from starting and we must eliminate these systems as causes one by one in order to locate the root cause.

A no-start diagnosis as it relates to mechanical failure is relatively simple. Begin by listening to the engine as it turns over. An engine in good condition should spin over smoothly and the starter should sound as if it has resistance to spinning. If the engine spins very fast and the starter does not sound as if it has a load against it, or it spins very unevenly, this can indicate a valve-timing problem. A stripped or broken timing belt or stripped cam gear on a chain drive system is the typical cause for these circumstances **(Figure 5)**.

Missing teeth

A

Missing teeth

B

**Figure 5.** (A) A stripped timing belt. (B) Stripped teeth of a cam gear.

If the engine spins freely, one of the first things you should check is to see if the camshaft is turning. This check can be accomplished on most vehicles by removing the oil fill cap, PCV, or fresh air tube from the valve cover. Observe one of the rocker arms or camshafts through the hole while an assistant turns the engine over. If the rocker arms do not move, this indicates a stripped or broken timing belt or a timing gear failure.

> **You Should Know** *Some engines have very close piston-to-valve clearance. If a failure occurs in one of these engines that affects cam timing, the valves can contact the pistons and valve or piston damage can occur.*

If the engine seems to spin over irregularly and the rocker arms or camshafts are moving, a jumped belt or chain might be indicated. A jumped chain or belt can be verified by observing the results of a compression test of all cylinders. An engine with incorrect valve timing will have erratic readings from cylinder to cylinder. Some can be normal, higher than normal, or lower than normal or a combination of these readings.

> **Interesting Fact** *Many times a gear or belt will strip immediately when the engine is turned off. As the engine slows, the compression pressure will resist turning the crankshaft and will sometimes actually kick the crankshaft backward. If a belt or gear is old and worn, this kickback effect can strip the cogs or teeth of the gear or belt.*

## TIMING BELT SERVICE

Timing belts, like any other rubber belt, are subject to wear and damage, and most manufacturers have specific belt replacement recommendations **(Figure 6)**. If a timing belt should break or strip the rubber cogs used to drive the belt, the engine will no longer start and run. If the engine is an interference design, a broken belt will result in several bent valves. It is critical that the timing belts be serviced per the manufacturer's maintenance interval. Most intervals will be at the 60,000- to 75,000-mile range. Vehicles that use timing chains do not have regular service intervals because timing chains are generally expected to last for the life of the vehicle.

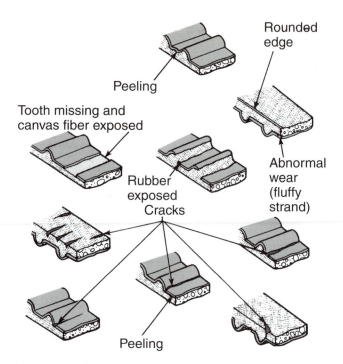

**Figure 6.** Common timing belt failures.

You will also find that many OHC engines will use the timing belts or chains to drive the water pump. Many engines use an eccentric water pump to adjust belt tension. Because a water pump and a timing belt have similar average life-spans, you should consider replacing the water pump and the belt at the same time for engines that use this method to drive the water pump. This rule is true even in the event that the water pump fails before the timing belt does. Because these two services overlap, it will usually save the customer a great deal of money in the long run by not having to perform virtually the same service twice.

> **You Should Know** *If a timing chain gear is found to be stripped, it is likely that the debris will find its way into the lubricating system. In this situation the oil pan should be removed and cleaned to remove any debris.*

In many cases, the timing belt replacement will be performed as a maintenance service. The specific steps for replacing a timing belt will vary greatly from vehicle to vehicle. The major differences will be usually in what has to be done to get the timing cover off. Once the cover is off, the steps are similar. Refer to a vehicle-specific service manual for specific steps to belt replacement. The general steps for replacing a timing belt are listed below. These steps are for SOHC engines; however, DOHC engines will be similar. You should refer to the specific service manual for specific cautions and timing mark locations.

1. Disconnect the negative battery cable to make sure that the starter cannot inadvertently engage while you are working in the engine compartment.
2. Remove the necessary components that will allow access to the timing cover.
3. Remove the timing cover.
4. Align the timing marks on the cam gear with the marks on the rear timing cover and cylinder head. If the belt has not slipped, the timing marks on the crank gear should line up as well **(Figure 7)**.
5. Release the tension from the belt and slide the belt off the pulleys. If the lower crankshaft pulley has not been removed, it might be necessary to do so at this time.
6. Inspect the tensioner for looseness in the springs, grooving of the pulley, and smooth operation of the pulley bearing. Replace as needed.

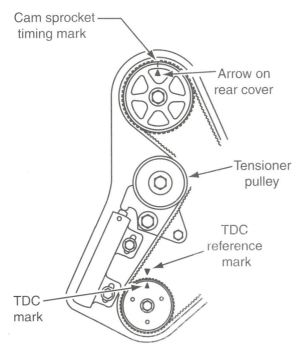

**Figure 7.** Typical locations for engine timing marks.

7. Inspect the cam and crankshaft gears and clean as needed to remove any debris.
8. If the timing belt drives the water pump, inspect or replace it at this time. When inspecting, look for any signs of leakage or a loose or rough pulley bearing. Replace as needed.
9. Make sure that all timing marks for the camshaft and crankshafts are lined up (Refer to Figure 7). Install the belt. The belt should be installed so that all of the slack in the belt is toward the tensioner side of the engine.
10. Tension the belt as outlined in the service manual. Note that some tensioners are self-adjusting.
11. Turn the crankshaft clockwise by hand two full revolutions and recheck the belt tension.
12. Turn the crankshaft by hand two additional times. All of the timing marks should be aligned. (Refer to Figure 7.) If the marks do not align, repeat steps 9–12.
13. Reinstall timing cover and engine-driven accessories. Reconnect the battery.
14. Check ignition timing if applicable.

## VALVE ADJUSTMENT

For many OHC engines, regular valve adjustment is a required procedure. Most OHV engines in use today do not require periodic valve adjustment and need attention only when valvetrain service has been performed. There are two methods of adjusting valve lash on an OHC engine and two methods for an OHV engine.

### OHV Engines

Overhead valve engines use an adjustable rocker arm, shaft-mounted rockers, or net lash rockers. Both the shaft-mounted assemblies and the net lash require only that the rocker arm bolts be torqued to specifications. No other adjustment is required. The adjustable rockers require a specific procedure. The basic steps are as follows:

1. Disable the fuel and ignition systems.
2. Remove the valve cover(s).

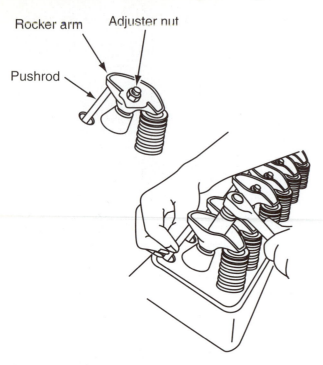

**Figure 8.** Valve lash adjustment on an engine with adjustable rocker arms.

3. Bring the number 1 cylinder to TDC on the compression stroke. Some engines might have a sequence of valves that can be adjusted in this position. If that procedure is available for your vehicle, follow those instructions. If a procedure is not available, each cylinder can be adjusted one at a time using the remaining steps of this procedure.
4. Loosen the rocker retaining nuts for each valve until a noticeable amount of play is present.
5. Using your thumb and index finger, grasp the pushrod below the rocker. Twist the pushrod with a back and forth motion **(Figure 8)**.
6. Tighten the rocker nut until a light drag is noticed on the pushrod. This is considered to be zero lash.
7. Tighten the nut an additional one-half to full turn. Refer to the service manual.
8. Once steps 3 through 7 are completed on this cylinder, bring the next cylinder in the firing order to TDC compression and repeat steps 3 through 7.
9. Reinstall valve covers.

## OHC Engines

OHC engines can typically be adjusted by changing the length of an adjustable rocker arm tip or by using shims between the camshaft and follower **(Figure 9)**. The procedure for adjusting OHC engines using selective shims is as follows:

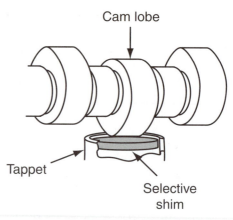

**Figure 9.** A typical OHC engine with selective shims.

1. Disable the fuel and ignition systems.
2. Remove the engine valve cover.
3. Bring the cylinder being adjusted to TDC on the compression stroke. Verify that the follower of the valve being adjusted is contacting the camshaft base circle.
4. Using a feeler gauge, measure the clearance between the cam lobe and the shim placed on the top of the follower. Compare measurement to that of the vehicle specifications.
5. If the clearance does not match the specifications, remove the old shim and measure the thickness. If the clearance is too tight, select a thinner shim to install. If the clearance is too loose, select a thicker shim to bring the adjustment into specifications. Install the new shim and remeasure the clearance.
6. Repeat steps 3 through 5 on all remaining cylinders.
7. Reinstall valve cover and restore vehicle.

> **You Should Know** *Two valve clearance specifications, hot and cold, are often given for an engine. These are direct references to the temperature of the engine. Failure to observe these specifications will result in improper valve adjustment. A valve that has excessive clearance will cause a ticking noise in the engine. A valve that is too tight will cause a loss of compression and result in an engine misfire.*

If the engine uses a rocker with an adjustable screw, use the following procedure. This procedure can also be used on OHV engines with this type of rocker. The procedure is as follows:

1. Disable the fuel and ignition systems.
2. Remove the engine valve cover.

3. Bring the cylinder being adjusted to TDC on the compression stroke. Verify that the rocker arm of the valve being adjusted is contacting the base circle of the camshaft.

4. Using a feeler gauge, measure the clearance between the valve stem and rocker tip. Compare measurement to that of the vehicle specifications.

5. If adjustment is required, using a screwdriver hold the adjustment screw while loosening the jam nut **(Figure 10)**.

6. Turn the screw until a feeler set to the proper specifications can be inserted between the valve stem and adjustment screw. The feeler gauge should have a light but noticeable drag when pulled.

7. Hold the adjustment screw with the screwdriver and tighten the jam nut.

8. Recheck clearance.

9. Repeat steps 3 through 8 on the remaining valves.

10. Reinstall valve cover and restore vehicle.

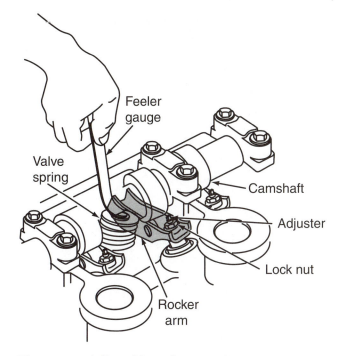

**Figure 10.**   Adjustable rocker arm tips.

## *Summary*

■ Engine performance concerns that result from engine mechanical problems are oil leaks, excessive smoke, misfire, and no-start conditions.

■ Oil leaks are almost always detected by a visual inspection. Aerosol foot powder and engine dye make oil leaks more visible. Excessive amounts of smoke from the exhaust can be caused by failures in the cooling system, fuel system, engine cylinders, and by valve guide wear.

■ Engine mechanical problems related to engine performance usually surface as a misfire or a no-start condition. It is imperative that operation of the fuel and ignition systems be confirmed before performing any in-depth engine mechanical diagnosis.

■ Vacuum leaks can cause various driveability concerns. Vacuum leaks can potentially appear as external or internal leaks.

■ Compression testing will determine the sealing ability of the pistons in the cylinders and if the valvetrain is operating. The cylinder leakage test is performed after the compression test to locate the cause of a low-compression concern.

■ Inaccurate valve timing caused by a broken or stripped timing belt or gear can cause a no-start condition. Observing the rocker arm movement while the engine is being turned over can easily identify if the timing components are still intact.

■ Timing belt service intervals should be strictly adhered to. Failure to properly maintain the timing belt can result in severe engine damage.

■ Valve clearance adjustment might be required on some engines as normal maintenance. When adjusting valve clearance, the follower or rocker arm should be on the base circle of the camshaft.

# Review Questions

1. Two technicians are discussing proper procedures for diagnosing oil leaks. Technician A says that some oil leaks might not be noticeable until the engine is allowed to run. Technician B says that the use of a chemical dye replaces the need for visual identification of a potential oil leak. Who is correct?
   A. Technician A
   B. Technician B
   C. Both A and B
   D. Neither A nor B

2. Explain what each of the following smoke colors indicates and give two examples each of what might cause these conditions.
   A. White
   B. Black
   C. Blue

3. Explain why an internal vacuum leak would not be possible on an in-line engine.

4. An engine has a misfire on the number 4 cylinder of a V6 engine. Both the fuel and ignition systems are performing to specification on that cylinder and the compression is within specifications. Technician A says that an internal vacuum leak can be causing the concern. Technician B says that the intake valve might have a severe amount of carbon buildup. Who is correct?
   A. Technician A
   B. Technician B
   C. Both A and B
   D. Neither A nor B

5. A vehicle with a single cylinder misfire is brought into your shop. Fuel and ignition system operation have been verified. A compression test shows that the cylinder in question has zero compression. A cylinder leakage test has found that the cylinder shows minimal leakage. Technician A says that a worn intake camshaft lobe could be the cause. Technician B says that a restriction in the exhaust system might be the cause. Who is correct?
   A. Technician A
   B. Technician B
   C. Both A and B
   D. Neither A nor B

6. Technician A says that failure to service the timing belt at the proper intervals can ultimately result in severe engine damage. Technician B says that improper alignment of the timing marks could result in severe engine damage. Who is correct?
   A. Technician A
   B. Technician B
   C. Both A and B
   D. Neither A nor B

7. In some instances, the condition of a timing belt can be determined by observing rocker arm or camshaft movement.
   A. True
   B. False

8. List and describe four methods used to adjust valve lash. This may include both OHV and OHC engines.

# Section 3

## Electronic Engine Management

## SECTION OBJECTIVES

After you have read, studied, and practiced the contents of this section, you should be able to:

- Describe and apply basic electrical principles.
- Describe the automotive computer and the role it plays in the modern automobile.
- Describe what an input sensor does and how various types of sensors operate.
- Describe what an actuator is and what each actuator is designed to do.
- Describe what OBDII is and how it affects the engine management system.
- Explain what is required of an OBDII-equipped vehicle.
- Explain how the engine management system controls and interacts with other vehicle systems.
- Describe and operate various pieces of electrical and electronic test equipment.
- Explain what a scan tool is and describe its many functions.
- Connect a scan tool and perform various data retrieval and bidirectional control functions.
- Perform various tests on the engine management system using specific tests and specialized equipment.

**Interesting Fact**

*Modern vehicles are equipped with many computers and electronic modules. These modules "talk" to each other through an on-board network similar in operation to the computer network in your school or business.*

# Chapter 12

# Electricity and Electronics

## Introduction

To be successful in the diagnosis and repair of modern vehicles, it is essential for the technician to have a fundamental understanding of electrical principles and relationships. There is no system in a modern vehicle that is not controlled or monitored by some type of electronic device. This chapter provides a basic introduction for those students who do not have any electrical background and a review for those students who are well versed in electrical systems.

## ELECTRICITY

Electricity is caused by the movement of electrons from one atom to another **(Figure 1)**. The movement of these electrons enables work to be done by an electrical circuit. We cannot see or touch these electrons, but we can observe and feel the work that electrons do. To have electrical flow, we must have an area that has extra electrons available and an area that has a deficiency of electrons. This is called potential difference. The two areas are connected together, causing the excess electrons to flow to the area of

deficient electrons. This is what happens in a battery; the battery supplies both the area with extra electrons (positive post, or power side) and the area that is electron deficient (negative post, or ground side). We connect the two posts together; electrons flow from the positive post back to the negative post. This is what occurs in the simple circuit depicted in **Figure 2**. There are four elements that you need to thoroughly understand to be successful in the diagnosis and repair of electrical systems: current, resistance, voltage, and voltage drop.

*There are two accepted theories of electrical flow, conventional theory and electron theory. Conventional theory states that the electrons flow from positive to negative; electron theory states that electrons flow from negative to positive. Either theory is accepted; however, conventional theory is observed in all vehicle systems.*

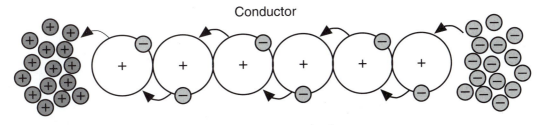

**Figure 1.** Electricity is the flow of electrons from one atom to another.

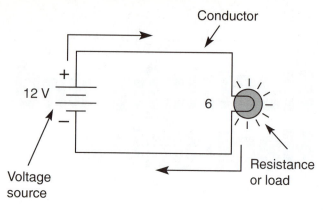

**Figure 2.**  A simple electrical circuit.

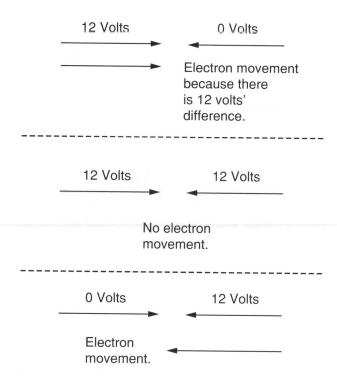

**Figure 3.**  Differences in voltage force the electrons to flow.

## CURRENT

**Current** is the cumulative flow of electrons from one atom to another. It is this movement of electrons that does the work in an electrical circuit. It takes the movement of 6.25 billion billion electrons moving past a single point in 1 second to measure 1A. To further put this into perspective, a simple dome light in a vehicle might draw 1–3 amps. Current, measured with a tool called an ammeter, is measured in a unit called amps and is often referred to as **amperage**.

The two types of electrical current are direct current (DC) and **alternating current (AC)**. Direct current means that the electrons can flow only in one direction and is the type most often used in vehicle electrical systems. Alternating current allows the electrons to flow from positive to negative and back to positive. The change of direction that takes place in an AC circuit takes place at a controlled rate called a **frequency**. Alternating current is used mostly in buildings and homes; however, a few components on a vehicle produce AC current. They will be discussed in later chapters.

## RESISTANCE

**Resistance** is the opposition to current flow. Resistance can be any situation or component that makes it difficult for the electrons to move from atom to atom. Some amount of resistance is naturally present in any material; however, external factors such as corrosion or conductor damage can increase resistance beyond acceptable levels. Any time that we force electrons through a resistance, some of this potential will be lost and heat will be created. The amount of heat will be dependent on the amount of current that we are trying to move and the amount of resistance that we are trying to move it through. The amount of resistance is measured in a unit known as ohms and is measured with an ohmmeter.

## VOLTAGE

**Voltage** is a difference of potential and is used to push electrons through a circuit. For example, a battery

has extra electrons at the positive post and a deficiency of electrons on the negative post. The difference between the number of electrons at the positive side of the battery and at the negative side is the difference in potential. When connected together, the buildup of excess electrons will push toward the side lacking in electrons. This creates a natural flow or pressure **(Figure 3)**. The pressure comes from all of these electrons trying to occupy the same space, much like a large crowd of people trying to move through a single door. The greater the number of people pushing, the greater the pressure. As the number of people dwindles, so does the pressure. As the number of electrons moves closer to being equal, we can say that voltage has dropped. When the amount of electrons is equal on the positive and negative sides, there is no difference in potential, and therefore no voltage is present. Voltage is measured in volts and measured with a voltmeter.

## VOLTAGE DROP

Voltage drop is the amount of potential that is lost when passing through a resistance. When you push electrons through a resistance, some of that potential will be turned into heat. For example, in a series circuit, if we have 12V going into a light bulb and 4V were consumed in the light bulb, we would have only 8V available for the remainder of the circuit **(Figure 4)**. Voltage drop is measured in volts using a voltmeter. Many computer circuits use con-

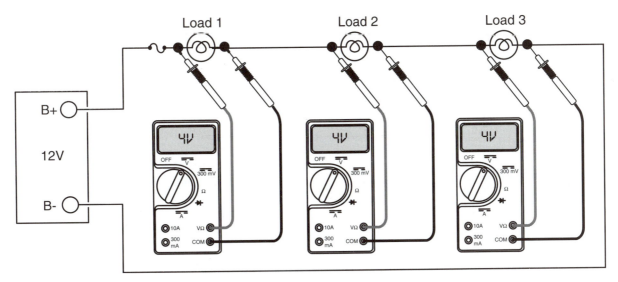

**Figure 4.** Voltage will drop each time it passes through a load.

trolled voltage drops as a means for monitoring a condition within a system.

## PRODUCTION OF ELECTRICITY

At this point, we have established a basis for how electricity travels within a conductor. Now we are ready to find out where electricity comes from. Electricity can be produced from two different methods: induction and chemical reaction. Both of these methods are used within vehicles.

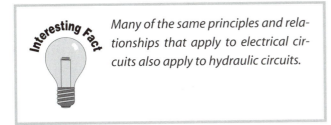

*Interesting Fact*

*Many of the same principles and relationships that apply to electrical circuits also apply to hydraulic circuits.*

### Magnetism

Magnetism and electricity are closely related—so closely that one can be used to produce the other. Any time that we push electrical current through a conductor, we create a magnetic field around that conductor. Likewise, any time that we move a conductor through a magnetic field, we induce electrical current into the conductor. Both of these properties are widely used in automotive electrical systems **(Figure 5)**. The principles of electromagnetism are used in the operation of relays and solenoids, and the principles of induction are used in the operation of generators and ignition coils.

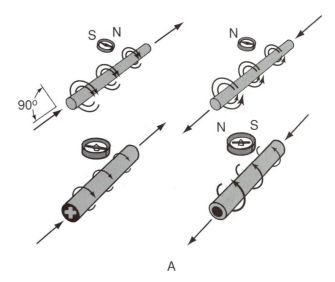

A

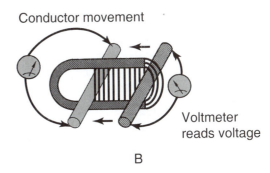

B

**Figure 5.** (A) Current flow through a conductor produces magnetic lines of force. (B) When a conductor moves through magnetic lines of force, voltage is induced into the conductor.

## Generator

A generator is an electrical component that produces electricity based on the principles of induction. A generator consists of a magnetic field and a conductor. As the conductor cuts the flux lines of the magnetic field, an electrical current is induced into the conductor; either the conductor can be moved through the magnetic field or the magnetic field can be moved across the conductor.

## Chemical Reaction

The second method of creating electricity is through chemical reaction, the only method by which DC voltage can be directly produced. Submerging two different metals in an acid solution produces electricity. The chemical reaction between the acid and the metals will promote the exchange of electrons between the different metals, creating an imbalance of electrons between the materials and therefore a difference in potential. A vehicle's battery uses two different types of lead-coated plates submerged in an acid solution. The acid solution promotes the transfer of electrons from one plate of lead to the other.

*Interesting Fact*

*A simple battery can be formed by placing a paper towel between a penny and a nickel and then saturating the components in an acid solution. Even a mild acid solution such as saliva will create a reaction sufficient enough to measure a small voltage. The voltage can be measured by touching each lead from a voltmeter to each of the metals.*

## OHM'S LAW

Ohm's Law is the basic theory we use to understand the direct relationship of voltage, amperage, and resistance. This theory must be completely understood before a technician can competently understand and diagnose electrical problems. Ohm's Law states that voltage in a circuit is directly proportional to the amount of current flow and the total resistance in the circuit. Here is what you must understand about Ohm's Law:

- Voltage, current, and resistance are directly related. If one of these values within a circuit changes, the other values will change as well.
- Assuming that voltage stays the same, if the circuit resistance goes down, the circuit current will go up. If the resistance goes up, the current goes down.
- Assuming that resistance stays the same, if the voltage goes up, the current will also rise. If voltage goes down, so does the current.

Understanding these three facts will form a foundation for understanding electrical circuits and their operation.

## CIRCUITS

An electrical circuit is the mechanism that allows us to harness and efficiently use electrical energy. Typical automotive circuits will consist of the following:

1. *Power source.* The power source provides a supply of electrons to introduce to the circuit. The primary power source for use in a vehicle is the battery. All electrical circuits will ultimately begin and end with the battery.
2. *Conductors.* **Conductors** are the wires and connections in which current flows. The conductors connect all of the elements of a circuit together.
3. *Loads.* Loads are those components installed in a circuit that consume electrical current to perform a function such as running a motor or light. The amount of current consumed is directly related to the load's resistance. All circuits must contain one or more loads.
4. *Controls.* Controls are devices that determine how and when a circuit operates. Controls are usually switches, relays, or electronic control modules. For a switch to control a circuit, the switch must be wired in series with the load that it controls.
5. *Circuit protection devices.* Circuit protection devices are used to limit the amount of current allowed to flow in the circuit. This is accomplished by creating an open in the circuit when current exceeds the rating of the protection device. The circuit protection devices are placed in the power side of the circuit and are rated to turn off the circuit well below the maximum capacity that the circuit was designed for. Common circuit protection devices are fuses, circuit breakers, and fusible links.

## TYPES OF ELECTRICAL CIRCUITS

Three types of circuits are used in vehicles today: series, parallel, and series-parallel.

The series circuit is the simplest of the three types. A series circuit supplies only one path for the current to flow **(Figure 6)**. In this type of circuit, an open at any point within the circuit will disable the entire circuit. The series circuit will always follow these laws:

- The total resistance is the sum of all resistances in the circuit.
- The same amount of current flows through each load regardless of its individual resistance.
- A voltage drop will occur across each load. The amount of the drop will be proportionate to the resistance of the load.
- The sum of all voltage drops within the circuit will always equal the source voltage.

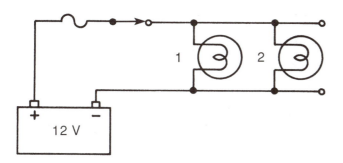

**Figure 6.**   A simple series circuit.

**Figure 7.**   A simple parallel circuit.

The parallel circuit provides a means for current to travel in more than one path **(Figure 7)**. In a parallel circuit, each path is called a **branch** and will be supplied with the same amount of source voltage. Each branch will also have

one or more loads and might have a separate control device. Unlike a series circuit, an open circuit in one branch will have no effect on the other branches of the circuit. The laws for a parallel circuit are:

- The current flow through each separate branch can be different depending on the amount of resistance in the individual branches.
- The total circuit current is the sum of the individual branch currents.
- Source voltage is supplied to each branch of a parallel circuit; each branch will drop the full amount of source voltage.
- The total resistance for the complete circuit is less than the value of the smallest resistance in any of the individual branches of the circuit.

The series-parallel circuit is a combination of both series circuits and parallel circuits. Usually the series-parallel circuit will have a load wired in series either before or after the parallel branches **(Figure 8)**. All three of these circuit types are widely used in vehicles and their use depends on a specific application.

## CIRCUIT COMPONENTS

Controls and circuit protection devices are some of the components in an electrical circuit that we have already examined. These particular items are found in almost any electrical circuit. Some of the more common components are:

- *Switches*. Switches are used to control how and when a circuit operates. Switches contain a set of contacts that are connected in series with the circuit. By closing these contacts, the circuit can then begin to work **(Figure 9)**. When the contacts are opened, the path of

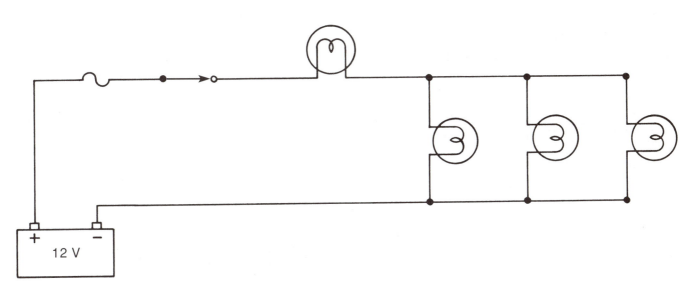

**Figure 8.**   A series-parallel circuit.

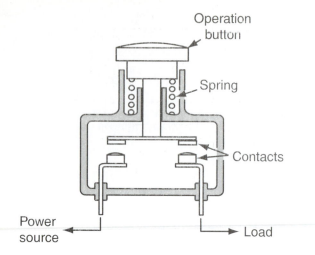

**Figure 9.** The switch is a simple form of circuit control.

electron flow is broken and the circuit stops working. Switches can be operated manually by the driver or automatically by pressure and/or temperature.

- *Relays.* Relays are used to allow the control of a high-current circuit using a low-current circuit. Inside the relay is a set of contacts; these contacts are closed by an electromagnet. A low-current switched circuit controls the electromagnet. When the magnet is energized, the contacts are drawn to close in order to complete the high-current circuit **(Figure 10)**.
- *Solenoids.* Like relays, solenoids use an electromagnet for operation. However, the core of the magnet is moveable. The moveable core is attached to a mechanical device and controls a mechanical action. This can be as

simple as applying pressure to a small steel ball to stop fluid flow or can be a strong mechanical lever used to engage the starter.

- *Resistors.* Resistors resist current flow. In some electrical circuits, it is desirable to restrict the amount of current flowing to a component. By restricting current, the speed of a motor or the brightness of a light can be controlled. Resistors have one connection on each end and are also wired in series in some circuits to create a voltage drop in order to deliver a specific voltage.
- *Potentiometers.* Potentiometers are variable resistors that are used as sensors. Potentiometers create a specific voltage drop based on a mechanical condition, such as throttle position. A potentiometer has three connections: a power wire is connected to one end of the resistor; the other end of the resistor is connected to a ground; the third connection is connected to the moveable arm, which is where the amount of voltage drop is measured and is usually sent back to a computer as an input signal **(Figure 11)**. By running a constant current through the resistor, the moveable arm reads only the voltage at a specific point across the resistor and is not a part of the circuit. Therefore, the voltage that can be sensed by the moveable arm is very stable.
- *Thermistors.* Another type of variable resistor is the thermistor, in which the resistance changes in relation to its temperature. Thermistors are used as fluid or air temperature sensors.
- *Diodes.* Diodes are semiconductors that allow current to flow in only one direction. These components are often used as one-way check valves. These types of

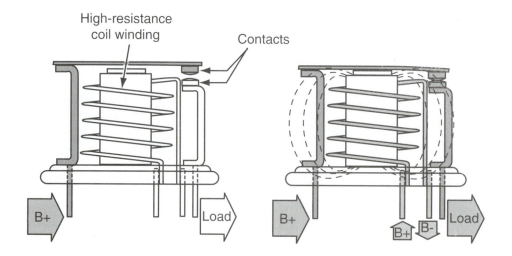

Note: High-resistance coil winding drawn much larger than actual size.

**Figure 10.** A relay controls high current using a low current.

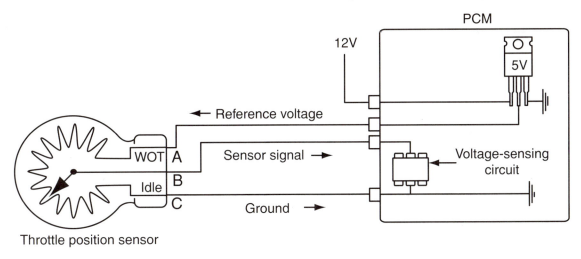

**Figure 11.**   Potentiometers are used to send signals to a computer base on a voltage drop.

diodes are used to protect electronic components from electrical surges or to rectify AC. A second type of diode is a **zener diode**. The zener diode will also allow current to flow in only one direction; however, if the voltage applied to the circuit reaches a predetermined level, voltage can be allowed to flow backward without damage. The third and most common diode to most of us is the **light-emitting diode (LED)**. The light-emitting diode again allows current to flow in

only one direction; however, this type of diode emits a light when current is flowing. Many times LEDs of different colors are used as indicator lamps in components such as radios.

- *Transistors.* Transistors are semiconductors that are used in many different applications but are mostly used inside of an electronic control module or a computer. The transistor in its simplest function can be thought of as an electronic relay.

## *Summary*

- Electricity is the movement of electrons from one atom to another. Electrical flow is established when excess electrons stored in one area flow to an area with a deficiency of electrons.

- Current is the cumulative flow of electrons through a given point within the circuit. Direct current allows electrons to flow in one direction only; alternating current allows the flow of electrons in either direction.

- Resistance is the opposition to electrical flow. All materials have some amount of natural resistance. Resistance is measured in ohms.

- Voltage is a difference of potential within an electrical circuit. It is the force used to push electrons through a circuit. Voltage drop is the amount of potential that is consumed within an electrical circuit.

- Electricity can be produced by magnetic induction or by chemical reaction. Any time a conductor cuts through a magnetic field, a voltage will be induced into the conductor.

- Ohm's Law defines the relationship of current, voltage, and resistance. Any change in one of these elements will affect the others.

- A circuit is the mechanism that allows us to use and control electricity. A complete circuit consists of a path from a power source through a load and to the negative side of the power source. Most circuits contain controls as well as a circuit-protection device.

- Three types of circuits are series, parallel, and series-parallel. Branches are alternate paths that allow electricity to flow in more than one intended direction.

# Review Questions

1. Explain what electrical flow is and how it is established.
2. All of the following statements about current are true *except:*
   A. Current flow is measured in amps.
   B. Current flow is the force used to push electrons through a circuit.
   C. Current flow is the cumulative flow of electrons in a circuit.
   D. Current can be measured with an ammeter.
3. Which of the following statements about resistance is true?
   A. Heat is created anytime electrons move through a resistance.
   B. Resistance is always the result of an external condition.
   C. Resistance is measured in units known as volts.
   D. Some amount of voltage will be created any time current is forced through a conductor.
4. Explain the difference between voltage and voltage drop.
5. Which of the following methods can be used to produce electricity?
   A. Chemical reaction
   B. Magnetic induction
   C. Chemical induction
   D. Both A and B
6. Ohm's Law states that 6.25 billion billion ohms flowing past a given point in an electrical circuit will produce 1 ohm.
   A. True
   B. False

7. All of the following statements about circuits are correct *except:*
   A. A series circuit allows only one path for electricity to flow.
   B. Some electrical circuits do not have loads.
   C. Some electrical circuits have components that place a load on the circuit but also perform a function.
   D. All vehicle electrical circuits ultimately start and end with the battery.
8. Technician A says that in a parallel circuit the total resistance of the entire circuit is less than the smallest resistor. Technician B says that an open anywhere in a parallel circuit will cause the entire circuit to stop functioning. Who is correct?
   A. Technician A
   B. Technician B
   C. Both A and B
   D. Neither A nor B
9. Technician A says that relays are usually found controlling low-current circuits. Technician B says that solenoids are electrical components that carry out mechanical actions. Who is correct?
   A. Technician A
   B. Technician B
   C. Both A and B
   D. Neither A nor B

# Chapter 13

# Automotive Computers

## Introduction

The automotive **computer** is an electronic device that gathers electronic data in the form of voltage signals, analyzes the information, and sends output signals to make changes to the actuators of a system. Computers and their associated sensors and actuators have taken over many tasks that were once handled solely by electromechanical or vacuum devices. The advantages of using on-board computers in vehicle applications are that they can make almost instantaneous adjustments with pinpoint accuracy. Additionally, they are equipped with diagnostic features that assist the technician in locating system problems with computer input and output circuits.

Today's modern vehicle can have as many as two dozen or more on-board computers or **modules (Figure 1)**. Computers and modules are similar in operation. Modules, however, usually are designed to handle one or two specific tasks, whereas computers are designed to handle multiple tasks simultaneously. Computers and modules can be found controlling everything from the fuel system operation to window operation. In most vehicles, each of these modules is linked to an on-board network that allows them to constantly communicate with one another.

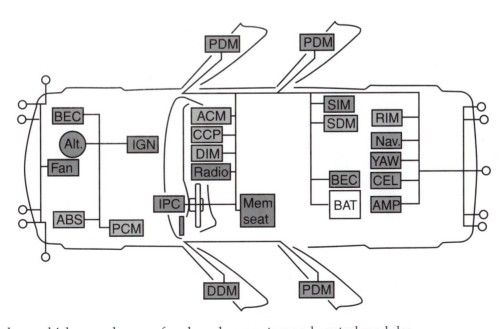

**Figure 1.** Modern vehicles use dozens of on-board computers and control modules.

# THE POWERTRAIN CONTROL MODULE

The **Powertrain control module (PCM)** is a microcomputer that is installed on a vehicle to monitor and control the actions of the powertrain systems, specifically the fuel, ignition, emission and transmission systems. The PCM is located inside of either the vehicle's passenger or engine compartment as shown in **Figure 2**. The PCM is connected to the vehicle wiring harness through the use of large connectors that house anywhere from 50 to 180 individual pins; each pin represents the circuits needed for the PCM to connect to the vehicle systems. Refer to **Figure 3**. The PCM uses voltage as a means of gathering information, analyzing information and sending information. These voltages are called **signals**. The way in which the PCM uses

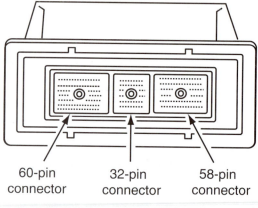

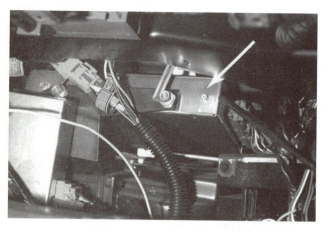

A

B

**Figure 2.** The PCM can be located (A) under the dash or (B) in the engine compartment.

60-pin connector   32-pin connector   58-pin connector

**Figure 3.** The PCM uses many different pins that connect it to various sensors and actuators.

these signals can be broken down into four functions, input, processing, storage, and output.

## INPUT

The computer receives input information from various **sensors** in the powertrain. A sensor is an electronic device that measures a mechanical condition and changes that condition to an electrical signal. Examples of mechanical conditions are coolant temperature, throttle blade position, or the amount of oxygen in the exhaust. The sensors are connected to the PCM by two or three wires. The PCM measures the amount of voltage on these wires and can sense when the voltage has changed. Sensors can provide this input to the PCM in three different ways:

- *Digital input.* A digital input signal will be one of two values: on or off. Digital signals are usually provided by a control component. The control component can be a contact switch, such as a brake switch, or it can be an output from another control module, such as an ignition module. Digital signals can be represented in two ways. In the first method, the PCM expects to see a specific external voltage level at a specific terminal. A control component is wired in series between the voltage source and the PCM. When the control component changes state from opened to closed or vice versa, the PCM notices the change in the level of voltage and knows that an action has taken place. The second method depends on the PCM to send out a reference voltage to a circuit. The PCM measures the amount of voltage supplied to the circuit. Again, the control component is wired in series; however, the component will supply a ground connection when closed. When the control component is open, the PCM will measure a high signal. When the switch is closed, the voltage is

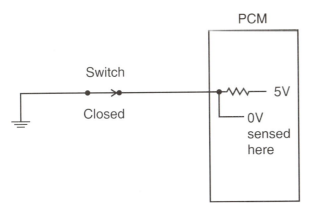

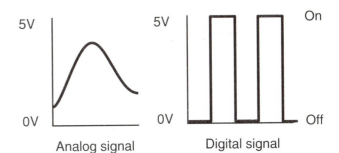

**Figure 6.** A comparison of analog and digital signals.

**Figure 4.** The reference voltage is pulled low when the switch is closed.

"pulled to ground" and the PCM measures a low signal. This concept is illustrated in **Figure 4**.

- *Voltage-modifying.* These signals originate in the PCM as a **reference voltage**. The PCM sends a known voltage, usually 5V, to the various sensors in the system. The sensor then alters the reference voltage based on the mechanical condition and returns it to the PCM as an **analog** signal **(Figure 5)**. An analog signal is one in which the voltage can be any value within a range. Based on the voltage of the signal, the PCM can calculate the status of the specific mechanical condition. Modified voltage signals can be analog, as described above, or digital in nature, as shown in Figure 4. An example of this type of sensor would be a **throttle position sensor (TP sensor)**. Digital and analog signals are shown in **Figure 6**.

- *Voltage-producing.* These sensors produce voltage that directly correlates to a mechanical condition. These sensors are most often used to detect engine speed,

vehicle speed, engine vibration, and engine fuel mixture. The signal produced by this type of sensor is an analog voltage.

## Input Conditioner

Input signals used as inputs to the PCM can be either digital or analog in nature **(Figure 7)**. Although the PCM uses both digital and analog signals as inputs, the processor section of the PCM can recognize only digital signals. Because of this, the PCM is equipped with special circuitry called a **signal converter** that converts analog signals into useable digital code that the processor can interpret. This is called signal conditioning. The signal converter is also responsible for increasing the strength of weak signals provided by some voltage-generating sensors.

## PROCESSING

The **microprocessor** is the component within the PCM that is in charge of making the calculations and decisions. It uses the information provided by the input sensors,

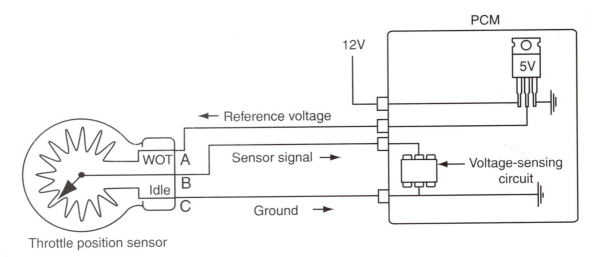

**Figure 5.** This type of sensor modifies the reference voltage and returns a signal to the PCM.

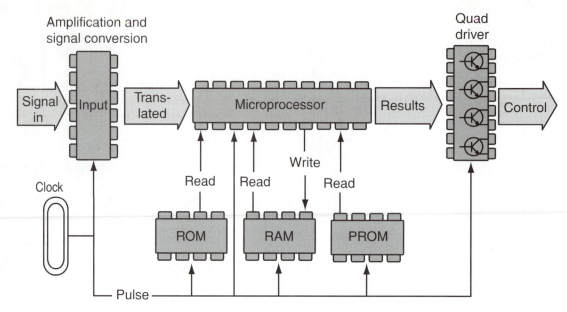

**Figure 7.** The signal converter changes analog signals to useable digital signals.

compares it with the parameters stored in the memory sectors, and sends commands to the actuators. The microprocessor is comprised of thousands of tiny transistors etched into one small microchip. The chip is then connected to the computer's main board with the other major components. (Refer to Figure 7.)

## STORAGE

For the PCM to operate, it must contain stored information about operating calibrations and parameters. This information, contained in "look-up" tables, is then used by the processor section of the PCM as a reference, which makes changes accordingly. Like any computer, the PCM uses various types of memory. The memory can store specific vehicle information, powertrain operating parameters, and temporary operating adjustments. The PCM uses four different types of memory:

- *Random-access memory (RAM).* Random-access memory is the memory block where frequently used information relating to PCM operation is stored. The PCM can read information from RAM as well as change the information stored there. The RAM is a volatile memory, meaning that when the key is turned off the information stored here is lost.
- *Read-only memory (ROM).* Read-only memory is the memory block in which the system operating calibrations are stored. These calibrations are always based on specific vehicle information such as engine size, gear ratios, and compression ratios. The ROM is a **nonvolatile** reference memory. This means that the infor-

mation is permanent and will not be erased when power to the PCM is discontinued and that the PCM can look at the information stored here but cannot change it. Because ROM is hardwired within the PCM, manufacturers that use ROM to store vehicle parameters must use a unique PCM for every vehicle model built.

- *Programmable read-only memory (PROM).* Programmable read-only memory is similar to ROM in that it is nonvolatile memory and stores specific vehicle information. The difference between PROM and ROM is that the PROM is a removeable memory chip that can be changed without changing the complete computer assembly. This allows the manufacturers a great deal of flexibility in the design and use of vehicle computers. Because the PROM is removeable, changes to vehicle calibrations can be made relatively inexpensively in the field by installing a different PROM.

 *Several aftermarket manufacturers produce ROM, PROM, and EEPROM upgrades for many vehicles. The upgrades are generally calibrated to improve performance for a specific application, such as towing or WOT operation.*

Electronically erasable programmable read-only memory (EEPROM), unlike PROM, cannot be removed from the PCM; however, EEPROM can be erased and reprogrammed in the field using a scan tool. This allows software changes to be made much quicker and with less expense. The manufacturers often update the EEPROM calibrations to fix minor glitches within the operating programs that can cause customer concerns.

- *Keep-alive memory (KAM).* Like RAM, the PCM can read from and write to the KAM. The main difference between the two types is that KAM will remain intact when the ignition is turned off but will be erased when the battery is disconnected. The PCM stores **adaptive strategy** adjustments within the KAM. The adaptive strategy is the PCM's ability to learn and record events that allow operating parameters to deviate from the program stored within the ROMs.

## OUTPUT

Once the processor has gathered information and made a decision, it will then send a command to a specific output **actuator**. The actuators are electromechanical devices, such as fuel injectors or control relays, that make the necessary changes to the operating systems to make the powertrain operate within the parameters set forth by the program. The actuators are connected to the PCM using wires. Typically, the PCM will exercise its control by completing the ground circuit to the component.

## OPERATION

The PCM is equipped with a program that outlines the operation of all PCM functions. This program is permanently stored in the ROMs, and within this program are a series of **control maps (Figure 8)**. The control maps are references used by the PCM to guide its actions. The PCM monitors information from the various sensors in the system, comparing the information that is received to the

information that is stored within the control map. If the information received is different from that of the control map, the PCM will adjust the desired actuators to make the parameters received from the input sensors match the control map. The PCM must constantly adjust to make the parameters match those found within the control maps of the program. For example, when the computer is signaled that coolant temperature has changed, the processor will refer to the control maps and find that particular temperature. For that temperature, there will be a recommended fuel system setting. The computer will make the recommended change to the fuel system and wait for feedback from the other sensors to determine additional changes. This gives you a very basic idea of how the PCM makes decisions. The scenarios themselves are much more complex because every sensor within the powertrain is giving feedback to the computer, and each piece of information goes into the final calculations, with some sensor signals receiving a higher priority than others.

## MODES OF OPERATION

The PCM has several different modes and strategies of operation that allow it to operate under many different conditions. The modes and strategies are as follows:

- *Open loop.* In the open-loop mode, the PCM ignores all signals from the oxygen sensors. Engine fuel and timing control are based solely on a limited number of sensor inputs. These signals are compared with a control map, and adjustments are made based on the reading of the sensors. Once the engine is running and idling in open loop, the PCM also takes MAP and rpm into consideration when making fuel and timing adjustments. The PCM will switch to closed-loop operation when the oxygen sensor has warmed up and begins to function and the engine has met specific run time and temperature criteria.
- *Closed loop.* The PCM has complete authority of fuel control and can make adjustments that are based on readings from the sensor readings, especially the oxygen sensor, and fuel map specifications. However, in closed loop the PCM is allowed to deviate from the amount of fuel that is prescribed by the fuel control map to meet immediate needs.
- *Clear flood mode.* The clear flood mode is activated when the PCM senses that the engine is cranking and the TPS signals that the accelerator is near WOT. When this occurs, the PCM restricts injector operation to produce a very lean air/fuel ratio. Once the TPS voltage drops back down or the engine starts, the PCM will return to normal fuel control. The driver can activate this mode in the event that the engine becomes flooded.
- *Deceleration enleanment.* When the PCM detects that the engine is decelerating at a rapid rate, the PCM will

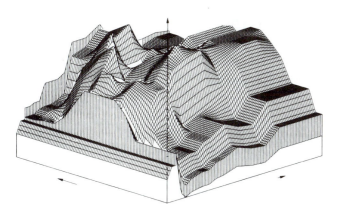

**Figure 8.**   A graphic representation of a control map.

act to lean down the air/fuel ratio. This is done to prevent raw fuel from entering the engine and exhaust systems, which could cause damaging backfire and increased emissions.

- *Adaptive strategies.* As stated earlier, this is the PCM's ability to learn. The PCM constantly monitors all of its input signals and actuator commands. When the signals and commands do not match the control maps, the PCM is allowed to make small adjustments that will enhance the accuracy of the control maps. When the PCM again operates in these modified cells of the control map, the enhanced calibrations are used. These enhancements will reside in KAM until they are overwritten or the battery is disconnected. There are adaptive strategies for many systems in the powertrain and each of these will be discussed in later chapters.

## SELF-DIAGNOSIS

Within the PCM operating program, there are self-diagnostic capabilities known as **on-board diagnostics (OBD)**. On-board diagnostics is simply the PCM's ability to monitor the input circuits for feedback and determine when a signal is out of an operating range as established by the program. The operating program determines how a sensor or actuator is monitored and establishes the sensitivity and frequency of the diagnostic test.

When a problem is detected, the PCM will set a **diagnostic trouble code (DTC)** into its memory. The DTC is an alphanumerical code that is stored in the PCM's memory; each monitored circuit will have one or more numeric codes specifying a condition. The technician can retrieve these codes by using a special piece of equipment called a **scan tool**. A scan tool is a hand-held computer that communicates with the PCM. The scan tool can read DTCs, read the data that is received from the sensors, and perform specific system override functions for testing purposes. The DTC can then be cross-referenced in the service manual to see what circuit(s) are affected and what diagnostic steps are used to pinpoint the cause of the failure. Depending on the particular failure, the PCM might request illumination of the **malfunction indicator lamp (MIL)** or service vehicle lamp. These lamps are used to alert the driver that a failure has occurred in the system.

## Circuit Monitoring

Circuit monitoring is a strategy in which the computer constantly monitors its active circuits. Any time the ignition is turned to the on position, the PCM will begin to monitor its input sensors and will expect to find certain voltages at specific PCM terminals. If the PCM does not see the expected voltage, it instantly knows that there is a problem

and will set a DTC and might illuminate the MIL. This is called **comprehensive component monitoring (CCM)**. For example, if a sensor was completely disconnected, the PCM will receive no feedback from the sensor and will interpret this to mean that a problem exists.

The PCM also has the capability of monitoring the integrity of a circuit or actuator. When a computer is used to control a relay or a solenoid, it will usually do so by providing a ground to the circuit. This means that voltage will be supplied to the power side of the circuit anytime the ignition is on. When the circuit is commanded off, the PCM should monitor battery voltage at the control terminal. This indicates that the circuit beginning at the power source and ending at the PCM is complete. If the PCM commands the actuator on and the voltage does not drop, it knows that the actuator has failed to operate.

*Interesting Fact*

*The PCM has the ability to assess what is logical and what is not. This is called a rationality check. For example, if the PCM assesses that the engine is under 85 percent load and engine rpm is at 4200 and the TPS is reading zero voltage, the PCM can detect this and determine that there is a problem with the TPS signal. This, in turn, would trigger the PCM to set a code and illuminate the MIL.*

## Active Monitoring

Active monitoring is similar to circuit monitoring but in an operating mode. This occurs when the PCM makes an actuator adjustment, expecting to have some measurable amount of response to the input sensors. If this response does not occur, the PCM can determine that a problem exists either in the actuator or in the sensor circuit.

## Intrusive Testing

Intrusive testing is similar to active testing. The major difference is that in an intrusive test the PCM commands one of the actuators on or off and will expect to see specific feedback from one or more input signals. These tests will take place under specific operating conditions. If the feedback does not match the expectations, the PCM can set a DTC and might choose to illuminate the MIL. The PCM has the ability to perform many different tests, and each manufacturer has its own set of tests as well as its own set of specifications for running these tests.

# Summary

- The automotive computer is an electronic device that gathers electronic data in the form of voltage signals, analyzes the information, and sends output signals to make changes to the actuators of a system. The advantages of using an on-board computer on a vehicle is that it can make almost instantaneous adjustments with pinpoint accuracy. In addition, it has its own set of diagnostic capabilities to assist the technician in locating system problems.
- The operation of the computer is broken down into four functions: input, processing, storage, and output. The PCM uses voltage as a means of gathering, analyzing, and sending information.
- The PCM stores specific information about the vehicle, engine, and fuel calibrations. It can also store running calculations on recent operating adjustments as well as any preferences that have been stored in memory. The four major types of memory are ROM, RAM, PROM, and KAM.
- The PCM receives information from various sensors within the system, compares input signals to the program, and sends commands to the actuators to match the input data to the control maps.
- The PCM has several operating modes to allow operation under many different conditions. Adaptive strategy is the computer's ability to learn.
- On-board diagnostics is simply the PCM's ability to monitor the input circuits for feedback and determine when a signal is out of an operating range established by the operating program and log the failure information. Circuit monitoring, active monitoring, and intrusive testing are methods used by the PCM to monitor the entire system for proper operation.

# Review Questions

1. Explain each of the four functions of the PCM.
2. Identify the four types of memory and explain their function.
3. Technician A says that information stored in ROM can be changed. Technician B says that EEPROM memory can be removed and replaced. Who is correct?
   A. Technician A
   B. Technician B
   C. Both A and B
   D. Neither A nor B
4. The PCM receives _____ from the _____ and sends _____ to the _____ so that the input information will match what is stored in the control maps.
5. The adaptive strategy is the PCM's ability to learn.
   A. True
   B. False
6. Which of the following statements is correct?
   A. The PCM is constantly monitoring the program and trying to make the program parameters match the sensor data.
   B. The PCM can use the data from all of its sensors any time that a signal is received.
   C. The PCM monitors one signal at a time.
   D. The PCM monitors multiple signals and makes corrections to the actuators in order to make the signals from the feedback match that of the operating program.
7. Technician A says that vehicle operating parameters can be stored in ROM. Technician B says that program adjustments cannot be made on vehicles that use ROM. Who is correct?
   A. Technician A
   B. Technician B
   C. Both A and B
   D. Neither A nor B

# Chapter 14

# PCM Input Sensors and Signals

## Introduction

A sensor is a device that has the ability to monitor a mechanical condition and change it into an electrical signal. The powertrain control module (PCM) uses a variety of sensors to provide it with necessary information about various operating conditions of the powertrain. Based on this information, the computer can make adjustments and corrections. In the case of the PCM, the sensor inputs are used to make corrections in the fuel, ignition, and transmission systems. The PCM itself can even act as an input to another computer. Each manufacturer might use any number or combination of inputs found here, depending on their specific needs and systems. Additionally, each sensor is given a priority in the decision-making process and the sensor priority can change, depending on the necessary corrections. It is always advisable to consult the manufacturer's specific service information when working on a particular vehicle. In this text, the sensors will be grouped as either voltage modifying or voltage producing.

Some of the sensors within any system may be used for diagnostic purposes only; this means that, although the sensor is operational and data is available, the data is ignored until the PCM program request that a specific test be performed. When a test is called for, the computer will then monitor what the sensor is relaying.

As we begin to look at the PCM inputs, we need to understand that not all of the signals are provided by a sensor, but some of the signals are shared between the PCM and another system. The PCM monitors several other systems just to keep track of what is going on in order to make minute adjustments to keep the powertrain operating properly.

## MODIFIED VOLTAGE INPUT SIGNALS

Modified voltage input signals may be analog or digital signals. These signals can be derived from a 5-volt reference that originates in the PCM and is modified by a sensor, a digital voltage input such as a brake switch, or an input from another control module or processor. The most commonly used inputs will be listed in alphabetical order.

## AIR CONDITIONER REQUEST

The air conditioner (A/C) request signal is monitored by the PCM to know when the A/C clutch has engaged. Because the A/C compressor adds a substantial parasitic load when operating, the computer needs to compensate by increasing the idle speed and in some cases might turn the A/C clutch off when extra power is desired, such as wide open throttle (WOT). This input signal is also used to activate the electric fans when the A/C clutch is engaged. This input signal can be obtained through a parallel connection from the dash switch, a direct input to the PCM, a vehicle body computer, or A/C programmer module **(Figure 1)**.

## BATTERY VOLTAGE (B+)

Because the operation of all system actuators is dependent on having the proper amount of voltage, it is desirable to monitor system voltage level. By monitoring the voltage levels, the PCM can make small adjustments to the operation of the fuel injectors, ignition system, and idle speed as the supply voltage changes due to the fluctuations caused by the parasitic loads from engine-driven

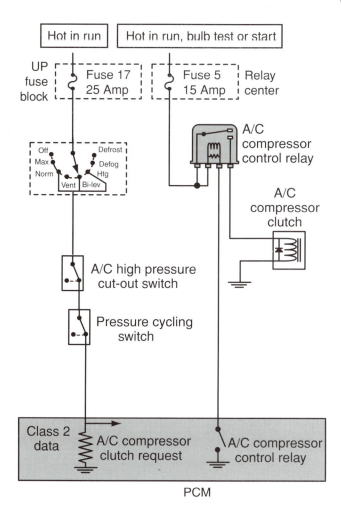

**Figure 1.** A typical A/C control circuit. Note that the PCM controls the A/C clutch.

accessories. This signal is further used in many vehicles to control the operation of the alternator field circuit.

## BRAKE PEDAL POSITION SWITCH

The **brake pedal position (BPP) switch** is located near the brake pedal. The BPP provides a 12-volt digital signal to the PCM. When the switch is open, the PCM will see zero voltage; when the switch is closed, the PCM sees 12V. The PCM uses this information specifically to control the operation of the torque converter clutch in the transmission. Input from this sensor can be used to allow or inhibit cruise control operation. Minimal idle speed adjustments can also be made based on BPP input.

The BPP switch in some cases will have two or three sets of switch contacts; the other contacts can control the brake lamps or cruise control or can be an **antilock brake system (ABS)** input. In addition to these functions, the BPP can have a vacuum switch built in. This is for the release of vacuum from a vacuum-actuated cruise control system.

## ENGINE COOLANT TEMPERATURE SENSOR AND INTAKE AIR TEMPERATURE SENSOR

The **coolant temperature sensor (CTS)** is a thermistor that is typically screwed into a coolant passage in the intake manifold or thermostat housing and is used to measure the coolant temperature of the engine. There are two types of thermistors: a **negative temperature coefficient (NTC)**, meaning that the internal resistance decreases as temperature rises, and a **positive temperature coefficient**, which means that the resistance rises as the temperature does. Most CTSs are of the **NTC** variety. The PCM uses these measurements to make fuel enrichment and timing adjustments, transmission shift, and converter engagement decisions. The typical CTS will have an operating range of −40–270°F (−40–133°C). Resistance at −40°F (−40°C) will be more than 100,000 ohms and can drop below 100 ohms at the high end of the operating range.

The CTS is connected to the PCM using two wires. One wire supplies 5V to the sensor through a resistor in the PCM and measures the voltage on that line. The other wire is a sensor ground. As the resistance within the sensor changes, the amount of voltage that is measured in the PCM is changed accordingly. **Figure 2** illustrates a typical CTS circuit. When the engine is cold, the measured voltage is high; as the engine warms up, the voltage drops.

The **intake air temperature sensor (IAT)** is mounted in the fresh air intake port of the engine. The purpose of this sensor is to relay to the PCM the temperature of the air entering the engine. The PCM uses this information in making fuel enrichment decisions. The operating information is the same as that of the CTS, with the only major difference being the location of the sensor. An IAT sensor and circuit are shown in **Figure 3**.

## EXHAUST GAS RECIRCULATION BACKPRESSURE TRANSDUCER

The EGR backpressure transducer is used to measure the flow of exhaust gas provided by the EGR valve. The EGR valve is installed on an engine to recirculate burned exhaust gases back to the intake manifold. A full operational description is provided in Section 6. The PCM uses feedback information to know if the EGR valve has opened and if it has opened enough. From this information, the PCM will make adjustments to EGR valve position, air/fuel ratio, and timing and serve as a diagnostic monitor. Through use of specific control maps, the PCM knows the specific amount that the EGR is supposed to be open for a given engine condition. If it does not detect the corresponding amount of flow, the PCM will assume that a problem exists.

The sensor is a three-wire sensor that receives a 5-volt reference from the PCM, has an internal ground in the PCM, and returns a signal back to the PCM. As the pressure that

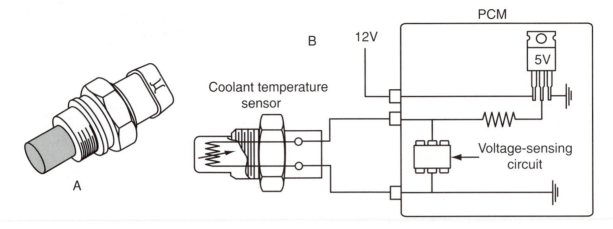

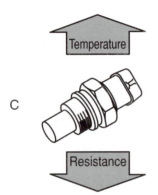

| Thermistor temperature to resistance valves | | |
|---|---|---|
| °F | °C | Ohms |
| 210 | 100 | 185 |
| 160 | 70 | 450 |
| 100 | 38 | 1,800 |
| 40 | 4 | 7,500 |
| 0 | -18 | 25,000 |
| -40 | -40 | 100,700 |

**Figure 2.** (A) A typical CTS. (B) A typical CTS circuit. (C) This chart shows the relationship between temperature and resistance.

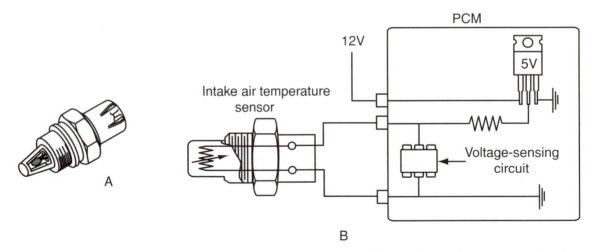

**Figure 3.** (A) A typical IAT sensor; the temperature-to-resistance relationship is the same as that for the CTS. (B) The typical IAT circuit.

the sensor is exposed to changes, the voltage returned to the PCM changes proportionately.

The sensor will be connected to the engine by the use of a flexible hose; the hose is connected to a port just below the inlet at the EGR valve. When the valve is closed, the pressure in the inlet tube will remain constant, and therefore the signal remains constant. As the valve is opened and the inlet is exposed to the lower pressure that is available at

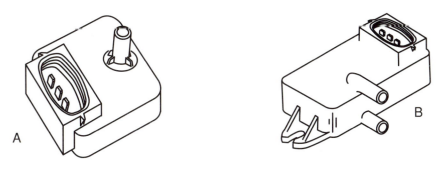

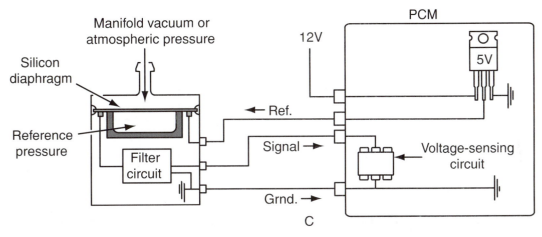

**Figure 4.**   (A) A typical EGR backpressure transducer sensor. (B) The differential EGR backpressure transducer sensor, which has the ability to compare two pressures. (C) The circuit for both sensors is very similar.

the outlet of the valve, the pressure in the inlet drops. The wider the valve opens, the more the pressure will change. As this takes place, the sensor changes the voltage signal to the PCM. A variation of this sensor will use a fixed orifice within the inlet tube. The backpressure sensor uses two hoses, one connected to each side of the orifice. The sensor then compares the amount of pressure on each side of the orifice. The placement of the orifice slows down the flow of exhaust gas; therefore, changes in flow are more exaggerated and the sensor gets a stronger pressure signal. This allows the sensor to detect smaller variations in the flow of gas **(Figure 4)**. This is called a differential EGR backpressure transducer.

## EXHAUST GAS RECIRCULATION VALVE POSITION SENSOR

Like the EGR pressure transducer the EGR position sensor provides the PCM with information about the operation of the EGR valve. The position sensor relays information about the actual position of the valve based on the pintle position inside of the valve. The flow rate of the valve is then calculated based on the position of the valve. The position sensor can detect only if the valve has opened. It cannot

detect that flow is occurring. The PCM uses information to make adjustments in the air/fuel ratio and timing specifications as well as perform self-diagnosis of the EGR valve.

The position sensor is a three-wire potentiometer circuit. One wire is a 5-volt reference from the PCM, the second is a dedicated ground within the PCM, and the third is a signal return back to the PCM. The sensor is mounted to the top of the EGR valve itself. As the pintle in the valve moves up or down, the effective resistance in the potentiometer changes, resulting in a change in the signal voltage. This sensor can be used on vacuum-actuated valves or can be an integral part of an electronic valve and is not serviceable. **Figure 5** illustrates a removable EGR position sensor.

## Exhaust Gas Recirculation Temperature Sensor

A third method of monitoring the operation of the EGR valve is by measuring the temperature of the EGR gases. By monitoring the temperature, the PCM can assume that a relative amount of flow has occurred. The **exhaust gas recirculation temperature (EGRT) sensor** is a thermistor that is mounted in the EGR passage. The information provided by this sensor is used primarily as a diagnostic test to ensure EGR operation.

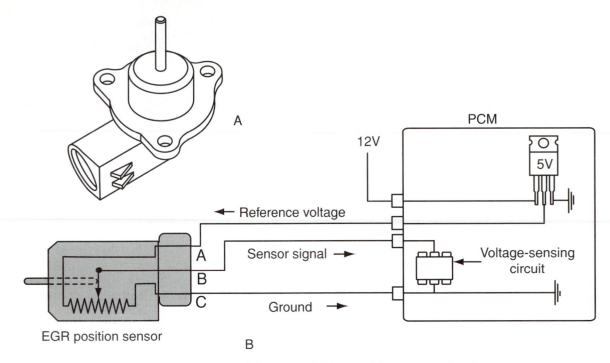

**Figure 5.** (A) An EGR pintle position sensor. (B) A typical EGR position sensor circuit.

## FUEL TANK PRESSURE SENSOR

As more stringent emissions regulations have been implemented, the need to better monitor the fuel storage system has increased. The **fuel tank pressure (FTP) sensor** is used to monitor the pressure within the fuel storage system. This is different from fuel system pressure in that the only thing being measured is the pressure within the fuel tank. The PCM actively monitors the fuel tank pressure through the use of the FTP sensor. Using this information, the PCM can request that the tank be evacuated of any excessive fuel vapors. In addition, the PCM can use the FTP sensor as a diagnostic sensor to ensure that the fuel system is airtight and will not leak vapors into the atmosphere.

The FTP sensor is mounted in the fuel tank; the sensor itself is a pressure-sensitive potentiometer. Three wires connect the sensor to the PCM: a 5-volt reference, a dedicated ground, and a signal wire. As the pressure within the tank changes, the voltage drop through the sensor changes as well, resulting in a signal of less than that of reference voltage to the PCM **(Figure 6)**.

## FUEL PUMP FEEDBACK CIRCUIT

This input can be used to alert the PCM that the fuel pump is receiving voltage. Additionally, as part of the enhanced diagnostic capability of the PCM, some systems may choose to monitor the available fuel pump voltage. This can primarily be used as a diagnostic tool in which the PCM will set a trouble code should the fuel pump voltage

become too low, usually resulting in a no-start condition. If the PCM detects that fuel pump voltage is low, it can assume that the relative fuel pump pressure and volume might have dropped. Based on this, the PCM can lengthen injector pulse to help compensate for the fuel delivery lost by the low voltage condition.

## IGNITION MODULE

In most systems, an ignition module is used to control the operation of the ignition coils. In this type of system, the ignition module will receive information about crankshaft and camshaft position. The module then feeds this information to the PCM, where it is used to determine engine rpm and help control ignition spark advance **(Figure 7)**.

### Ignition Switch

The ignition switch provides one or more direct inputs to the PCM for its operation. In addition, the switch provides the input voltage for other sensors, switches, and modules that provide other crucial information to the PCM.

## MASS AIR FLOW SENSOR

The **mass air flow (MAF)** sensor is located in the air inlet of the engine. All of the air that enters the engine has to pass through the MAF sensor, allowing the sensor to effectively measure the amount of air that the engine is using. By knowing the exact amount of air passing into the

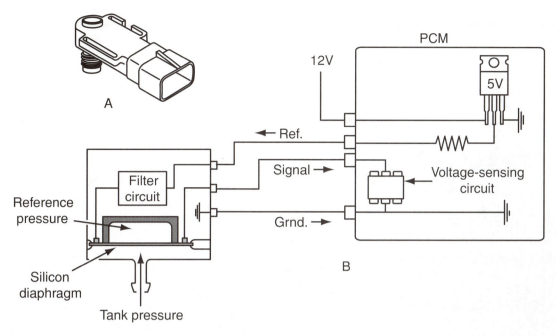

**Figure 6.**   (A) A fuel tank pressure sensor. (B) A typical electrical circuit for the FTP sensor.

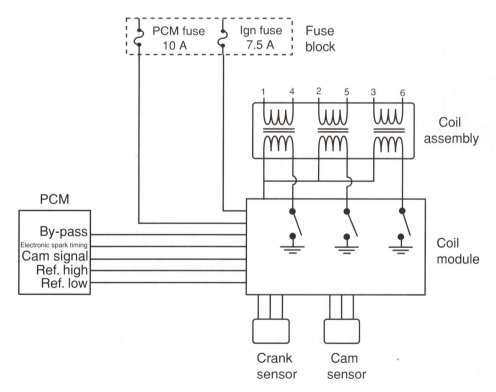

**Figure 7.**   The ignition module receives the camshaft position (CMP) and crankshaft position (CKT) signals and passes them on to the PCM.

engine, the PCM can precisely control the amount of fuel needed to maintain a 14.7:1 air/fuel ratio. The MAF sensor data is used to calculate air/fuel ratio, ignition timing, EGR operation, and some transmission calculations.

There are two basic types of MAF sensors in use today:
1. *Heated Wire and Heated Film*. The heated wire and heated film MAF sensors operate in similar fashion. The basic premise for both of these types of sensors is that

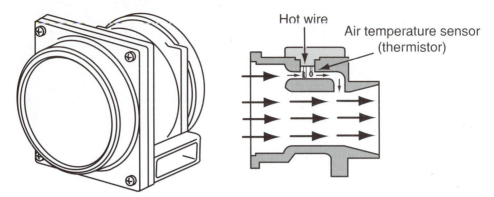

**Figure 8.** The air flows over the heated element in this type of sensor.

a resistor or wire is placed in the incoming air stream. The MAF electronics send enough current through the wire or resistor to keep it at a predetermined temperature. As air is passed across the wire or resistor, the rush of the air cools the resistor. The electronics respond by sending additional current to the resistor to bring it back up to the desired temperature **(Figure 8)**. The amount of current required to keep the sensor at its desired temperature is relayed back to the PCM in a signal proportionate to the airflow. Depending on the type of sensor used, the signal relayed back to the PCM can be either in an analog voltage or it can convert to a digital frequency.

2. *Vane Airflow.* The second type of MAF sensor that is commonly used, the vane airflow meter, is more of a meter than a sensor. The meter is placed directly in the intake air stream, as the others are; however, the vane-type meter uses a swinging door that is attached to a potentiometer **(Figure 9)**. As air begins flowing into the engine, the door moves accordingly. As airflow increases, so does the amount that the door is opened. As the door is moved, so is the potentiometer. With the movement of the potentiometer, the reference voltage is changed to directly correspond with the amount of air flowing into the engine. The sensor is connected like the other potentiometer-type sensors, using a 5-volt

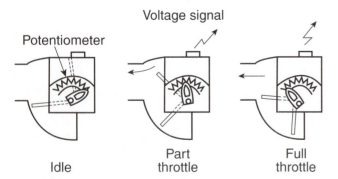

**Figure 9.** This sensor depends on the movement of a small door to detect airflow.

reference, a dedicated ground, and a signal wire. The heated wire sensor is more widely used in late model vehicles than is the vane-type sensor.

## MANIFOLD ABSOLUTE PRESSURE, BAROMETRIC PRESSURE SENSOR

The **manifold absolute pressure (MAP)** sensor is installed to monitor the relative load that is placed on the engine. This is accomplished by monitoring the vacuum of the engine. As engine load increases, the vacuum decreases. As load decreases, the vacuum increases. By knowing the relative engine load, intake air temperature, and engine rpm, the PCM can accurately determine how much air is flowing into the engine and can estimate the amount of fuel needed within the engine. The MAP readings are heavily factored in the air/fuel ratio adjustments, ignition timing adjustments, and EGR operation.

The MAP sensor changes the vacuum reading into a voltage signal. The sensor is typically located on or near the intake manifold or the firewall of the engine compartment. The sensor is connected to the engine using a vacuum hose. The sensor is connected to the PCM through the use of a 5-volt reference, a dedicated ground, and a signal wire **(Figure 10)**. There are two different types of MAP sensors used. One type is a pressure-sensitive potentiometer that sends a modified reference signal back to the PCM based on the amount of vacuum inside the sensor. The second type converts the 5-volt reference signal into a frequency-based signal. This provides the PCM with extremely accurate information.

The MAP sensor is also able to calculate base barometric pressure readings. When the ignition switch is turned to the on position before the engine is turned over, the PCM will take a sample reading of the pressure. Because the engine is not spinning over, the pressure within the engine is the same as that of the atmosphere. By measuring this, the sensor can observe a relative pressure reading of the atmosphere. The barometric pressure reading is used to obtain base fuel system adjustments. With this sensor reading, the system is able to automatically compensate for

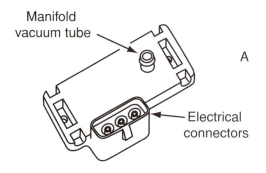

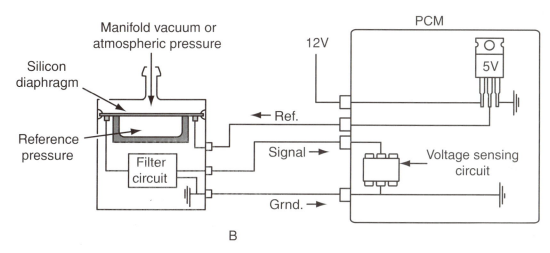

**Figure 10.** The MAP sensor is connected to the intake manifold using a small hose. (A) The MAP sensor also produces the barometric pressure reading. (B) A typical MAP circuit.

drastic changes in pressure or altitude. Some vehicles use a separate sensor to measure barometric pressure. However, its operation is similar to a MAP sensor. These are called barometric pressure sensors.

A variation of the MAP sensor used by some manufacturers is called the **temperature/manifold absolute pressure (TMAP) sensor**. The TMAP combines the functions of the IAT sensor and the MAP sensor into one unit.

## POWER STEERING PRESSURE SWITCH

The **power steering pressure switch (PSPS)** is a normally open switch that is placed in the high-pressure steering hose. As the steering is moved toward the lock position, the pressure within the system rises, creating additional strain on the power steering pump. The additional strain causes the pump to create a larger parasitic drag on the engine. The excess drag on the engine can cause the idle speed to be drastically lowered and can cause the engine to stall. The PSPS signal is used to notify the PCM when these conditions occur. The PCM supplies a signal wire to the switch. When the switch is open, the PCM senses 12V on the signal wire. When the pressure in the steering system is increased to the threshold that is built into the switch, the

switch closes and grounds the input wire from the PCM, causing the voltage that is sensed by the PCM to be less than 1V. When the PCM receives this high-pressure signal, it can modify engine speed and ignition timing and in some cases can turn the A/C compressor off.

## THROTTLE POSITION SENSOR

The throttle position sensor (TP sensor) is fastened to the throttle body, where it monitors the position of the throttle. By doing this, it can relay information to the PCM about how much and how fast the operator is opening or closing the throttle. This information is used to make fuel enrichment and timing adjustments and it is taken into account for shift points and torque converter operation. The TP sensor is a potentiometer that receives a 5-volt reference from the PCM and effectively lowers the voltage in direct relation to how far the throttle is opened. The sensor has an operating range of 0.5V at closed throttle to 4.5V at WOT.

Typically, there are three wires connecting to the TP sensor, each of them originating at the PCM. One wire is the reference voltage, one is the sensor ground, and the last one attaches to the wiper. The wiper produces the signal and is effectively connected to the throttle blade. The 5-volt

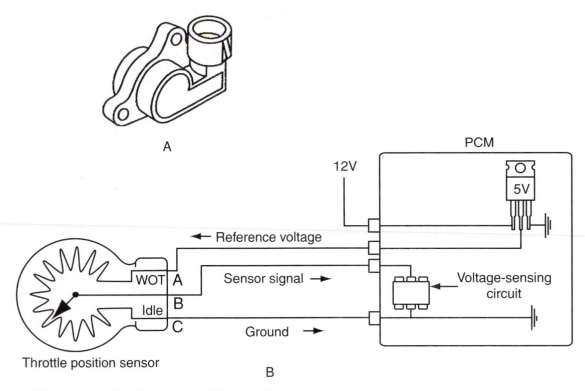

**Figure 11.** (A) A typical TPS. (B) A typical TPS circuit.

reference is taken to ground through the internal resistor within the sensor; this provides a constant voltage drop across the reference signal. The movable wiper arm moves across the resistor and samples the amount of voltage at a given point. This information is sent to the PCM as position data **(Figure 11)**.

## TRANSMISSION RANGE SENSOR

The transmission range switch is located on the exterior of the transmission at the point that the shifter shaft protrudes from the transmission case. The **transmission range sensor (TR sensor)** provides the PCM with information regarding the specific transmission range that the operator has selected. Most switches house several switches in one housing. The switches open and close in different combinations. The PCM detects and interprets these combinations of digital signals into a gear position and uses this information to make idle speed adjustments. The transmission range switch also has an additional switch to interrupt the starter circuit to keep the vehicle from starting in gear and a switch to operate the reverse lights. The PCM further uses TR sensor input in controlling transmission functions.

## METHODS OF VOLTAGE GENERATION

Voltage-generating sensors are sensors that produce a voltage relative to a mechanical condition. This type of sensor is typically used to measure the relative position of a rotating object, speed differential between two components, engine vibrations, and oxygen content of the exhaust. There are several different methods that a sensor can use to produce voltage, and there are several different sensors that send signals based on this method.

## HALL EFFECT

Hall effect switches do not produce a voltage but modify an existing voltage; however, Hall effect switches and voltage-generating devices can be interchanged in particular sensor applications, depending on the needs of the manufacturer. Hall effect switches are electronic switches that produce a digital voltage signal. This type of sensor is supplied with a 12-volt ignition feed, a ground circuit, and a 5-volt reference signal.

The switch itself is comprised of an electronic circuit that when exposed to a magnetic field turns on a transistor that, in turn, pulls the reference voltage low (0V). When the magnetic field is blocked, the voltage then again rises back to 5V (high) **(Figure 12)**. The constant switching from high to low produces a digital voltage signal.

The electronic circuit is placed on a small microchip; the magnet is placed across from the chip and the two are separated by an air gap. An **interrupter ring** mounted to a gear or shaft rotates through the air gap in the sensor. The interrupter ring is typically a thin metal ring that has windows or gaps cut out of it. These windows allow for the magnetic field to pass and activate the chip. As the shaft or gear spins,

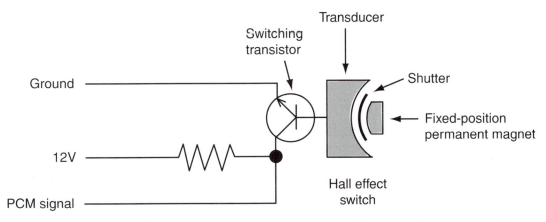

**Figure 12.**   A Hall effect switch creates a digital signal by pulling the reference voltage to ground.

the interrupter ring passes through the gap between the magnet and the chip of the sensor. As the windows pass into the air gap, the magnet activates the chip and the reference voltage is pulled low. As the window closes, the voltage again rises. Any number of windows can be placed in an interrupter ring to meet the demands required by the system. It is not uncommon to have multiple Hall effect switches and interrupter rings installed in one sensor, allowing a corresponding number of signals to be obtained at once. These sensors are typically used as cam and crank sensors.

## Optical Sensors

Optical sensors are similar in nature to a Hall effect sensor. The optical switch, like the Hall effect switch, produces a digital signal by pulling a reference voltage low. A LED faces toward a **photocell**, a component that produces a voltage when it is exposed to light. When the photocell is exposed to the LED, it creates a small voltage that turns on a transistor. The transistor then completes the circuit for the reference voltage to go to ground **(Figure 13)**.

The optical sensor is mounted stationary and uses a rotating interrupter ring with windows cut out in various locations. The interrupter ring is placed between the LED and the photocell. As the windows move in-line with the LED and the photocell, the photocell produces a voltage and completes the circuit. Optical sensors are most often found in ignition distributors.

## VARIABLE RELUCTANCE

Variable reluctance sensors rely on the principles of magnetic induction to create a signal and are often referred to as permanent magnet generators. There are two variations that have been used.

The first style and the one most common today consists of a magnet that is wrapped with many coils of copper wire. Increasing the number of copper coils strengthens the electrical signal. This type of sensor is mounted at a stationary point and uses a conductive wheel to compress the magnetic field, effectively forcing it back into the magnet and through the coils, thus increasing the voltage.

The conductive wheel is called a reluctor wheel. The reluctor has gaps cut in the outer circumference of the wheel; this leaves a series of gaps and teeth. As the reluctor wheel passes across the end of the sensor, the teeth move

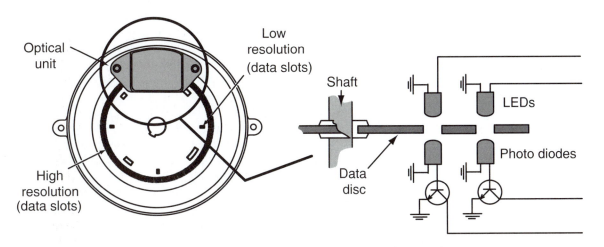

**Figure 13.**   An optical sensor uses LEDs and photodiodes to create specific signals.

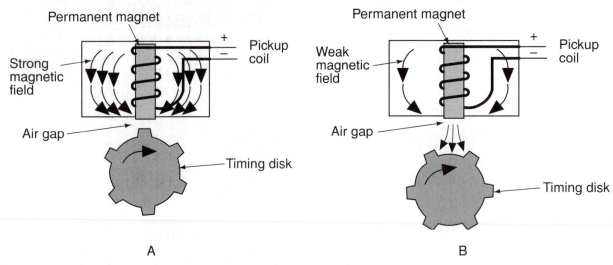

**Figure 14.** (A) As a tooth of the reluctor nears the magnetic pole, magnetic energy is forced into the sensor and voltage is induced into the windings. (B) As the tooth passes the end of the magnet, the energy expands and the voltage drops.

closer to the end of the sensor. The magnetic field at that point is compressed and a voltage builds within the sensor. The voltage continues to build until one of the reluctor teeth moves to its closest point at the end of the sensor. As a tooth of the reluctor moves away from the sensor, the magnetic field begins to expand and the voltage decreases until the gap is at its deepest point in the wheel. As the wheel turns, the cycle starts over again. An illustration of this is shown in **Figure 14**. These sensors produce an A/C voltage and are widely used as crank sensors, cam sensors, and wheel speed sensors. Additionally, these sensors can have the sensor and the reluctor housed in one unit, typical of a wheel speed sensor, or can be separate components, typical of most crank sensors.

The second type of sensor relies on a circular conductor wrapped in copper coils; this is called a pickup coil. The conductor has sharp points that point to the center of the circle. A circular magnet called a pole piece rotates within the pickup coil. The pole piece has a corresponding point for each point on the pickup coil. The pole piece is connected to a shaft and rotates so that the points from the pickup coil and the pole piece actually move in and out of alignment as the shaft spins. This type of arrangement is typically used in electronic ignition distributors, and the number of points corresponds directly with the number of cylinders that the engine has. The principles of the collapsing and expanding magnetic field in the previous description apply to this sensor as well, with the difference being that the magnet moves rather than the conductor. Both of these types of sensors produce an A/C voltage.

## Crystal-Based Sensors

Crystal-based sensors contain a piezoelectric crystal installed within a threaded housing. The crystal is of specific

properties that when it vibrates within a given frequency range it produces a small A/C voltage. These sensors are typically used as knock detection sensors.

## Reaction-Type Sensors

Reaction-type sensors rely on a material that is reactive to oxygen to either produce a voltage or support a change in resistance. Typical materials are zirconia and titania. These sensors are used as exhaust gas oxygen sensors.

## VOLTAGE-GENERATING SENSORS

Voltage-generating sensors are the sensors that are used to produce a signal based on the previously described methods of voltage generation.

## CAMSHAFT POSITION SENSOR

The **camshaft position sensor (CMP sensor)** signal is used by the PCM to determine the number-one firing position. Based on this position, the PCM can start sequential fuel injection. Depending on the particular system in which it is installed, this sensor may send data directly to the PCM or it may go to an ignition module, where it is relayed to the PCM. The sensor is often located in the front cover of the engine, where the signal is taken by monitoring the cam gear. This CMP sensor design uses a Hall effect switch with either an interrupter ring or a magnet attached to the cam gear.

Other CMP designs use a Hall effect switch and interrupter contained in a sealed housing. The assembly is driven by the camshaft and is located in a typical distributor location **(Figure 15)**. An inductive-type sensor with a reluctor

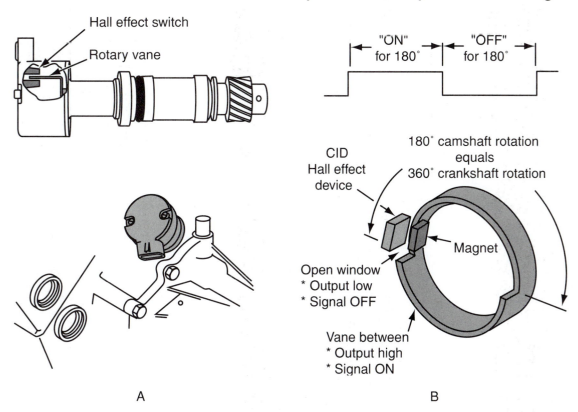

**Figure 15.** (A) This type of sensor contains the Hall effect switch and the reluctor. (B) A signal is created when the window of the reluctor passes through the switch.

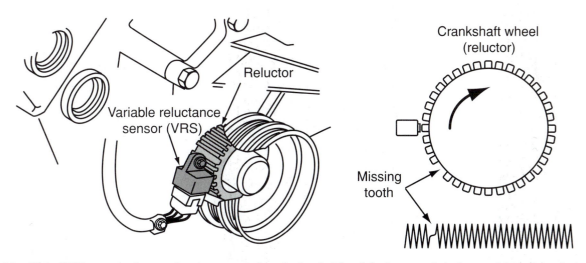

**Figure 16.** This CKT uses a reluctor that is mounted to the backside of the harmonic balancer. Note the missing tooth.

wheel attached to the cam or gear may also be used in some models. The operation is similar to the crank sensor arrangement depicted in **Figure 16**.

## CRANKSHAFT POSITION SENSOR

The crankshaft position sensor (CKP sensor) is used by the PCM to determine engine rpm and firing order and to calculate ignition timing positions. The CKP sensor can be a Hall effect switch or an induction-type sensor. The Hall effect switch is typically located at the front of the engine. The interrupter wheel is fastened to the backside of the harmonic balancer. As the balancer spins, the windows of the interrupter open and close in the sensor, producing a digital signal. It is typical for a dual Hall effect switch to be installed and produce two distinct signals that can be used for precise ignition timing control. Induction-type sensors are equally common; these sensors can be located virtually anywhere near the crankshaft. When this type of sensor is used, the reluctor wheel is typically part of the crankshaft or harmonic balancer. (Refer to Figure 16.)

The number of windows or gaps in the sensor wheels of both types of sensors is calibrated to give the PCM specific information about crankshaft position. For instance, a wide gap or two very close single gaps will usually indicate the firing position of the number-one cylinder. Each vehicle manufacturer will use designs with a varying number of teeth or gaps; these are precisely spaced and will produce distinct electrical signals. By observing these signals, the PCM can detect crank position, engine speed, and crankshaft acceleration. These precise signals make it possible for the PCM to provide extremely quick starting and pinpoint accurate spark timing adjustments. Additionally, the PCM can monitor the acceleration and deceleration of the crankshaft, a strategy widely used to detect an engine misfire. Like the CMP sensor, the CKP sensor can be a direct input to the PCM or signals may go to the ignition module first.

## Distributor

On vehicles that use distributor-type ignition systems, the functions that are carried out by the CMP and CKP sensors are handled within the distributor. The distributor is driven by the camshaft using a helical gear that contacts a mating gear built as part of the camshaft, or the distributor can be driven directly off of the front of the cam gear. The distributor may use a Hall effect switch, a pickup sensor and pole piece, or an optical sensor.

> **You Should Know**  *An optical sensor distributor that has been exposed to moisture can build condensation within the distributor. When this happens, the LED and photocell might become cloudy and fail to accurately transfer a signal. It is important to make sure that the vent hose on these distributors remains clear.*

## KNOCK SENSOR

The **knock sensor (KS)** uses a piezoelectric crystal installed within a threaded housing, which is screwed into the engine block. Locations range from near the crankshaft, in the valley of the engine, or in some cases, at the rear of the cylinder head **(Figure 17)**. The KS is used to inform the PCM when the engine is detonating. When an engine detonates, it creates a frequency-specific vibration that is detected by the crystal. The vibration causes the crystal to create a small AC voltage. The signal produced by the KS is weak and must be amplified in the input conditioner within the PCM before it can be processed.

When the PCM detects detonation, it will retard ignition timing until detonation is no longer detected; when detonation is no longer detected, the PCM will then advance timing until detonation is detected again. This cycle is repeated as long as the engine is running. The range of this activity is generally only a few degrees of timing and will not be noticed by the operator. The KS can provide data

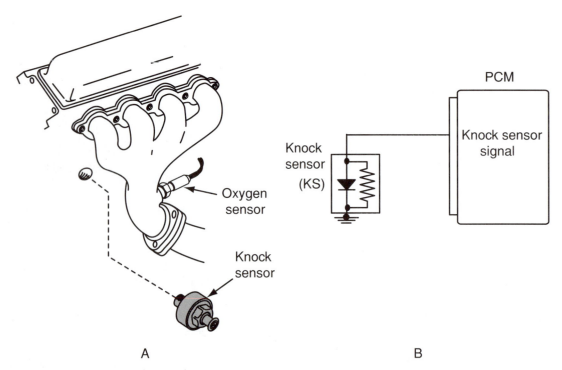

A                                   B

**Figure 17.**   (A) The knock sensor is usually mounted in the block or cylinder head. (B) A typical knock sensor circuit.

directly to the PCM or it can be directed to a separate electronic spark control module.

> You Should Know
>
> *The KS is very sensitive to the torque at which it is tightened. If the sensor is overtightened, it can become too sensitive. If the sensor should become too sensitive, the timing might be retarded too much and an engine performance condition will ensue; if the sensor is left loose, it might not be sensitive enough and the timing might be advanced too much, stay at a point of severe detonation, and cause severe engine damage.*

## OXYGEN SENSOR AND HEATED OXYGEN SENSOR

The **heated oxygen sensor (HO₂S)** is the most important fuel control sensor found on the automobile today. This sensor is a feedback sensor that tells the computer what the relative air/fuel ratio is, based on the amount of oxygen in the exhaust gases. The sensor is located in the exhaust system and detects the amount of oxygen remaining in the exhaust gases. By monitoring oxygen content, we judge how **rich** or **lean** the fuel mixture is. If there is a large amount of oxygen in the exhaust, we know that there was not enough fuel in the engine to use all of the oxygen; this is called a lean condition. If there is little or no oxygen in the exhaust, we know that there was more fuel than was needed to use the supplied amount of oxygen, indicating a rich condition.

The sensor produces a range of voltage: from 0.1V up to 0.9V. Anything that is less than 0.450V is considered to be a lean condition; anything over 0.450V is considered to be a rich condition. The PCM will constantly adjust fuel, causing the sensor to toggle between a voltage greater than 0.450V and less than 0.450V. The sensor will typically cross 0.450V approximately two times per second. Should the voltage become fixed on either side of 0.450V, this is an indication that the PCM cannot control the fuel any longer. When the sensor is cold, it produces little or no voltage. When this is the case, the computer operates in open loop because the PCM does not know if the mixture is rich or lean. (The open loop mixture is typically rich.) Once the sensor warms to 600°F (318°C) or more, the sensor will begin fluctuating above and below 0.450V. The PCM sees this and interprets this to mean that the sensor is functional; at this point, the PCM will go into closed loop.

Vehicles that you are likely to work on can have from one to four **oxygen sensors (O₂Ss)**. Vehicles produced after 1996 are likely to have three to four sensors: a sensor in each exhaust pipe for each engine bank, a sensor directly in front of the catalytic converter, and one directly behind the catalytic converter. By using this large number of sensors, the PCM can more specifically control the fuel to each bank of cylinders and monitor the effectiveness of the catalytic converter. These sensors are equipped with heaters to enhance their overall performance. An HO₂S is shown in **Figure 18**.

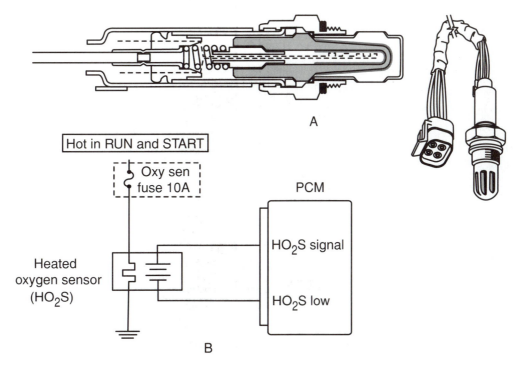

**Figure 18.** (A) A typical HO₂S. Note that this sensor has four wires. (B) A typical HO₂S circuit.

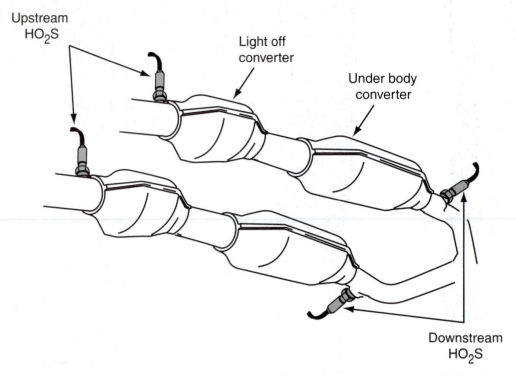

**Figure 19.**   All vehicles produced after 1996 have at least two HO$_2$Ss and some can have as many as four. The location of the sensors is standardized.

> **You Should Know**   *As a part of OBDII, the terminology of O$_2$S locations is standard. The sensors will be numbered according to which bank of the engine they serve and what order in each bank they are positioned in (**Figure 19**). Bank one will always be the bank housing the number-one cylinder and will always contain the pre- and postcatalyst sensors. Bank two, if the engine has one, will be opposite of bank one and will house only bank two sensor one. The exception to this rule is if a vehicle has dual exhaust. Then the bank will contain its own separate catalytic converter and postconverter sensor.*

## Materials

The O$_2$Ss are made from two common materials, zirconia and titania. Each of these materials has different properties and therefore reacts differently to oxygen.

The zirconia sensor is built from a steel shell that uses 18-mm threads to screw into the exhaust system. Inside the shell is a ceramic tube that is closed at one end. Both the inside and outside of the tube are coated with a thin film of zirconia. The inner portion of the cone is attached to a wire that sends a signal to the PCM. The outer portion of the cone is connected to the protective shell. The inside of the cone is exposed to atmospheric air by drawing in air through the electrical wiring connected to the cone. The outer portion of the cone is exposed to the exhaust gases. When oxygen content is high on both the inside and outside of the cone, the sensor produces little or no voltage, usually less than 0.1V. As the oxygen content in the exhaust becomes lower, a differential of oxygen is created between the inside of the cone and the outside of the cone, and a voltage is produced, up to a 0.9V maximum. For this reaction to begin, the sensor must be at a temperature of at least 600°F (318°C). Below this temperature, this sensor has little or no activity.

In an effort to increase the efficiency of the sensor at cold startup, sensors are now equipped with heaters. The heater is activated as soon as the ignition is turned on. By artificially heating the element, the sensor can be raised to operating temperature must faster, and it is not prone to cooling off during long periods when exhaust flow is low, such as when the engine is idling. The O$_2$Ss are equipped with one, two, three, or four wires.

1. One-wire sensors have one wire going to the PCM. The outer surface of the element is attached to the shell and grounded to the exhaust.

2. Two-wire sensors have a center wire going to the PCM, and the shell has a dedicated ground circuit.
3. Three-wire sensors are equipped with a PCM signal wire, a dedicated ground, and a 12-volt feed directly from the ignition switch.
4. Four-wire sensors have a PCM signal wire; a signal low wire to the PCM, in place of the previously dedicated ground wire; a dedicated ground for the heater element; and a 12-volt feed for the heater operation.

Titania sensors perform the same function as the zirconia sensors; however, the operation is slightly different. Titania does not produce a voltage as zirconia does. Operation of the sensor depends on the conductive resistance of titania. The resistance changes proportionately to the amount of oxygen that it is exposed to. The sensor consists of a sensing element housed in a steel case that is screwed into the exhaust system. The sensor element is not exposed to outside air but only to the exhaust. A 1-volt reference voltage is supplied to the element. As the resistance within the sensor changes, so does the voltage that the PCM senses on the signal wire. As for the zirconia sensor, a voltage reading above 0.450 is considered to be a rich condition and a voltage below 0.450 is lean. A relatively small sensing tip and 14-mm threads can identify the titania sensor.

> **You Should Know** *Contaminants, such as silicon, that are introduced into the fuel system or intake system can form a coating on the sensor's element and render the sensor inoperative. Any sealers or fuel additives used should be marked as "$O_2S$ safe."*

## VEHICLE SPEED SENSOR

The **vehicle speed sensor (VSS)** signal is used by the PCM to detect the speed of the vehicle. This information is used to control transmission shifting, torque converter engagement, cruise control, and cooling fan operation. Other systems that use the VSS signal are the ABS and traction control. The VSS uses an induction-type sensor located near the output shaft of the transmission or near the final drive on a transaxle. The external reluctor ring is typically pressed onto the output shaft or final drive. Some sensors can be completely self-contained and driven by a gear **(Figure 20)**. The sensor generates an AC voltage.

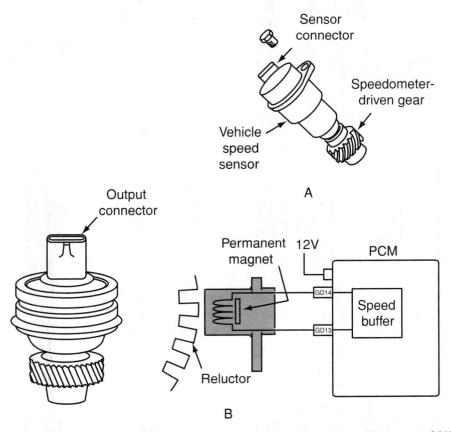

**Figure 20.** (A) This type of VSS sensor is self contained and is driven by a gear. (B) A typical VSS electrical circuit.

# Summary

- Sensors are devices that monitor mechanical conditions and change them into electrical signals that are relayed back to the PCM. Sensors can produce analog or digital signals.
- The PCM uses sensors that modify supplied voltages and those that produce their own voltages.
- The A/C signal provides a digital signal to air the PCM to help determine idle speed.
- The PCM changes how an actuator functions based on system voltage.
- The BPP sensor provides a digital signal, allowing the PCM to detect if the brake pedal is applied.
- The PCM monitors EGR system operation by monitoring flow, valve position, or EGR gas temperature.
- The CTS and IAT sensors operate virtually the same, providing a voltage drop at the PCM. When temperatures are cold, the resistance within the sensor is high. This produces a small voltage drop across the sensor, and the PCM detects a high voltage.
- The FTP sensor monitors fuel storage pressure and helps test the sealing ability of the fuel storage system.
- The PCM can increase injector on time based on the fuel pump voltage.
- The MAP sensor provides information about the amount of engine load that is being placed on the engine. The MAP sensor also has the ability to provide the barometric pressure reading to the PCM. Signals can be provided by an analog signal or a digital signal, depending on sensor design.
- The MAF sensor detects the amount of engine load by measuring the amount of air flowing into the engine.
- The TP sensor provides the PCM with an analog signal, relaying information about how far and how fast the throttle is opened.
- Voltage-generating sensors produce voltage based on the principles of induction, vibration, or reaction. Hall effect and optical sensors provide digital signals by pulsing a reference voltage high and low. Induction, piezoelectric, and reaction sensors all produce analog voltages.
- The PCM controls the fuel injector firing and ignition timing by using references from the CMP and CKP sensors. In distributor-equipped vehicles, the CMP and CKP signals are provided by the distributor.
- The PCM uses KS information to advance or retard ignition timing.
- The $O_2$Ss measure the oxygen content of exhaust gases. The $O_2$S and $HO_2$S are the most important sensors on the vehicle. The $O_2$Ss are located in the exhaust and can both produce and modify voltage. The signals provided by the $O_2$S and $HO_2$S are analog. Voltages below 0.450V are considered lean; voltages above 0.450V are considered rich.

# Review Questions

1. Explain how sensors relay a mechanical condition to the PCM.
2. Describe how modified voltage signals are produced.
3. All of the following about the A/C input signal are correct *except*:
   A. The A/C input signal can be used to modify engine speed.
   B. The A/C input is used to control the operation of the cooling fans.
   C. The power steering can be turned off when the A/C is turned on.
   D. The A/C input is monitored because the A/C compressor has a high parasitic load.
4. Technician A says that the PCM can change how long the fuel injector is opened based on the battery voltage input. Technician B says that the PCM can change alternator output based on the battery input signal. Who is correct?
   A. Technician A
   B. Technician B
   C. Both A and B
   D. Neither A nor B
5. Which of the following is true of the BPP signal?
   A. The BPP sensor provides a 5-volt reference signal.
   B. The PCM can control the operation of the torque converter clutch using the BPP signal.
   C. The BPP switch is used to control only the brake lights.
   D. The PCM sees 12V when the switch is opened.
6. All of the following are correct about CTS and IAT sensors *except*:
   A. The internal resistance of an NTC CTS sensor increases as the temperature increases.
   B. CTS information is used to make fuel and ignition timing adjustments.
   C. The CTS is a voltage-modifying sensor.
   D. Most coolant sensors are of the NTC variety.
7. Which of the following methods can be used to monitor EGR valve operation?
   A. A backpressure measurement
   B. EGR valve position
   C. EGR gas temperature
   D. All of the above

8. Describe how both the heated wire and the vane air flow meter can measure the amount of air that is flowing into the engine.

9. Technician A says that the MAP sensor can provide the PCM with a barometric pressure reading. Technician B says that the MAP sensor must also use air temperature and rpm information to determine the amount of air flowing into the engine. Who is correct?
    A. Technician A
    B. Technician B
    C. Both A and B
    D. Neither A nor B

10. On some vehicles, when high power steering pressure is present, the PCM can turn the A/C compressor off.
    A. True
    B. False

11. Describe how the TP sensor sends a signal to the PCM.

12. Which of the following about a TP sensor is true?
    A. The TP sensor can measure the rate at which the operator is closing the throttle.
    B. The voltage drop across the TP sensor is 5V only at WOT.
    C. The TP sensor voltage is low at WOT.
    D. A defective TP sensor will not affect how the transmission shifts.

13. Explain the operation of the Hall effect switch.

14. Describe the operation of a variable reluctance sensor.

15. Technician A says that Hall effect switches produce a voltage when exposed to a magnetic field. Technician B says that reaction-based sensors produce a digital signal. Who is correct?
    A. Technician A
    B. Technician B
    C. Both A and B
    D. Neither A nor B

16. The PCM uses information from the _____ _____ to determine engine _____, _____ _____ and to calculate ignition _____ .

17. All of the following statements about a HO$_2$S are correct *except:*
    A. Too much oxygen in the exhaust indicates a rich condition.
    B. High HO$_2$S voltage indicates that too much fuel is present in the exhaust.
    C. The PCM ignores HO$_2$S input when operating in open loop.
    D. A voltage of 0.300V indicates an abundance of air in the exhaust.

# Chapter 15

# PCM Outputs

## Introduction

So far, we have looked at the computer operation and how it makes decisions, and we have also looked at the various PCM input sensors and signals. Now we consider the various output actuators. Actuators are those components that use electrical signals provided by the PCM and turn them into mechanical actions. The PCM uses actuators to carry out its requested actions. Like the PCM inputs, not all of these actuators will be used on every vehicle; application will depend on the manufacturers and their demands. Specific manufacturers may use their own variation of any actuator. Listed here are brief generalized operational overviews to allow the student to grasp to what extent the PCM does control the operation of the vehicle. It is always advised to refer to the specific manufacturer's service information when servicing a particular vehicle.

## CONTROL

The PCM will control most actuators by supplying a ground to the component. This means that voltage must already be present at the actuator. The PCM can control the ground circuit in three different ways: simply by supplying a constant ground, **pulse width modulation (PWM)**, or **duty cycle**. The simplest method for a computer to control a component is by supplying a constant ground until the actuator is no longer needed. This method is used for many actuators that are run constantly, such as fuel pumps and A/C compressors. However, when controlling actuators charged with making precise adjustments, such as fuel

injectors, this method is neither fast enough nor accurate enough; in this case, PWM or duty cycle is used.

PWM and duty cycle are methods of controlling an actuator by rapidly turning it on and off. These methods are used to control many components on a vehicle, including fuel injectors, blower motors, solenoids, and lights. Resistors were once used to control the amount of electrical energy that was supplied to a component and did a good job of controlling the speed and brightness of motors and lighting; however, much of the electrical energy was turned to heat. When a component is PWM controlled, there is no loss of energy and the component is switched off and on so rapidly that the operation of the component is seamless and cannot be noticed by the operator. This works well for components such as fuel injectors because adjustments can be made almost instantly with exceptional accuracy.

PWM occurs at a given frequency, meaning that the component will cycle from off to on and back off a specific number of times in 1 second. Frequency is measured in a unit called hertz (Hz). If a signal has cycled ten times in 1 second, that would be a measurement of 10 Hz **(Figure 1)**. When an actuator is PWM controlled, we measure how long the actuator has been energized within the cycle. This

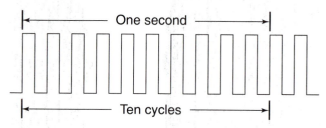

**Figure 1.** A component operating at a 10-Hz cycle.

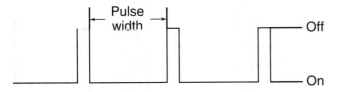

**Figure 2.** Pulse width is the amount of time that a component is turned on within a cycle.

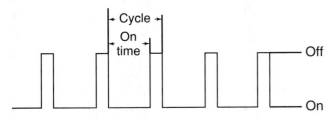

**Figure 3.** A 75 percent duty cycle. Note the component is turned on 75 percent of the cycle and is turned off for 25 percent.

is the **pulse width** and is illustrated in **Figure 2**. The PCM has the ability to adjust the frequency of operation as needed.

Duty cycle is also variable and is a measure expressed as a percentage of the time that the component is turned on within a cycle. For example, a 75 percent duty cycle at a 10-Hz frequency would mean that for every one cycle the signal would be on 75 percent of the time and off for 25 percent **(Figure 3)**. What makes duty cycle different from PWM is that the frequency of a duty cycle–controlled component will generally remain fixed. The accuracy with which both of these methods operate is dependent on the fact that changes can be made to pulse width or duty cycle in a fraction of a second.

## RELAYS

The PCM is able to control many components directly. Many components, however, require PCM control but draw a large amount of current to operate. Because the PCM is not equipped to handle high-current loads, a **relay** is used. A relay is an electrical switch that can be controlled from a remote location, in this case by the PCM. The use of a relay allows two important functions to occur. First, it allows the PCM to control high-amperage components by completing the ground circuit to a low-amperage electromagnet within the relay. The second benefit is that the relay can be placed in close proximity to the power source and the component. This minimizes the amount of high-current wiring that is routed throughout the engine and passenger compartments, reducing the chance for an electrical fire.

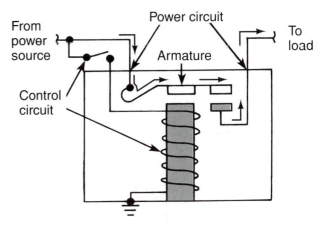

**Figure 4.** A typical relay.

A relay can contain one or two sets of switch contacts and can operate in a normally open or normally closed manner. A relay with one set of contacts operates as follows. The relay contacts and electromagnet are contained in a small box. One contact point is attached to a power source, and the other contact is connected to the component that is being controlled. The electromagnet is used to control the relay contacts. One end of the magnetic coil is connected to a power source, whereas the opposite side of the magnet's coil is connected to the control device, in this example the PCM. The PCM completes a ground circuit when operation of the component attached to the relay is requested. When the coil is energized, a magnetic field is created and the switch contacts are pulled together. A relay is illustrated in **Figure 4**. In the case of a normally closed relay, the contacts are pulled apart to break the circuit rather than complete it.

A relay with two sets of contacts operates in fundamentally the same way as a single-contact relay. The difference is that the movable contact arm has a contact on each side of the arm and two stationary contacts. This allows one relay to act as either a normally open or a normally closed relay. Normally open relays are used to turn components on, whereas normally closed switches are used to turn components off.

## ACTUATORS

The actuators are the components that are responsible for doing all of the work in a powertrain management system. Actuators are required to make very rapid and precise adjustments to operating systems to ensure efficient operation and exceptional driveability qualities. The actuators in this chapter will directly affect the performance of the engine. Chapter 16 discusses some other PCM-controlled systems that may or may not directly affect engine performance.

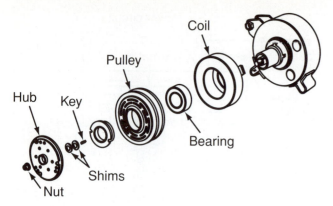

**Figure 5.** The typical components found in an A/C clutch assembly.

## A/C CLUTCH

The A/C clutch is a large electromagnetic drive mechanism located on the front of the A/C compressor that is used to turn the compressor off and on. A compressor clutch and its related components are shown in **Figure 5**. Because the A/C compressor induces a large parasitic load on the engine, the PCM is allowed to control the compressor. This is accomplished using a relay. The PCM uses an A/C input signal, provided by the A/C control assembly or another control module, to signal that an A/C request has been made. When a request has been processed, the PCM will provide a slight increase in engine speed milliseconds before a ground is supplied to the control side of the relay to engage the clutch. This helps to eliminate stumble and surge created when the compressor is engaged. If engine load should become too great, the PCM can temporarily override the A/C request and disable the clutch. Two common occurrences include excessive power steering pressure and heavy acceleration.

Some A/C systems supply refrigerant pressure information to the PCM to monitor system pressure conditions. If refrigerant pressures should fall outside the acceptable range, the PCM may inhibit compressor operation to prevent severe system damage.

## ACTIVE CYLINDER CONTROL

Active cylinder control is used by some manufacturers to allow variable engine displacement operation. Variable engine displacement allows the engine to have the power and acceleration of a large displacement engine and the economy of a small displacement engine. Fuel economy savings of up to 25 percent or more are possible with these systems. The principle of operation is that during acceleration or high load modes all cylinders are operational. However, at steady speed cruising, only a fraction of the power is needed to maintain speeds, and half of the cylinders

are disabled. Reactivation of the cylinders can be made almost instantly and the transitions are seamless. Variations of the variable displacement system have been around for more than 20 years, but operation was not spectacular. The advancement in technology has made the latest versions of these systems acceptable.

The active cylinder control system has the ability to enable and disable cylinders at will to produce the desired amount of engine power. This is accomplished by deactivating the valve operation of particular cylinders. In one particular version used by Mercedes, rocker arms can be engaged or disengaged by oil pressure. The oil pressure to the rocker arms is controlled by a PCM-controlled solenoid. When the PCM determines that a decrease in displacement is desired, it will activate the solenoid. When the solenoid is active, it will allow oil pressure to exert pressure on an internal piston within the rocker arm and effectively disengage the rocker. If the system should fail, the engine will operate using all available cylinders. Another variation used by General Motors employs a collapsible lifter assembly. Again, the operation of the lifter is activated by oil pressure that is controlled by PCM solenoids. When pressure is applied to the lifter, the valve is deactivated; when pressure is released, the cylinder returns to operation.

## COOLING FANS

Most modern automobiles and some light-duty trucks use electric cooling fans. In these applications, the PCM uses the input from the CTS and the VSS to control the operation of the cooling fans. When the engine coolant temperature rises above a predetermined temperature, usually between 225 and 235°F (108–114°C), the cooling fans are turned on. When the temperature falls down below specifications, usually 210–225°F (99.7–113°C), the fans are turned off. As the vehicle speed increases, so does the amount of air flowing across the radiator. Because of this, the fans may be turned off once the vehicle reaches a predetermined road speed.

When the A/C compressor is engaged, the temperature of the A/C condenser is extremely high. Because of this, the fans are generally commanded on to draw air through the condenser and radiator to remove heat from both components, thus improving A/C efficiency and preventing overheating. The PCM uses a relay to control the operation of the fan(s). A typical fan control circuit is shown in **Figure 6**. The fans may also be commanded on when certain DTCs are set. This is done to prevent engine overheating.

## EXHAUST GAS RECIRCULATION VALVE

The EGR valve is used to recirculate burnt exhaust gases back into the engine to control combustion temperatures,

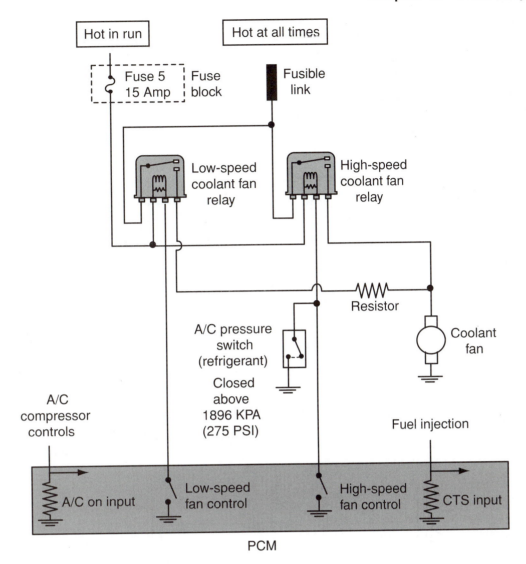

**Figure 6.**   A typical cooling fan control circuit.

thus reducing harmful engine emissions. A complete explanation of the formation and prevention of the formation of these gases will be discussed in Section 6. However, because the EGR valve is an integral part of the computerized engine management system and many complaints that are associated with a malfunctioning EGR surface as driveability concerns, we will observe the EGR value's operation as an actuator.

Either vacuum or electrical current can be used to actuate the EGR valve. Those valves that operate from engine vacuum are controlled directly by the PCM using an in-line vacuum solenoid. The solenoid uses a small electromagnet with a movable core. The core serves as a pintle that seals against the seat, thus blocking the flow of vacuum when de-energized or allowing flow when energized. The vacuum is then allowed into a vacuum chamber, which is connected to a movable pintle that either

blocks or allows EGR flow. This system is illustrated in **Figure 7**.

Two types of electrical EGR valves are in use: digital and linear. The digital EGR valve typically uses three separate solenoids. Each solenoid uses a movable core that doubles as an EGR pintle. Each separate solenoid controls a different-sized pintle: small, medium, and a large. In this type of valve, the solenoids allow the pintle to be either opened or closed. However, to achieve proper EGR flow, any combination of the individual solenoids can be used to achieve the desired amount of flow **(Figure 8)**.

The linear EGR uses a large electromagnet; the magnet uses a movable core that is attached to the pintle of the valve. The pintle will either rise or fall in direct proportion to the duty cycle applied to the coil. This type of valve has an infinite number of positions. The electrical EGR is controlled directly by the PCM. The PCM uses input signals from the

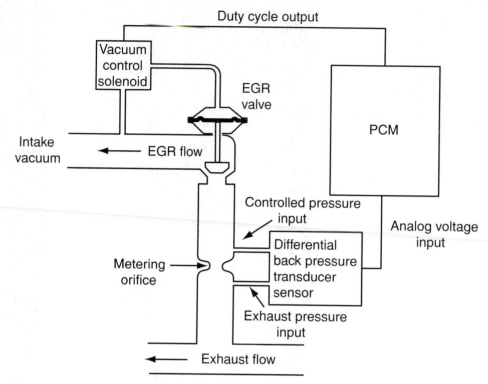

**Figure 7.** A vacuum-actuated EGR circuit using PCM control.

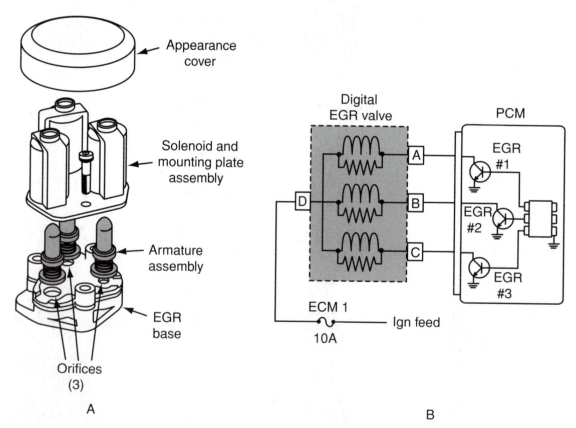

**Figure 8.** (A) A digital EGR valve using three solenoids. (B) A typical control circuit for a digital EGR valve.

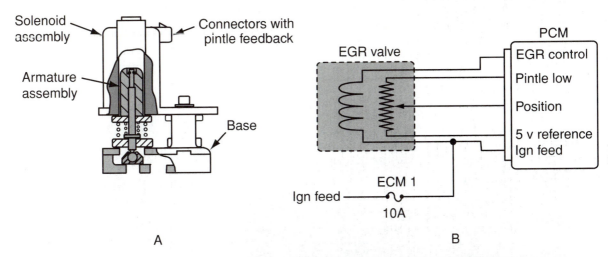

**Figure 9.** (A) A typical linear EGR valve. (B) A typical control circuit for a linear EGR valve.

TPS, VSS, MAP, MAF, CTS, and either an EGR position or an EGR backpressure signal to determine the operating position of the valve. This type of valve is shown in **Figure 9**.

## EMISSION CONTROLS

The PCM directly controls many of the emission control devices found on the vehicle. By monitoring these systems, the PCM can both control and in many cases will have specific diagnostic tests for these components. These components may depend on direct control or they may be controlled by a relay. These components, which will be described in depth in Section 6, are:

- Air injection reaction (AIR) pump
- AIR management valves
- Evaporative emissions purge valve (EVAP)
- EVAP vent valve

## FUEL INJECTORS

The fuel injectors are the most critical PCM output in any system. The fuel injector not only makes the engine start and run, but it also affects power output, idle quality, and emission output. The fuel injector is an electrical solenoid. Fuel enters the injector through the top or sides of the injector and flows across the coils of the solenoid. The fuel, in addition to making the engine run, cools the injector coil and dampens the noise created from operation. A nozzle is located at the end of the injector. The pressurized fuel is ultimately pushed through the nozzle to create a highly atomized fuel mixture. Separating the injector body and the nozzle is a valve and seat. This valve can be either a needle-style pintle or a ball-type valve. The movable core or armature of the solenoid acts on the pintle or ball. When the injector coil is energized, the armature is lifted up toward the top of the fuel injector, lifting the pintle or ball off its seat and allowing fuel flow. When the sole-

noid coil is de-energized, spring pressure acting on the pintle or ball forces its seal against the seat assembly, thus cutting off fuel. Battery voltage is supplied to one terminal of the injector coil anytime the ignition is turned on. The PCM controls injector operation by controlling the injector ground circuit by PWM. Typically, operating values are displayed on a scan tool as pulse width and displayed in milliseconds. A typical fuel injector and control circuit is shown in **Figure 10**.

## FUEL PUMP

The fuel pump is used to supply fuel to the engine with sufficient pressure and volume needed to make the engine run. Most fuel-injected engines have the pump located in the fuel tank. Some systems, however, will have the pump located in-line between the tank and the fuel rail. The fuel pump itself is composed of a DC electric motor. The armature of the motor drives the pump portion of the unit. The pump also contains a pressure by-pass valve that allows the fuel to bypass the fuel outlet if outlet pressure should become too high. Located within the pump is a check valve that prevents the fuel lines from draining when the pump is turned off. This feature allows the engine to receive fuel instantly and provides faster starting. All of these components are built into one compact unit and cannot be serviced separately. A modern modular fuel pump and sender is shown in **Figure 11**.

Traditionally, systems with an electric fuel pump have used a fuel return line. By returning unused fuel to the fuel tank, we can maintain constant fuel supply pressure and prevent pump damage. If a pump is allowed to run with excessive pressure, its life will be significantly shortened.

Pumps that use a return line are generally controlled by the PCM using the relay. In most circuits, the PCM provides a priming pulse to the relay as soon as an ignition "on" signal is detected by the PCM. This allows the relay to close

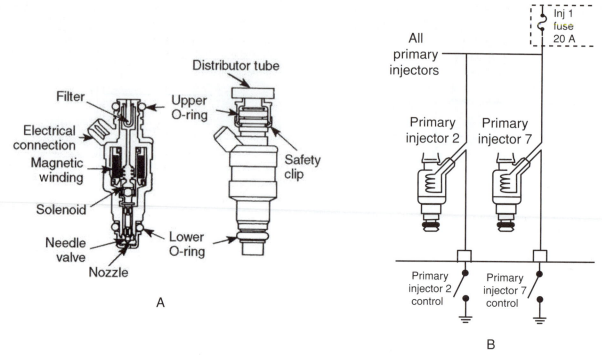

**Figure 10.** (A) A typical top-feed fuel injector. (B) A typical control circuit for a fuel injection system.

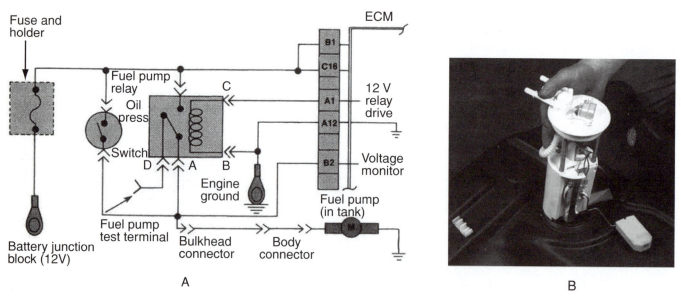

**Figure 11.** (A) A typical fuel pump control circuit. (B) A combination fuel pump and fuel gauge sending unit.

for 2 seconds to allow a brief buildup of pressure, which helps the engine to start quicker. Once the PCM detects an rpm signal provided by the CKP sensor or the ignition module, it will provide a constant ground to the relay until the ignition is turned off. Some systems use an inertia switch located in the passenger compartment or in the trunk to disable the fuel pump in the event of a collision. A typical fuel pump circuit is shown in Figure 11.

Recently, some systems are using a returnless fuel system. This system uses a separate fuel pump module in place of the relay. The PCM will send signals to the module based on the TPS, MAF, CTS, and HO$_2$S. The module then uses PWM to control the pump. By using PWM, the control unit can control the speed of the pump, which in turn will affect the pressure and the volume present in the supply system. This system does not require a fuel return-to-tank line.

## GENERATOR

The PCM on many late-model vehicles directly controls the generator output. This allows the PCM to make up for small changes in system voltage. In addition, the PCM also knows what demands on system voltage are about to be made; it knows this by monitoring input signals or by information received by another computer communicating a future request. An example of this would be when the PCM has received an A/C request, and, based on this signal, the PCM knows that the clutch is about to need a specific amount of current and that the cooling fans will be required, drawing even more current. Using this information, the PCM can command the alternator to ramp up current production milliseconds before it is needed.

The PCM does this by controlling the field circuit of the generator. In a generator, the field circuit is the magnetic field. The stronger the magnetic field is, the more current the generator is able to produce. When the PCM detects an impending demand on the electrical system, it will increase the field current to the generator. Likewise, if it detects that current demand is about to decrease, the PCM can decrease the field current. The PCM controls the field by providing a PWM ground circuit to the field circuit of the generator **(Figure 12)**.

## IDLE AIR CONTROL VALVE

Most modern fuel systems use an **idle air control valve (IAC valve)** to control the engine idle speed. The IAC valve, sometimes called an idle by-pass valve, is a variable position valve that allows air to bypass the throttle blades when the throttle is closed in order to control the idle speed of the vehicle. The operation of the valve allows the engine to be started without depressing the accelerator on fuel-injected engines. The PCM can adjust the position of the valve depending on operating conditions. The PCM uses rpm, TPS, TRS, A/C, and engine load signals to determine valve positions.

There are two common types of IAC valves: a **stepper motor** type and a digital type. The stepper motor uses a small DC motor; the armature of the motor has a hollow center that is threaded. The IAC valve pintle is threaded into the armature. The other end of the pintle has a tapered head that seals against a corresponding seat in the throttle body. The motor uses four wires connected to separate brushes on the motor. The PCM alternately provides power and ground to alternate brushes. Each time current is applied or removed, the motor will move small increments or steps. As the motor moves the pintle, it moves closer to or farther from its seat, which in turn changes the amount of airflow. Each time the engine is turned off, the PCM drives the pintle all the way until it seats. The PCM then retracts the IAC pintle back a predetermined amount of steps. Doing this prepares the system for the next starting sequence. Once the key is turned on for the next sequence, the PCM quickly reads the pertinent sensor data, specifically CTS, and moves the pintle to the necessary position for those conditions. The number of steps that the PCM uses is determined by the adaptive strategy. This means that the computer has been constantly measuring and recording how many steps were required to make the engine idle under specific conditions. The adaptive strategy is used because, as the valve becomes dirty, the amount of air available at a specific step has been reduced. By monitoring the IAC positions, the PCM can compensate for this.

Once the engine has started, the PCM constantly monitors the data and changes the IAC pintle position in the same direction as the throttle angle is changed. What this means is that as the throttle is opened, the IAC valve will open as well; as the throttle is closed, the IAC valve will also begin to close. This is done for two reasons. The first is to prevent a closed throttle stall. If the throttle is suddenly closed and inadequate air is provided to the engine, it will subsequently stall. If the IAC valve is allowed to open as the throttle opens, if the throttle were suddenly closed it gives the engine enough airflow to prevent stalling and the PCM can then very quickly return the engine to a safe idle speed. The second reason that the IAC valve is opened in relation to that of the throttle is that the amount of airflow at WOT can be increased slightly.

We have referred to the second type of valve as a digital valve because it has two positions, opened and closed. Unlike the stepper motor that can infinitely vary the amount of airflow, the digital type has one preset maximum airflow

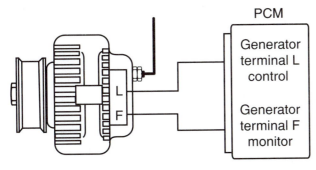

**Figure 12.** A control circuit for a PCM-controlled generator.

Typical Ford Motor Company
intake-mounted control
(digital)

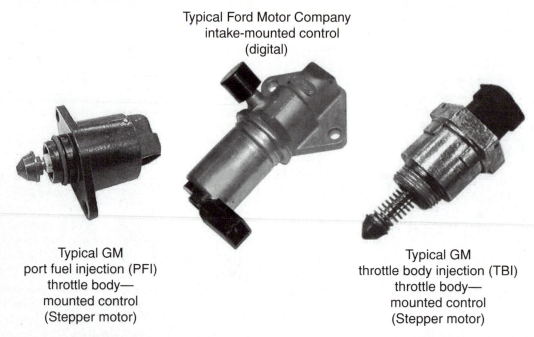

Typical GM
port fuel injection (PFI)
throttle body—
mounted control
(Stepper motor)

Typical GM
throttle body injection (TBI)
throttle body—
mounted control
(Stepper motor)

**Figure 13.** Three of the most popular designs of IAC valves.

amount and the valve is either opened or closed. However, the PCM is able to vary the exact amount of airflow by using PWM or duty cycle to control the valve. By controlling the amount of time that the valve is open within a given cycle, the airflow becomes infinitely variable. Like the stepper motor, the digital valve can become dirty, and the amount of adjustment to the operation has to constantly change. The basic operation and logic of valve operation remains the same. Only the type of valve and how it is controlled have changed. The commonly used IAC valves are shown in **Figure 13**.

> You Should Know
>
> *Many of the IAC systems that use adaptive strategy have special procedures that must be adhered to when power to the PCM is lost or valve service has occurred. These procedures are used to help the PCM relearn its base settings.*

## IGNITION COILS

The ignition coil supplies the voltage and current that is required to ignite the incoming air/fuel mixture. The coil uses the principles of magnetic induction to amplify battery voltage into several thousand volts at the spark plugs. Coil construction and operation are discussed

in greater depth in Section 4. For many years, the coil was controlled by an electronic ignition module, which uses transistors and signals from the PCM to control when the coil is pulsed. Two variations of these systems existed. One used one single coil to provide spark to all cylinders; the second type of system used one coil for every two cylinders. The most recent version uses one ignition coil for every cylinder, with the coil being located very close to the spark plug, using a very short secondary cable; this is called either a **coil-near-plug (CNP)** ignition or a **coil-on-plug (COP)** ignition. These types of systems circumvent the use of an ignition module. Each coil contains a transistor for switching the coil off and on. The internal transistor is connected directly to the PCM. The PCM assumes all of the functions that were commonly controlled by the ignition module, such as coil dwell and spark timing advance. The typical COP coil will have a PCM connection, a ground connection, a battery feed from a switched ignition source, and the secondary wire leading to the spark plug **(Figure 14)**.

## INTAKE MANIFOLD RUNNER CONTROL

The intake manifold runner control (IMRC) is used on some port fuel-injected engines. On these engines, the **plenum** is divided into separate passages. The plenum is the upper part of the intake system in which the air enters the throttle body and is distributed into different tubes that feed individual cylinders. The separate passages allow for

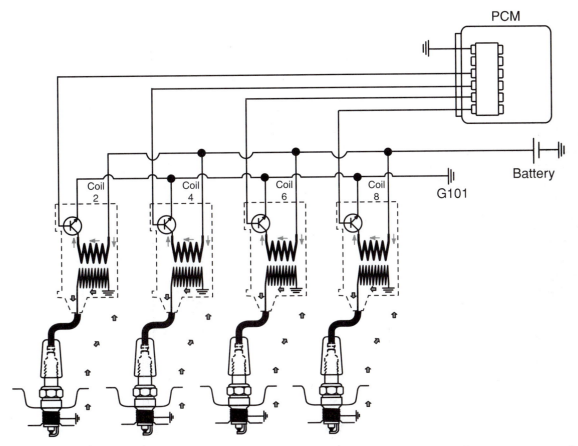

**Figure 14.** This is a coil-on-plug or coil-near-plug system schematic. Note the absence of an ignition control module.

the length and volume of the tube to be changed. This changes the amount and the speed of the air moving into the cylinders. The plenum uses a movable flap that opens or closes the auxiliary passages. This valve can be controlled by a vacuum valve and solenoid or by a DC motor. Both of these methods are controlled by the PCM. The solenoid is grounded by the PCM and the motor is controlled by a PCM-activated relay.

The valve blocks the auxiliary passages at low speeds and opens at higher engine speeds. At low engine speeds, a large amount of air is not needed. Closing the valve at low speeds allows air velocity to be maintained, allowing for a better atomized air/fuel mixture. At higher engine speeds, the additional air is desirable to increase engine performance. Although the amount of air is increased, the additional engine speed increases the velocity of the air.

## MALFUNCTION INDICATOR LIGHT

The MIL is a critical component in the operation of the engine management system. The MIL is used to communicate system faults to the vehicle operator. These faults can range from emissions-related problems that cause

the engine to emit excessive pollutants to serious system failures that could disable the vehicle. Without the use of a system warning light, the vehicle operator might continue to operate the vehicle until it became inoperable. Many consumers still have misconceptions about what the MIL really means. Many consumers are still under the impression that the MIL might illuminate because of a low oil or coolant condition. It is commonplace to have a customer explain that when the MIL illuminated, they had checked the oil and coolant and did not find a problem. Although this is not bad in practice, the MIL will illuminate only in the event of an emission or powertrain control circuit failure. The MIL is located in the instrument panel in direct view of the operator **(Figure 15)**.

When the ignition is first turned on, the MIL will go through a bulb check procedure. The bulb check is performed each time the ignition is turned on. The MIL and many other instrument panel lamps are illuminated for 2–3 seconds when the ignition is first turned on. This is just to show the operator that the lights are in operating condition. The MIL is a simple series bulb circuit; one side of the bulb is connected to a switched ignition feed that is powered anytime that the ignition is on. (Refer to Figure 15.) The ground side

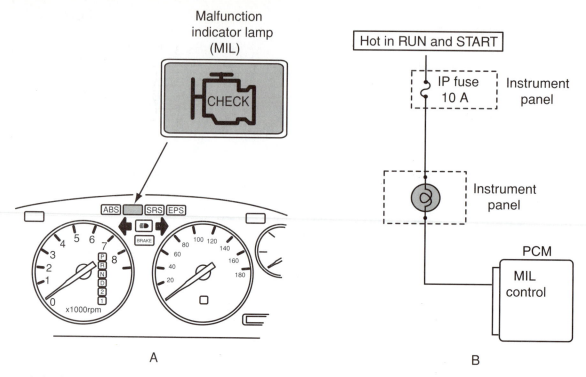

**Figure 15.** (A) The MIL lamp is located in the instrument panel. (B) The MIL is controlled directly by the PCM.

of the circuit is connected to the PCM. The PCM will illuminate the MIL whenever a system parameter falls outside of the specifications that are set forth by the operating program.

## IGNITION MODULE

The ignition module is an electronic device that uses transistors to directly control the operation of the ignition coil(s). The module controls when and how long the coil is pulsed. Spark timing in an engine will always occur before the cylinder reaches top dead center (BTDC). The number of crankshaft degrees BTDC that the air/fuel mixture is ignited is called **spark advance** or spark timing. The PCM has ultimate control over the desired amount of spark advance, but the module makes the actual adjustment based on the PCM control signal. The PCM will use engine load, TPS, rpm, KS, and CTS signals to make timing decisions. The PCM searches the control maps to determine the necessary amount of spark advance and sends signals to the ignition module requesting it to make the desired adjustments. The PCM will also monitor the KS to determine if the spark timing needs to be **retarded**. Retarded is a term that is used to describe that the spark timing has been moved closer to TDC. The ignition module can be located in or on the distributor and control the ignition coil remotely. In systems that do not use distributors, these are called **distributorless ignitions** or **electronic ignitions (EI)**. The coils are fastened directly to the module and this assembly can be located anywhere on the engine block or on an accessory

bracket. A typical ignition control module control circuit can be seen in **Figure 16**.

## THROTTLE ACTUATOR CONTROL

Many modern vehicles are using **throttle actuator control (TAC)**, also known as **drive by wire**, to control the movement of the engine throttle. The system uses a DC bidirectional motor that is attached directly to the throttle body of the engine. Using PWM, the PCM can control the motor directly. Other variations use a separate TAC module to control the TAC motor. Vehicles equipped with TAC still use an accelerator pedal; however, the pedal has no mechanical connection to the engine and is now called an **accelerator pedal position (APP) sensor**. The APP can use two or three variable resistors to send signals to the TAC module or PCM in relation to throttle position and rate of change. The PCM uses the TPS to determine if the commands are being completed as directed. Because the PCM directly controls the throttle opening, a separate cruise control system is no longer needed. In addition, the PCM can readily reduce throttle opening in traction control situations if so desired **(Figure 17)**.

## VARIABLE VALVE TIMING

**Variable valve timing (VVT)** is a method of changing the relationship between intake valve opening/closing events and exhaust valve opening/closing events. The VVT

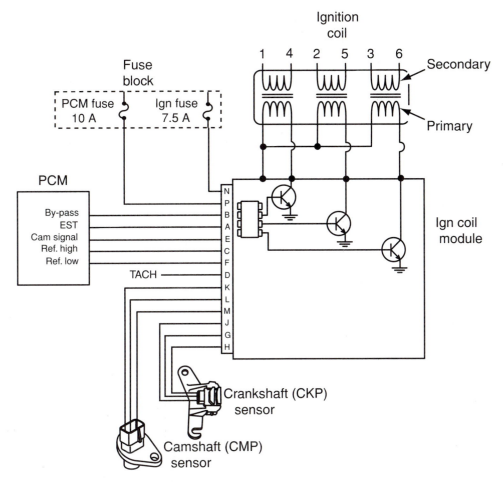

**Figure 16.**  The ignition control module controls the ignition coils based on input from the PCM.

can be only on those engines that use separate intake and exhaust cams. Traditionally, the valve timing is a specification that is built into the camshaft, meaning that valve timing could not be changed once the camshaft was installed. Because timing could not be changed, the manufacturers

had to grind the camshaft with compromise in mind. They had to strike a delicate balance between performance, economy, and tail pipe emissions. The VVT system allows the engine management system to change engine valve timing "on the fly." The valve timing system changes the rel-

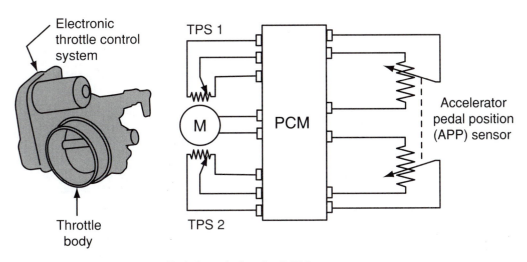

**Figure 17.**  A TAC system can be controlled directly by the PCM.

ative position between the exhaust camshaft and the intake camshaft.

The ability to change valve timing allows the engine to assume two slightly different personalities. At low engine speeds, the engine can idle smoothly and provide extra torque, but at higher speeds can provide additional horsepower and better fuel economy. Doing this provides increased performance and decreased emissions. Because increased valve overlap allows extra exhaust gas into the cylinders, providing a cooling effect to the combustion process, the ability to increase valve overlap under engine load eliminates the need for an EGR system in some applications.

Changing the valve timing is done in two different ways: changing the relationship between the camshaft and the drive gear or changing the relationship between the intake and exhaust cams by manipulating the drive chain.

The most popular method uses a piston assembly that is mounted to the cam gear **(Figure 18)**. The piston has internal helical splines and the camshaft has matching external splines; the contact between the splines drives the camshaft

in one direction in relation to the cam gear. Oil is directed to the piston using a PCM-controlled actuator; the PCM uses PWM to precisely control the amount of oil traveling to the piston. As the piston moves fore and aft within its bore, the helical splines cause the camshaft to twist in relation to the gear assembly. These systems generally allow 25–45 degrees of camshaft movement. Variations of this method use one or two CMP sensors to allow the PCM to monitor the relationship between the intake and exhaust cams.

The second method uses the slack within the timing chain to control the intake-to-exhaust cam relationship. This system uses a secondary chain that connects the intake and exhaust cams together. Between the two cams is a variable oil charged chain tensioner. The tensioner is able to move the slack between the top of the gears and the bottom of the gears. As the slack is moved, the relationship between the intake and exhaust cams changes. The PCM controls the oil flow with a PWM-controlled actuator.

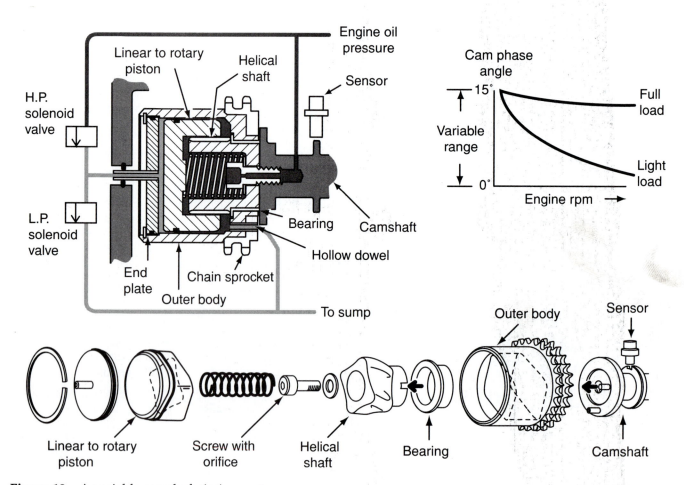

**Figure 18.**   A variable camshaft timing system.

# Summary

- Actuators are the components that convert electrical signals provided by the PCM and convert them to mechanical actions. Individual systems require the use of specific actuators.

- The PCM will control actuators by supplying a ground to complete the circuit. The PCM may supply a constant signal or it may supply a PWM signal. PWM is the act of cycling a component at a specific frequency.

- A relay contains a small electromagnet that magnetically closes a set of switch contacts. By controlling the operation of the electromagnet, the PCM can effectively control a device with high current demands remotely.

- The PCM uses a relay to control the operation of the A/C clutch. The PCM control of the clutch is desirable because the compressor places large loads upon the engine.

- Active cylinder control allows the PCM to turn off one-half of its cylinders under certain conditions. The PCM activates the cylinder control system using solenoids to control pressurized oil.

- The EGR system can be controlled by the PCM with a solenoid-controlled vacuum supply or directly by using an electrical valve. Three types of EGR valves are vacuum, linear, and digital.

- The PCM controls the fuel injector directly; by doing so, it controls the amount of fuel allowed into the engine. The PCM uses PWM to control the operation of the injector.

- The fuel pump is the heart of any fuel injection system. The PCM controls the fuel pump operation directly by a relay or indirectly with a separate fuel pump control module used in returnless fuel systems.

- The PCM controls the field current of the generator to adjust system voltage. The PCM can ramp up voltage in anticipation of impending loads.

- Idle air control valves are used by the PCM to directly control the idle speed of the engine. Two types of valves are the stepper motor valve and the digital-type solenoid.

- The COP technology allows the PCM to directly control each engine coil and eliminates the need for a separate ignition control module.

- The MIL is one of the most critical components of any engine management system. The MIL is used to alert drivers to a system failure and alert them to take their vehicle in for service. The MIL is controlled directly by the PCM.

- The ignition module receives input from the PCM on the desired amount of spark timing advance the engine requires. The module can also transfer CKP and CMP signals to the PCM.

- VVT is being used on a growing number of vehicles. The use of VVT allows the manufacturers to boost performance and economy. The VVT systems are hydraulically controlled by the PCM using solenoids supplying oil to a gear-mounted piston assembly. This system can change valve timing 25–45 degrees.

# Review Questions

1. Technician A says that all actuators that are controlled by the PCM are controlled by electrical signals. Technician B says that some PCM-controlled actuators may be vacuum operated. Who is correct?
   A. Technician A
   B. Technician B
   C. Both A and B
   D. Neither A nor B

2. Technician A says that most actuators have battery voltage present whenever the ignition is turned on. Technician B says that the PCM may apply a constant state ground to the circuit or it may be a PWM signal. Who is correct?
   A. Technician A
   B. Technician B
   C. Both A and B
   D. Neither A nor B

3. All of the following facts about PWM are true except:
   A. We are only concerned with the amount of time that the component is turned on.
   B. Using PWM, a component can be controlled by turning it on and off.
   C. Duty cycle is the amount of time a component is turned off expressed as a percent of cycle time.
   D. PWM modulation helps to conserve energy.

4. Technician A says that the use of relays allows the PCM to control high-amperage circuits by controlling a low-current electromagnet. Technician B says that if a short to ground occurred in the circuit of the component controlled by the relay, the PCM could be damaged. Who is correct?
   A. Technician A
   B. Technician B
   C. Both A and B
   D. Neither A nor B

5. The PCM can turn off the A/C compressor when the power steering pressure is too high.
   A. True
   B. False

6. Technician A says that active cylinder control allows the PCM to deactivate the cylinders one at a time. Technician B says that all of the cylinders are needed when the vehicle is cruising at steady speeds. Who is correct?
   A. Technician A
   B. Technician B
   C. Both A and B
   D. Neither A nor B

7. Technician A says that problems in the EGR system rarely surface as driveability concerns to the customer. Technician B says that the linear EGR valve usually has three specific positions. Who is correct?
   A. Technician A
   B. Technician B
   C. Both A and B
   D. Neither A nor B

8. Technician A says that the ball in a ball-type injector is used to push the fuel through the nozzle. Technician B says that the fuel passing through the injector serves to cool the solenoid coil. Who is correct?
   A. Technician A
   B. Technician B
   C. Both A and B
   D. Neither A nor B

9. Technician A says that the fuel pump runs anytime the ignition is turned on. Technician B says that if the PCM does not receive a CKP signal, the fuel pump may fail to operate. Who is correct?
   A. Technician A
   B. Technician B
   C. Both A and B
   D. Neither A nor B

10. Technician A says that it is desirable to have the PCM control generator operating in order to assist the technician in diagnosis of the charging system. Technician B says that the PCM can decrease alternator output when a high-current accessory is turned off. Who is correct?
    A. Technician A
    B. Technician B
    C. Both A and B
    D. Neither A nor B

11. Technician A says that a dirty IAC can cause the engine to stall while operating at WOT. Technician B says that as the throttle is opened, the IAC valve closes to prepare the engine to idle. Who is correct?
    A. Technician A
    B. Technician B
    C. Both A and B
    D. Neither A nor B

12. The COP ignition is connected to the PCM through the secondary cable.
    A. True
    B. False

13. All of the following facts about the MIL are true except:
    A. The MIL will always illuminate for a few seconds when the ignition is first turned on.
    B. The MIL alerts the driver to an emissions failure.
    C. The MIL will alert the driver to a powertrain control circuit failure.
    D. The MIL will illuminate when the engine is low on engine oil.

14. Technician A says that the ignition module acts as both an input and an actuator for the PCM. Technician B says that the ignition coil may be fastened directly to the module. Who is correct?
    A. Technician A
    B. Technician B
    C. Both A and B
    D. Neither A nor B

15. Technician A says that the amount of camshaft advance or retard is adjusted using a piston with external straight splines fastened to the camshaft. Technician B says that when the slack in the timing chain is present, the valve timing can be altered. Who is correct?
    A. Technician A
    B. Technician B
    C. Both A and B
    D. Neither A nor B

# Chapter 16

# Related Systems

## Introduction

In this chapter, we learn about some additional systems that are controlled by the PCM but are not directly related to the operation of the engine. Each of these systems has an important role in the overall operation of the vehicle, and allowing the PCM to control these systems greatly enhances the overall driveability, performance, and quality of the vehicle. Some of the systems covered in this chapter have been mentioned in other chapters because a malfunction in the system will affect vehicle operation, but a more thorough examination is in order.

### AIR CONDITIONING SYSTEM

Anytime that the compressor is engaged, a heavy load is placed on the engine. This load can often be enough that the engine may try to stall. At the very least, the engine idle must be raised up to compensate for the additional load. Because of the dramatic effect that compressor operation has on the operation of the engine, compressor control has been assigned to the PCM. The PCM has the ability to adjust engine parameters milliseconds before the compressor is engaged. This enables almost seamless operation of the A/C compressor.

### Air Conditioning Control Input Signals

In order to control the operation of the A/C clutch, the PCM uses several different input signals. These sensors and switches provide input signals from the refrigeration system and share input signals with the engine management system. The most common input signals are as follows:

- *A/C request signal.* The A/C request signal is a digital signal provided to the PCM that indicates that the vehicle operator has requested A/C operation. This signal can be received directly from the A/C control switch or can be transferred by a separate A/C control module.
- *Pressure cycling switch.* The pressure cycling switch is used to control the evaporator temperature. A normally closed pressure switch is used in this application. When the pressure at the evaporator outlet drops below a predetermined level, the switch opens, which results in the compressor being turned off. This switch may interrupt a PCM request signal to the A/C relay or it may be supplied to the PCM as a direct input that the PCM would use to modify the request status to the relay.
- *High-pressure monitoring.* High-pressure monitoring is used to turn the A/C compressor off in the event that high side refrigerant pressure exceeds a predetermined level. This can be accomplished through a switch that is placed in series with the clutch. Used in this fashion, the switch will interrupt the clutch control circuit. Other systems accomplish the same thing by making the switch or a sensor an input to the PCM. In this arrangement, the PCM would sense the high-pressure condition and disable clutch operation.
- *Engine rpm.* Engine rpm is monitored so that the PCM can make idle speed adjustments in relation to the operation of the compressor.
- *Throttle position.* The PCM monitors the TPS to disable clutch operation when the TPS signal indicates a WOT condition. Disabling the clutch under WOT is done to provide the engine with the maximum amount of power.

- *Power steering pressure.* Power steering pressure is monitored by the PSPS to disable the A/C clutch in situations when the power steering system is causing an additional engine load. This usually occurs during full lock parking lot maneuvers.

## Outputs

The PCM has only one output circuit for operation of the A/C system, the A/C clutch relay. The PCM controls operation of the relay by supplying a ground to the control circuit of the relay. The clutch is supplied with a permanent ground circuit. When the relay is energized, 12V is supplied to the coil through a fused circuit. This causes a magnetic field to develop in the coil. When the magnetic field is present, the clutch plate that is splined to the compressor shaft is drawn to the pulley assembly. This action causes the compressor to spin at a speed relative to that of the crankshaft.

## Control

When the vehicle operator moves the A/C control switch to either the A/C or defrost position, a signal is then sent to the PCM to indicate that compressor operation is desired. The PCM then compares input signals provided by various sensors to decide if the conditions are right for clutch engagement.

If the conditions are all met, the PCM will modify engine operation to accommodate the additional engine load. This may include increasing the IAC counts, increasing injector pulse width, and/or increasing ignition timing. This preparation takes place milliseconds before it supplies a ground to the A/C relay. By controlling the clutch in this manner, much of the typical engine surge associated with clutch engagement and disengagement can be avoided. A typical A/C clutch control circuit is shown in **Figure 1**.

## ELECTRONICALLY CONTROLLED TRANSMISSION

Most of the automobiles manufactured today make use of an electronically controlled transmission. What this means to you is that the PCM has complete control of when a transmission shifts, how a transmission shifts, when and how the **Torque Converter Clutch (TCC)** is applied, and the line pressure within the transmission.

In the same way that the engine management system uses control maps to control fuel and ignition systems, the PCM also has specific control maps to control the transmission. The PCM relies on various inputs and actuators to control the operation of the transmission. In addition to controlling transmission operation, the PCM is also able to provide specific transmission diagnostic tests and set specific DTCs relating to these problems. The technician has access to a wide variety of transmission data through the PCM. This data can be viewed by using a scan tool.

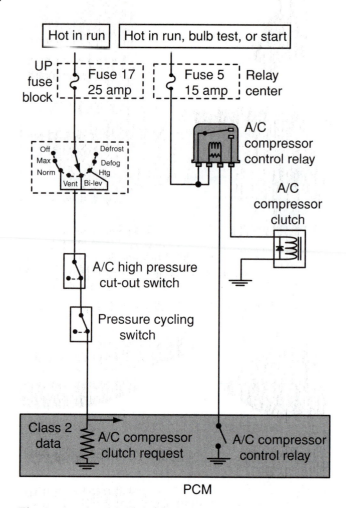

**Figure 1.**   A typical A/C circuit.

## Input Signals

The PCM uses various input signals to control the operation of the transmission. Some signals are shared by the engine management system and some are specific transmission inputs **(Figure 2)**. These are the most common transmission input sensors and signals used by the PCM:

- *Coolant temperature.* Coolant temperature information is obtained from the coolant temperature sensor (CTS) and is specifically used in determining TCC operation.
- *Intake air temperature.* Intake air temperature is obtained from the intake air temperature sensor (IAT sensor). This information is used as an input to line pressure boost control.
- *Transmission fluid temperature.* The transmission fluid temperature is provided to the PCM by a **transmission fluid temperature sensor (TFT sensor)**. The TFT sensor is a thermistor that is located within the transmission pan. The sensor measures fluid temperature and is used only by the transmission in determining proper shift scheduling and TCC operation.

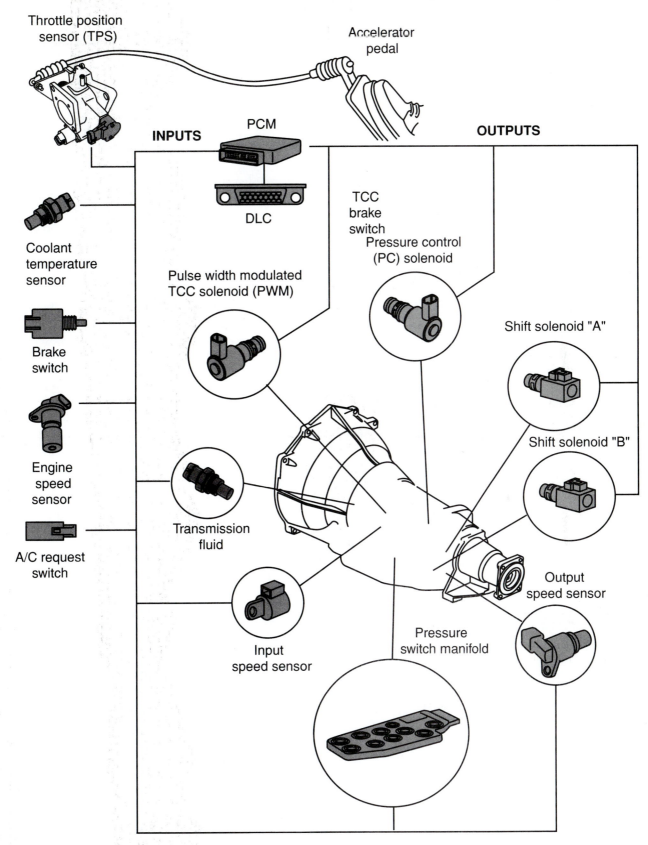

Throttle position sensor (TPS)

Accelerator pedal

**INPUTS**

PCM

DLC

**OUTPUTS**

Coolant temperature sensor

Brake switch

Engine speed sensor

A/C request switch

Pulse width modulated TCC solenoid (PWM)

Transmission fluid

Input speed sensor

TCC brake switch

Pressure control (PC) solenoid

Shift solenoid "A"

Shift solenoid "B"

Output speed sensor

Pressure switch manifold

**Figure 2.**   The PCM controls all of the major transmission functions.

- *Transmission range.* The TR input signal is provided by the TR sensor. The TR sensor is bolted to the side of the transmission and is activated as the shifter is moved. The information provided by the TR sensor is used to increase line pressure and send a signal regarding the desired TR.
- *Brake pedal position.* The BPP information is gathered from the BPP switch. The information provided by the BPP is used to disengage the TCC when the brake pedal is applied.
- *A/C clutch request.* The A/C clutch request signal is used to boost transmission line pressure when the A/C is turned on to compensate for additional engine load.
- *Mass airflow.* The MAF provides information about engine load to adjust shift timing, shift feel, and line pressure.
- *Output shaft speed.* **Output shaft speed (OSS)** information can be provided by the vehicle speed sensor or a separate OSS sensor. The OSS sensor is a permanent magnet generator that requires the use of a reluctor wheel mounted to the output shaft of the transmission. Information from this sensor is used to control shift timing, shift feel, TCC operation, and line pressure adjustments.
- *Engine rpm.* Engine rpm information is gathered from the ignition system and is used in part to control TCC operation, line pressure control, and shift timing and feel.
- *Transmission input speed.* Transmission input speed information is received from a dedicated sensor. This sensor is located near the input shaft of the transmission and reads from a reluctor wheel mounted to a transmission component that spins at the same speed as the input shaft. Transmission input speed information is used to control TCC control, shift timing, shift feel, and pressure control. Information provided by this sensor can be combined with information from the OSS sensor to determine if internal transmission components are slipping. When combined with engine rpm information, the PCM can determine if torque converter slippage is within acceptable range.
- *Throttle Position.* The PCM receives TP information from the TP sensor. This signal is used to control the operation of the TCC as well as an input to change throttle pressure, line pressure, and shift calibrations.
- *Pressure switches.* Various fluid pressure switches are used throughout the transmission to allow the PCM to monitor operation of the hydraulic systems. By monitoring these switches, the PCM can specifically determine if the actuators have carried out specific commands.

## Output Actuators

Once the PCM has retrieved all of this valuable information from various input sensors, it can send commands to various output actuators throughout the transmission.

These actuators control major transmission functions, shift timing, shift feel, TCC operation, and line pressure control. These are the common actuators used within the transmission:

- *Shift solenoids.* Shift solenoids are mounted on the transmission valve body and are used to control the fluid flow to the various shift valves within the transmission. When the solenoid is commanded closed, all fluid pressure flows in the particular shift circuit. This forces specific valves within the transmission to shift and redirect fluid flow. When the solenoid is commanded opened, fluid is bled from that shift circuit. This results in spring pressure overcoming fluid pressure and closing the valve, thus redirecting the fluid to another location. The number of shift solenoids in a particular transmission varies from two to four. The solenoids are operated in several different combinations to derive specific gear ranges.
- *Pressure control.* Pressure control is typically carried out by a solenoid/valve combination unit. The PCM uses various input sensors to control the position of the solenoid, resulting in specific valve positioning. The solenoid is controlled using PWM.
- *TCC.* The TCC is controlled by a solenoid similar to that of a shift solenoid. When the solenoid is commanded open, fluid pressure leaks down, no pressure is applied to the TCC valve, and the TCC does not apply. When commanded closed, pressure is allowed to flow to the TCC control valve and the TCC is applied. The TCC solenoid can be a continuous signal or controlled by PWM, which allows the torque converter clutch to apply gradually for smoother application.

## ADAPTIVE LEARNING

Like many other powertrain systems on the modern vehicle, the transmission operating program is equipped with an adaptive learning strategy. Some manufacturers refer to this as a "shift adapt." This strategy allows the PCM to compensate for wear in transmission components. This is accomplished by learning the modifications to the transmission line pressure. This allows shift feel to remain consistent throughout the life of the vehicle. Additionally, the adaptive strategy has the ability to adjust to the operator's driving habits. Like other adaptive strategies, when power to the PCM is interrupted the strategy must be relearned.

*Interesting Fact*

*A customer might complain of harsh shifting after the battery has been disconnected. This is due to the "learned" parameters being lost. It will usually take several miles of driving for the PCM to relearn the operator's driving habits.*

## ANTITHEFT SYSTEMS

Most factory-installed antitheft devices work in conjunction with the PCM. Many of the systems will have stand-alone modules that monitor the vehicle conditions but will interact directly with the PCM. The duty of the PCM is usually to disable the fuel system or the starting system to prevent the vehicle from starting. The PCM may also store some DTCs associated with such features.

## INSTRUMENT PANEL CLUSTER

The instrument panel cluster has taken on many functions in the modern vehicle. What once was just a display for vehicle information has assumed many operational functions. In many vehicles, the PCM sends powertrain information to the instrument panel cluster using a serial communication link. This information is displayed to the driver as gauge or indicator information. However, in many modern vehicles the role of the instrument panel cluster has expanded to be a clearinghouse for many vehicle functions. In addition to the traditional functions such as speedometer or temperature gauge, some of the many additional vehicle functions are to:

● Display live PCM data and stored DTCs.
● Assume the functions of the antitheft module.
● Provide reset functions for many auxiliary warning systems, such as oil change monitor or low tire pressure.
● Provide the input controls for many personal choice settings.

## CRUISE CONTROL

The cruise control is a unit used to hold the vehicle at a constant speed. The cruise control uses either a vacuum-operated servo or an electronic stepper motor connected to the throttle. A typical cruise control unit is shown in **Figure 3**. A stepper motor is a small DC motor in which the input voltage is pulsed, rather than maintained at a steady flow. By pulsing the current, the armature spins in small steps each time a voltage is applied. The stepper motor design also allows the current to be reversed; therefore, the motor can back up as quickly as it can move forward. This is important in any stepper motor function.

The cruise control can have a separate module to command the servo or stepper motor or can be completely controlled by the PCM. For the cruise control to operate properly, information must be received about vehicle speed, TP sensor, BPP switch, and servo or motor position. Based on these inputs, the PCM will issue the commands to the servo or motor to increase or decrease throttle position.

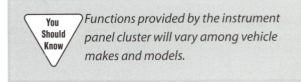

You Should Know *Functions provided by the instrument panel cluster will vary among vehicle makes and models.*

## AUDIO SYSTEMS

In many newer vehicles, the audio system has become the control point for many systems. Like the instrument cluster, the radio may display data or access other modules to reset data. In many cases, special provisions have to be made when an aftermarket radio is installed because many systems will fail to function if the radio is removed. Additionally, some audio systems may use a VSS input from the PCM to increase or decrease volume as vehicle speed changes. A scan tool may be used in some systems to test the operation of audio system functions. Some of the common auxiliary functions that are provided by the radio are oil life monitor reset, tire pressure monitor reset, and a means to program personal security and choice features as shown in **Figure 4**.

**Figure 3.** A typical PCM-controlled stepper motor cruise control.

**Figure 4.** The audio system is a control point for the operation of many vehicle systems.

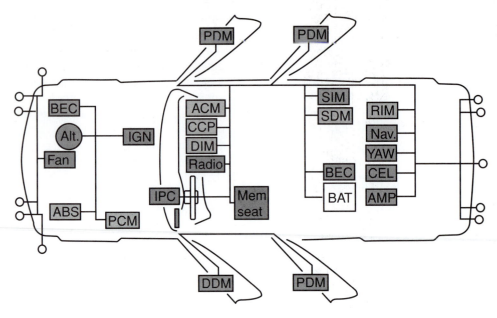

**Figure 5.** Vehicle networks include most electronic vehicle systems.

## VEHICLE NETWORK SYSTEMS

Because an increasing number of computers and modules within a modern vehicle share information, a communication network exists in most vehicles. This allows for increased speed of communication between various modules in the vehicle. These computers may communicate with each other using a specific communication protocol. Through this communication protocol, various modules are able to share data as well as act as input sources to each other. Several different protocols are used and may be different from what is used by the PCM. With all of these modules connected, they can typically be accessed using a scan tool connected to the DLC. A basic network is shown in **Figure 5**.

The most common components of a vehicle network system are:

- PCM
- ABS
- **Body control module (BCM)**
- DLC
- Heating, ventilation, and air conditioning (HVAC)
- Four-wheel-drive (4WD) controls
- Audio system
- Instrument panel cluster

## *Summary*

- Because of the effect that compressor operation has on the engine, compressor control has been assigned to the PCM.
- The PCM has complete control over transmission operation. The PCM supports DTC diagnostics and adaptive strategies for the transmission.
- The PCM actively controls many vehicle antitheft systems.

- The instrument panel cluster may control many functions as well as display data.
- The PCM controls the function of the cruise control.
- An on-board network allows many vehicle modules to communicate with one another.

# Review Questions

1. Technician A says that if the IAC system is inoperative the engine may stall when the compressor is engaged. Technician B says that the PCM may adjust the ignition timing before the clutch is engaged. Who is correct?
   A. Technician A
   B. Technician B
   C. Both A and B
   D. Neither A nor B

2. All of the following concerning transmission electronic controls is true *except:*
   A. The PCM controls transmission operation using solenoids.
   B. Much of the sensor data used to control the engine management system is also used to control the transmission.
   C. The PCM has the ability to perform diagnostic tests on the transmission.
   D. The addition of electronic controls eliminates the need for a mechanical valve body.

3. Technician A says that diagnostic codes may be available for the antitheft system. Technician B says that the instrument panel cluster may control the functions of the antitheft device. Who is correct?
   A. Technician A
   B. Technician B
   C. Both A and B
   D. Neither A nor B

4. Technician A says that some instrument panel clusters have taken the place of the PCM. Technician B says that some instrument panel clusters may display PCM codes. Who is correct?
   A. Technician A
   B. Technician B
   C. Both A and B
   D. Neither A nor B

5. Technician A says that the radio is such an integral part of the vehicle system that if the original radio is removed some systems may not operate properly. Technician B says that the radio is the control point for changing the fuel injector pulse width. Who is correct?
   A. Technician A
   B. Technician B
   C. Both A and B
   D. Neither A nor B

# Chapter 17

# On-Board Diagnostics II

## Introduction

On-board diagnostics II (OBDII) is the second generation of vehicle diagnostic systems. The intent of OBDII is to reduce vehicle emissions by detecting system and component failures in the early stages before vehicle emissions have the opportunity to exceed 1.5 times the federal standards. In addition, OBDII also requires standard terminology and service procedures. This allows any technician with the proper specialized but generic equipment to diagnose and repair any vehicle emission failure on any make of vehicle. All vehicles sold in the United States since 1996 are equipped with OBDII.

### VEHICLE REQUIREMENTS

The implementation of OBDII required that all manufacturers selling vehicles in the United States meet the following requirements:

- *Nomenclature.* All manufacturers are required to use the same name for components and systems that perform similar tasks.
- *MIL.* All OBDII-equipped vehicles must be equipped with a warning indicator located on the instrument panel that will illuminate any time the PCM detects that emissions levels are above 1.5 times the federal standards.
- *Protocol.* All vehicle PCMs must be capable of communicating with generic diagnostic equipment using class 2 serial data language.
- *Diagnostic link connector.* All vehicles must have a standard **diagnostic link connector (DLC)** that is a specified size and shape, has specified pin locations, and is

visibly located under the dash on the left side of the vehicle.
- *DTCs.* Vehicles must use standard DTCs, component monitoring procedures, and test modes. The system must have the ability to allow DTCs to be erased using a generic scan tool.
- *Freeze frame.* The PCM must be able to record specific data parameters at the time of a detected emission failure.

### COMMUNICATION AND THE DIAGNOSTIC LINK CONNECTOR

The first order of standardizing the diagnostic process was to establish a standard communication **protocol**. A protocol is the language that computers use to communicate with one another and diagnostic equipment. The PCM in all OBDII-compliant vehicles must be able to communicate with diagnostic equipment through a protocol called a class 2 data link. Class 2 data is transferred via a single wire that transfers a 7-V digital signal. The 7-V reference is PWM and the pulse width varies, thus sending what could be called a computer "Morse code" **(Figure 1)**. Through this code, the PCM may communicate with diagnostic equipment or it may form a network with the other computers or

**Figure 1.** Class 2 serial data is a toggling 7-V digital signal.

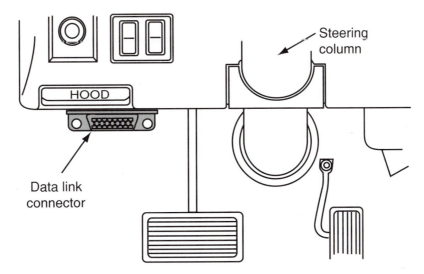

**Figure 2.** This is the standard OBDII DLC location.

modules within the vehicle. Class 2 data is transferred at a rate of 10.4 kilobits of information per second. Class 2 data links have the ability to manage several data transfers simultaneously; data is automatically transferred by order of greatest importance. This means that if two messages need to be transferred at the same time, the message with the highest priority will automatically be transferred first. Although some vehicle systems may use a different data transfer protocol for communications with accessory control modules, the PCM must be able to communicate with Class 2 diagnostic equipment.

To further standardize diagnostic process, a standard DLC was established. All OBDII-compliant vehicles must use a standard DLC that contains 16 pins and is located under the instrument panel on the left side of the vehicle near the steering column **(Figure 2)**. The DLC has a specific shape, and each pin within the DLC has an assigned task. Seven of the pins have specific designations as required by OBDII; this means that these terminals must provide the same function on all vehicles. Terminal location is shown in **Figure 3**. The remaining pins are reserved for the manufacturer to use as desired. This allows the manufacturer the flexibility to interface other system modules with diagnostic equipment.

 *Vehicles with the OBDII-style connectors began appearing in 1994. These vehicles are not OBDII equipped and are referred to as "pull ahead models." With these models, the manufacturers began implementing many of the OBDII features that would soon become required.*

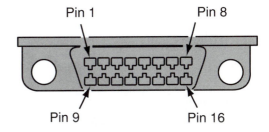

Pin 1: Manufacturer discretionary
Pin 2: J1850 bus positive
Pin 3: Manufacturer discretionary
Pin 4: Chassis ground
Pin 5: Signal ground
Pin 6: Manufacturer discretionary
Pin 7: ISO 1941-2 "K" line
Pin 8: Manufacturer discretionary
Pin 9: Manufacturer discretionary
Pin 10: J1850 bus negative
Pin 11: Manufacturer discretionary
Pin 12: Manufacturer discretionary
Pin 13: Manufacturer discretionary
Pin 14: Manufacturer discretionary
Pin 15: ISO 9141-2 "L" line
Pin 16: Battery power

**Figure 3.** OBDII specified pin-out locations.

## ON-BOARD DIAGNOSTICS II MONITORING CONDITIONS

OBDII uses specific definitions to identify the conditions in which the vehicle is operated. These definitions help to define when system **monitors** are performed, when DTCs are set and cleared, and the conditions in which

## DIAGNOSTIC TIME SCHEDULE FOR I/M READINESS
### (Total time 12 minutes)

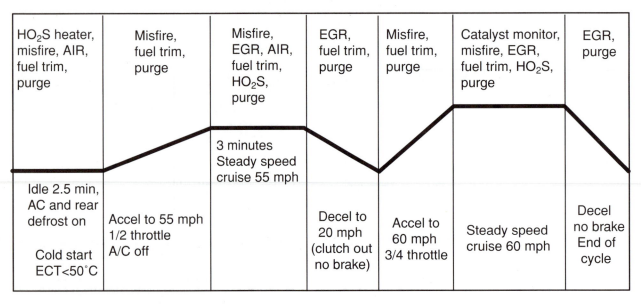

| HO$_2$S heater, misfire, AIR, fuel trim, purge | Misfire, fuel trim, purge | Misfire, EGR, AIR, fuel trim, HO$_2$S, purge | EGR, fuel trim, purge | Misfire, fuel trim, purge | Catalyst monitor, misfire, EGR, fuel trim, HO$_2$S, purge | EGR, purge |
|---|---|---|---|---|---|---|
| | | 3 minutes Steady speed cruise 55 mph | | | | |
| Idle 2.5 min, AC and rear defrost on <br><br> Cold start ECT<50°C | Accel to 55 mph 1/2 throttle A/C off | | Decel to 20 mph (clutch out no brake) | Accel to 60 mph 3/4 throttle | Steady speed cruise 60 mph | Decel no brake End of cycle |

**Figure 4.**   Typical sequence for completing a drive cycle. Vehicle manufacturers may vary.

the MIL is illuminated. A monitor is a diagnostic test that is performed in a specific sequence and is used to monitor the operation of a specific system. Understanding the following definitions will help the technician understand the logic behind when and how decisions are made.

- *Drive cycle.* A drive cycle occurs when a vehicle is driven in such conditions that all of the diagnostic monitors have been run. As a technician you will often want to create an environment that re-creates a drive cycle. This will help you diagnose a failure and verify repairs. The steps for the completion of a typical OBDII drive cycle are provided in **Figure 4**.
- *Enable criteria.* Enable criteria are the conditions that must be met for a monitor to run. The status of the enable criteria can be viewed on the scan tool.
- *Trip.* A trip is similar to a drive cycle. A trip, however, occurs when the enable criteria for a specific monitor are met and the test is run. The scan tool lists all of the specific OBDII monitors and if the enable criteria have been met and the test run. The result will be displayed as yes or no next to the monitor. Additionally, the scan tool should indicate whether the system passed or failed. This information is used by the technician when duplicating a concern or verifying a repair in a monitored system.
- *Warm-up cycle.* A warm-up cycle occurs when engine temperature increases at least 40° Fahrenheit (4.5°C) from the startup temperature and reaches a minimum of 160° Fahrenheit (72°C). The specific temperatures in which a warm-up cycle occurs may vary slightly between manufacturers.

## DIAGNOSTIC TROUBLE CODES

A DTC is an alphanumeric code, stored in the PCM, relating to the failure of a specific monitor. The DTC provides the technician with specific information regarding the failure that has occurred. The vehicle service manual provides diagnostic information, including a DTC description and specific diagnostic information for all DTCs. Almost all vehicle systems that are operated or monitored by a computer have failure codes available.

All OBDII systems use a standard five-character DTC. The first digit of the code is a letter designating in which system the failure occurred. The DTCs that begin with the letter "P" indicate a powertrain failure; a code that begins with the letter "B" indicates a body system failure; the letter "C" indicates a chassis system failure; and a "U" indicates a communication (network) failure.

The second character is a number and it will indicate whether the code is an OBDII generic code or a manufacturer-specific code. OBDII codes are indicated by the numeral zero as the second character; these codes have the same description and meaning for all vehicle manufacturers. Codes with a numeral one in the second position indicate a manufacturer-specific code. These codes are proprietary to the manufacturer. For example, two different manufacturers might choose to use a P1000 code but, because this is not an OBDII code, manufacturer A might use this to designate a transmission failure, whereas manufacturer B might choose this code to indicate a fuel system failure.

The third digit will always be a number and will indicate a subsystem failure. There are eight different subsystem

Example: P0137 Low-voltage bank 1 sensor 2

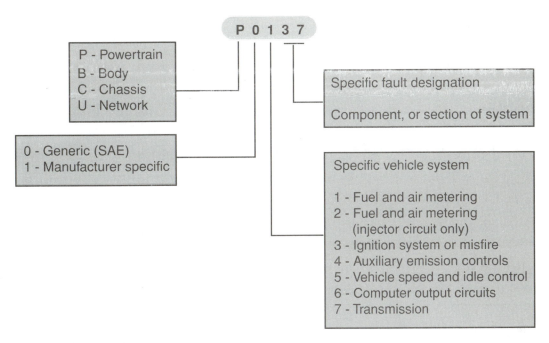

**Figure 5.** DTC identification chart.

codes for the powertrain. The fourth and fifth digits indicate the specific failure that has occurred in the systems **(Figure 5)**.

## Diagnostic Trouble Code Types

System failures and DTCs are arranged in a specific hierarchy in the determination of how the PCM reacts when a DTC is set into memory. The DTCs are classified as type A, B, C, or D. There are additional types but these four are the types that you will deal with on a daily basis.

A type A DTC indicates a failure that will not only cause an increase in vehicle emissions but will also cause damage to the catalytic converter. A type A failure will result in a DTC being set and the MIL being illuminated after only one failure. In some cases, such as serious engine misfire, the MIL may flash.

A type B failure will cause the emissions level to rise above the specified 1.5 times the allowable amount. When a type B failure occurs, it will result in the DTC being set and the illumination of the MIL after a monitor has failed on two consecutive trips. The first failure is logged by the PCM, but no DTC is set. This is called a pending DTC. If the failed monitor should pass on the next trip after the first failure, the system is reset and it will again require two consecutive failures for PCM action.

Non-emission altering failures are classified as type C or D. These failures will result in the recording of a DTC and may illuminate a separate service vehicle lamp if so equipped.

## Clearing Diagnostic Trouble Codes

After repairs are made to a vehicle, it is desirable to clear out the existing trouble codes. There are two methods available for clearing trouble codes. A scan tool can be used to clear the trouble codes from the PCM memory. Trouble codes may be cleared from the PCM when the vehicle has completed 40 warm-up cycles without a subsequent failure of the same test. The exception to this rule is any code indicating a misfire condition; it may take up to 80 warm-up cycles for this code to be cleared.

> **You Should Know** *Not all driveability concerns will be accompanied by a DTC. For a DTC to set, a failure must meet very specific criteria, which may include specific operating temperatures, time limits, or voltage thresholds. If a failure occurs outside of the criteria, the customer might experience a driveability concern without the system setting a DTC.*

## MALFUNCTION INDICATOR LIGHT OPERATION

The operation of the MIL is directly connected to the setting of a DTC. The MIL will illuminate anytime that a type

A or B DTC is set. In the event that a specific failure exists that could damage the catalytic converter, the PCM will command the MIL to flash. Once illuminated, the MIL will remain on until the code is cleared or until the vehicle completes three consecutive trips in which the failed monitor is run without a recurring failure.

## FREEZE FRAME

Freeze-frame records are stored to aid the technician with vehicle diagnosis. Freeze-frame data is recorded whenever a type A or B DTC is stored. When a DTC is triggered, the PCM will record all of the pertinent data parameters when the failure occurred. The technician can view freeze-frame data using a scan tool. The technician can use this data to observe the readings of a failed sensor. Freeze-frame information is also used to operate the vehicle under the same conditions in which the DTC was set to duplicate the concern or verify a repair.

OBDII requires that the PCM be able to store one freeze-frame event. If more than one DTC has been set, the PCM will store the data of the DTC with the highest priority. For example, if a type B code occupies the freeze frame and a type A code is set, the type A data will overwrite the type B data. In the event of a repeat failure of one monitor, the most recent data will be stored. Freeze-frame data is directly tied to the DTC and will be cleared when the DTC is cleared. DTCs and freeze-frame data should not be cleared until you have recorded the information or corrected the problem.

## FAILURE RECORDS

Some manufactures provide additional storage of system failure data in a different file called failure records. The data that is stored in the failure records is the same as that which is stored in the freeze frame. The difference is that a failure record will be stored for any DTC that the manufacturer has chosen. The PCM is often programmed to store up to five failure records compared with one freeze-frame event. The data in the failure record is replaced on a first in first out basis. What this means for the technician is that in addition to having information for non-emission–related codes, the failure record may have a backup copy of the freeze-frame data for a code in which the data had been replaced.

## SYSTEM MONITORS

A monitor is a miniature program that is designed to test the operation of specific components within the system. The monitor is performed under specific vehicle operating conditions. This is required in order to maintain the consistency of the results and properly indicate a pending failure. By closely controlling the conditions in which a monitor is performed, a repeatable, predictable result can be obtained with great consistency. The operating conditions are so critical to monitor operation that a monitor can be stopped and restarted if the vehicle operating conditions change. Once the test is completed, the PCM will log the test as either pass or fail. This can be viewed with the use of a scan tool. A failure of a system monitor can result in a DTC and the illumination of the MIL. Specific monitors and the method in which these monitors are carried out can vary among manufacturers. The PCM may monitor the system by using either passive or active methods.

Using the passive method, the PCM constantly monitors the signals provided by the sensors and some actuators. This is called comprehensive component monitoring (CCM). Using CCM, the PCM compares the data received from the sensors and the actuators to what is stored in the control maps of the operating program. If a data parameter should fall outside of the specifications, a DTC may be set; this is called a rationality check.

Using the active method, the PCM initiates an action and monitors the response on a particular sensor. For example, an EGR monitor might require the PCM to open the EGR valve a certain percentage under specific operating conditions. When the EGR valve is opened, the PCM would expect to see a change in MAP sensor data. If it did not, it could safely assume that an EGR problem existed. This mode may alter the operation of the vehicle and may affect the tailpipe emissions. A failure of one monitor can result in the suspension of another monitor. Typical OBDII monitors include the following.

### Air Injection Reaction Monitor

The AIR monitor in many cases is both a passive test and an active test. The passive test is performed anytime that the AIR is switched to the exhaust. The PCM monitors the $HO_2S$ operation. When air is switched to the exhaust, the $HO_2S$ reading should be relatively low. When air is diverted from the exhaust, the $HO_2S$ signal should rise. By comparing the data during these two tests, the PCM can determine if the system was activated.

If the passive test should fail or prove inconclusive, the PCM will then run the active test. The active test monitors the $HO_2S$ in much the same way as the passive test, however in a more controlled environment. In a low-load closed-loop situation, the PCM will command that air be pumped into the exhaust system. The PCM then monitors the $HO_2S$ and expects to see a specific amount of change in the sensor reading. If the $HO_2S$ does not register the proper amount of change, a DTC will be logged.

### Catalyst Efficiency Monitor

The catalyst efficiency monitor is a passive test in which the PCM uses two $HO_2Ss$ to monitor catalyst efficiency. One sensor is located in front of the converter and is called a precatalyst $HO_2S$. In some applications, the PCM may use this sensor for fuel control as well. The other sensor is located behind the catalyst and is called the postcatalyst $HO_2S$. With the engine operating in closed loop, the PCM compares the data of the two sensors. When the catalyst is operating as designed, the precatalyst sensor should produce a constantly varying signal, while the postcatalyst signal should be relatively stable. If the postcatalytic signal should fluctuate beyond what is allowed by the monitor, the PCM can assume that a problem exists.

## Exhaust Gas Recirculation Monitor

The EGR monitor, like the AIR monitor, can be both a passive test and an active test. The passive test may involve monitoring EGR position and EGR temperature or may measure the EGR flow. The passive test is continually monitored. The active tests involve turning on the EGR valve under certain conditions. The introduction of EGR gases will result in a change in the MAP signal. The active test is typically performed once per trip.

## Evaporative Emissions Control System Monitor

The EVAP monitor is an active test that occurs once per trip. There are several methods for monitoring EVAP system performance. Some systems may be equipped with a flow control valve. Under controlled conditions, the system is activated and the flow is measured by the PCM using sensors plumbed into the system. Another method involves pulling a vacuum on the entire fuel supply system and measuring the amount of vacuum that is lost in the system.

**Interesting Fact** *A loose gas cap can cause a failure of the EVAP monitor. This will result in a DTC being set and the illumination of the MIL.*

## Fuel System Monitor

The fuel system monitor, called fuel trim, is an adaptive strategy that is monitored by the PCM. There are two fuel trim values, short term and long term. The baseline point for fuel trim is the calibration programmed into the PCM operating program. As the vehicle operates, deviations from the program are needed to compensate for changing conditions and component wear. Perfect fuel trim values would be identified on a scan tool as 0 percent. If fuel were added to the system, values would be listed as positive percentage; subtractions in fuel are viewed as negative percent. The short-term fuel trim makes short-term corrections to

the fuel calibration necessary to keep the $HO_2S$ toggling above and below 450 mV. The long-term fuel trim reacts to short-term fuel trim and makes corrections to keep the short-term fuel trim near zero. Both long-term and short-term fuel trim have a maximum authority to change fuel control baseline between −20 percent and +20 percent; however, manufacturers may differ slightly. If both the long-term and short-term fuel trim show consistent values at or near these maximums, the PCM will trigger a DTC.

## Heated Oxygen Sensor Monitor

The $HO_2S$ monitor is used to determine the efficiency of the $O_2S$ heater circuit. This monitor is run only after a cold engine start, as defined by the engine coolant temperature, and is run a maximum of one time per ignition cycle. A cold engine start is usually defined as a startup when the engine temperature is below 95° Fahrenheit but this may vary among manufacturers. The PCM will monitor how long the $HO_2S$ takes to become active. If elapsed time that the sensor took to become active was too long, a DTC will be stored. The PCM determines the maximum allowable amount of heat-up time based on startup engine temperature and engine load.

## Oxygen Sensor Monitor

The oxygen sensor monitor is used to determine the efficiency of the sensor. The PCM performs a diagnostic test on the sensors shortly after the system goes into closed loop. The PCM will monitor the number of times that the sensor transitions from rich to lean in a predetermined amount of time, usually 100 seconds. Because the PCM continually monitors the $HO_2S$ to obtain rich-lean feedback, the PCM can also detect if the sensor stops working anytime that the engine is running. A failure of either of these monitors will generate a DTC.

## Misfire Monitor

The misfire monitor is used to determine when an engine misfire occurs. Misfire is determined by measuring the acceleration of the crankshaft. When the engine is operating, each cylinder provides a certain amount of crankshaft acceleration. Crankshaft acceleration is monitored with a high-data crankshaft sensor. When a cylinder misfires, the acceleration of the crankshaft at a given point is slowed.

There are three types of engine misfire: type A, B, and C. Type A misfire is the most serious and will cause catalyst damage if ignored. Type A failures will store a DTC and cause the MIL to flash. It is possible for a misfire to change in severity. When this occurs, the MIL could change from flashing to steady. Type B failures will cause the emissions to be raised beyond the 1.5 times threshold. A type C misfire is one that is sufficient to a point that the vehicle might fail an emissions inspection. Type B and C failures will result in a DTC being

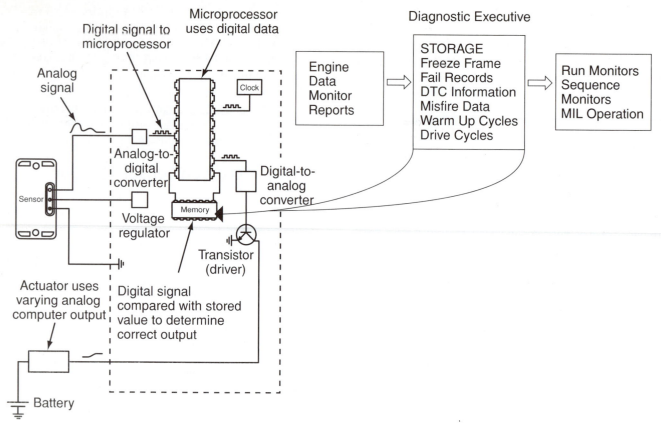

**Figure 6.** The diagnostic executive.

stored and the illumination of the MIL. DTC P0300 codes indicate random engine misfire. When a number other than zero follows the three in a P03XX code, the specific cylinder is indicated. For example, P0306 indicates a misfire in cylinder number 6. The PCM closely monitors the number of times that a particular cylinder misfires. This information is stored in the PCM in addition to freeze-frame records.

The PCM monitors engine rpm in 200-rpm segments; the most current 16 segments are stored. If misfire in any 10 segments exceeds 2 percent, a DTC will be recorded and the MIL will be illuminated. If the misfire is severe, the PCM will trigger during the first 200-rpm segment, set a type A DTC, and flash the MIL. During type A failures, the PCM may turn off the injectors to the misfiring cylinders to prevent catalyst damage.

## DIAGNOSTIC EXECUTIVE

The diagnostic executive (DE) is the portion of the PCM program that manages all of the operations discussed in this chapter **(Figure 6)**. The DE coordinates the operation of the following functions:

- Running the monitors
- Sequencing the monitors
- Determining if a DTC should be set
- Requesting MIL operation
- Tracking both drive and warm-up cycles
- Managing freeze-frame information
- Erasing DTC and freeze-frame information as needed
- Managing the misfire counters

# *Summary*

- OBDII was introduced in 1996. The primary intent of OBDII was to reduce vehicle emissions by detecting failures in their early stages.
- All OBDII-equipped vehicles are required to meet certain requirements applying to the detection and service of emissions failures.
- All OBDII-equipped vehicles use a standard communication protocol and DLC.
- All OBDII-specific codes have the same meaning among all manufacturers. All manufacturers use a five-digit alphanumeric code to identify system failures.

- There are four common types of DTCs. The PCM uses a different protocol for each type of DTC. Each type of DTC has specific criteria for the operation of the MIL. All DTCs are a direct result of a failed system monitor.
- DTCs may be cleared using a scan tool or the PCM can clear trouble codes after 40–80 trips without a subsequent failure.
- Freeze-frame and failure records store specific data parameters as they existed at the time of a failure. The PCM is required to be able to store one freeze-frame event. Events are stored according to priority.
- System monitors are programs designed to test the operation of a specific sensor or system. Monitors can be passive or intrusive in nature.
- The PCM can monitor systems by passive or active methods.
- The DE is in charge of coordinating all OBDII functions.

## Review Questions

1. Explain the requirements that all vehicles must meet concerning OBDII.
2. Explain what a communication protocol is.
3. Technician A says that all vehicle codes must be accessible with a generic scan tool. Technician B says that all OBDII-equipped vehicles use a standard seven-pin connector. Who is correct?
   A. Technician A
   B. Technician B
   C. Both A and B
   D. Neither A nor B
4. Technician A says that OBDII stipulates that only OBDII information can be transferred through the DLC. Technician B says that a vehicle can be equipped with the standard OBDII connector and not be OBDII compliant. Who is correct?
   A. Technician A
   B. Technician B
   C. Both A and B
   D. Neither A nor B
5. Technician A says that an OBDII drive cycle has occurred when all of the monitors have run. Technician B says that a trip and a drive cycle are the same. Who is correct?
   A. Technician A
   B. Technician B
   C. Both A and B
   D. Neither A nor B
6. All of the following statements about OBDII are correct *except:*
   A. All OBDII DTCs must be five digits.
   B. The first letter in a DTC will signify which system has failed.
   C. The MIL will illuminate anytime a DTC is stored.
   D. A type D DTC will not illuminate the MIL.
7. Which of the following statements about OBDII is correct?
   A. Freeze-frame records are updated on a first in first out basis.
   B. The freeze-frame records are useful for replicating a concern as well as verifying a repair.
   C. OBDII mandates that the PCM must be able to store up to five freeze-frame records.
   D. Freeze-frame data must be cleared separately from DTC information.
8. Technician A says that CCM is a form of passive testing. Technician B says that intrusive testing by the PCM initiates an action expecting to see a specific result. Who is correct?
   A. Technician A
   B. Technician B
   C. Both A and B
   D. Neither A nor B
9. Technician A says that some diagnostic monitors may use both intrusive and passive tests. Technician B says that a 7 percent short-term value means that fuel is being added above what is recommended by the baseline calibration. Who is correct?
   A. Technician A
   B. Technician B
   C. Both A and B
   D. Neither A nor B
10. Technician A says that a DTC will be set if short-term fuel trim is reading +20 percent and long-term fuel trim is at +1 percent. Technician B says that P0305 indicates a misfire for the number 5 cylinder. Who is correct?
    A. Technician A
    B. Technician B
    C. Both A and B
    D. Neither A nor B

# Chapter 18

# Diagnostic Equipment

## Introduction

Today's engine performance technician has more in common with an electrical engineer than with a mechanic. In order to diagnose the operation of various sensors, actuators, and computers you will have to be more familiar with a voltmeter than with a screwdriver. The test equipment used by the engine performance technician varies from a one-cent paper clip to a $5,000 hand-held scanner. In this chapter, we familiarize ourselves with the most common equipment used in the automotive shop today.

### JUMPER WIRE

A jumper wire is simply a length of wire with two terminals or alligator clips attached. Jumpers with a variety of terminals attached are particularly useful to front-probe terminals and to connect two connectors together without damage. To prevent severe circuit damage, jumper wires should be equipped with fuses **(Figure 1)**. Jumper wires can be used to:

1. Supply an external power source.
2. Supply an external ground source.
3. Substitute for a switch.
4. Substitute for a sensor.

### CONNECTORS AND TERMINALS

The proper connectors and terminals are indispensable when making electrical tests. Used connectors can be obtained from old components or wiring harnesses. They can be purchased from recycling yards very economically.

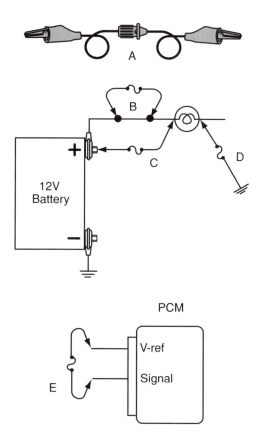

**Figure 1.** (A) Fused jumper wire. (B) Substituting a jumper wire in place of a switch. (C) Using a jumper wire to supply an external power source. (D) Using a jumper wire to supply an external ground. (E) Using a jumper wire to substitute for a sensor (only when instructed by the appropriate service manual).

As a technician, you will often find the need to build special jumper harnesses or specialized test leads, and using the proper connectors makes safe connections easy.

> **You Should Know**
>
> *When testing electrical circuits, probes or leads should not be forced into a terminal. Doing so will spread out the contact surfaces of the terminal. This can cause an otherwise operable circuit to have an open or intermittently open circuit. When front probing is necessary, only matching terminals should be used. As many technicians will tell you, a problem that is created while testing is the most difficult to diagnose.*

Repair terminals are found in all dealerships and most repair shops. These specific original equipment terminals are used to make terminal replacements in factory quality connectors and are not of the multicolored crimp-on variety. These connectors come in the exact dimensions as the factory terminals. These are useful when front probing a terminal is necessary; the mating terminal can be used without damage to the terminal being tested. You may also desire to build special jumper leads that will directly fit a specific terminal type. These specialized leads are also available premade from some vendors. All of these items are identified in **Figure 2**.

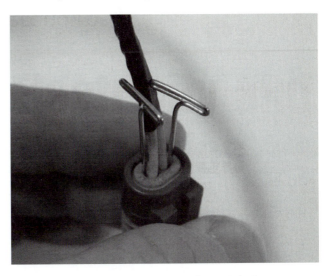

**Figure 3.** Back probing a weather-sealed connector using "T" pins.

## SEWING PINS

Sewing pins have become a staple in the tool box of electrical and driveability technicians. The pins are extremely useful when **back probing** a circuit. Back probing is a method of testing a circuit by touching connectors' terminals on the backside. This prevents damage to the terminals' mating surfaces. Because most connectors that are exposed to the elements have seals that prevent water intrusion, the thin, sharp design allows the pin to slide between the wire and the seal without damaging the wire or the seal **(Figure 3)**. The pins come in two designs: a straight pin or a "T" pin. The "T" pin is preferred because it is easier to make meter lead connections. Either of these pins can be purchased at any store where craft or sewing materials are sold.

## TEST LAMP

The test lamp consists of a sharp probe, similar to that of an ice pick, attached to a clear screwdriver-type handle. A light bulb is placed within the clear handle with a short cord attached to the bulb at one end and an alligator clamp on the other. The probe acts as the second contact for the bulb. When the probe and clamp are connected to power and ground, respectively, a complete circuit will be formed and the bulb will illuminate. A test lamp is shown in **Figure 4**.

The test lamp is one of the most versatile pieces of equipment in your toolbox. For many years, engine performance technicians have shunned the test lamp, but when used properly it can be a most valuable asset. The test lamp applies a load to the circuit being tested and will uncover many faults caused by high resistance that a voltmeter would not uncover. Many tests can be performed with the

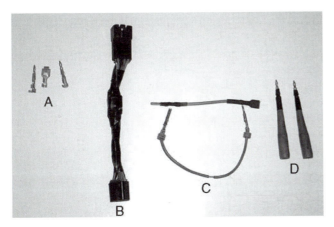

**Figure 2.** (A) Factory repair terminals. (B) A custom jumper harness made from old connectors. (C) Jumpers constructed from factory terminals. (D) High-quality factory-made jumper wires.

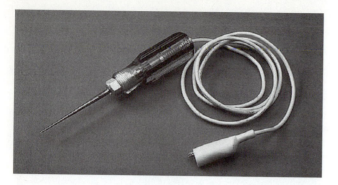

**Figure 4.** A typical test lamp.

test lamp. When testing with a test lamp, a few things must be observed:

1. Do not pierce any wiring.
2. Do not front-probe wiring terminals.
3. Do not use a test lamp to test very low–current PCM circuits, such as a TPS circuit.
4. Do not use a test lamp for testing low-voltage circuits.

Tests that can be performed with a test lamp are as follows:

- Testing for ground:
1. Connect the clamp to a known voltage source.
2. When the probe is touched to a ground circuit, the light will illuminate.
- Testing for voltage:
1. Connect the clamp to a known ground source.
2. When the probe is touched to a voltage source, the light will illuminate.

- Substituting a test lamp for a component: a test lamp may be used to substitute a high-current component such as a motor or solenoid.
1. Disconnect the component.
2. Using jumper wires, connect the test lamp across the connector in place of the component. Connect the probe end to one side and the clamp end to the other. This will create a series circuit.
3. Turn on the circuit.
4. If the lamp illuminates, an operable circuit is indicated. These tests are shown in **Figure 5**.

## LOGIC PROBE

A logic probe is similar in construction to a test lamp. Instead of using a light bulb, the logic probe uses three LEDs: red, yellow, and green. Unlike the test lamp, the logic probe uses two leads. For the logic probe, both leads must be connected. The red lead will connect to a known, stable voltage source. The black lead will be connected to a known good chassis ground. Both of these connections are needed. The power lead is responsible for supplying the needed voltage to power the probe. The ground probe is provided as a ground and a reference for the probe. A logic probe is shown in **Figure 6**.

When the probe is touched to a ground circuit, the green LED will illuminate. The red LED will illuminate when the probe is touched to a power source. Touching a pulsed voltage, such as a fuel injector circuit, will result in the illumination of the yellow light.

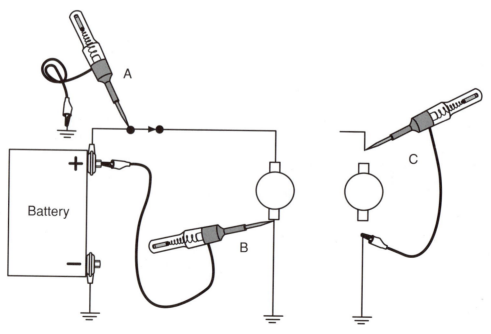

**Figure 5.** (A) Testing for voltage with a test lamp. (B) Testing for ground with a test lamp. (C) Substituting a test lamp in place of a component.

**Figure 6.** A typical logic probe.

## DIGITAL MULTIMETER

The **digital multimeter (DMM)**, also known as a **digital volt-ohm meter (DVOM)**, has become a staple in almost every tool arsenal **(Figure 7)**. The analog or needle-type meters are still found in some shops, but because these meters have low internal resistance, the analog meter can actually modify the operation of the circuit being tested. This can lead to false test readings and can cause damage to sensitive electrical circuits. For this reason, the use of an analog meter is highly discouraged. Most DMMs are able to function as a DC or AC voltmeter, ammeter, ohmmeter, and diode tester and will be able to measure signal frequency, duty cycle, and pulse widths. Other functions may include a backlit display, minimum/maximum

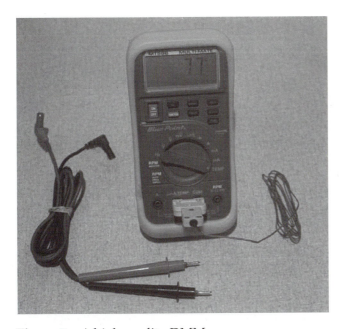

**Figure 7.** A high-quality DMM.

> **Interesting Fact**
> The DMMs come in many different price ranges and are equipped with many different features to meet specific industrial needs. To make the best possible use of your equipment it is imperative that you familiarize yourself with all of the functions, operations, and limitations for your particular meter.

signal recording features, temperature measurement capabilities, continuity buzzers, and the ability to read rpm. The DMMs have several different input terminals. To utilize all of the functions of your meter, the meter leads must be connected in the proper locations on the front of the meter. The meter controls vary greatly among models. Some meters require you to select a specific unit and range of measurement, whereas others might only need you to select a unit of measure; this type of meter will automatically select the desired range of measurement. These are called self-ranging meters.

> **You Should Know**
> Self-ranging multimeters automatically select the range of measurement within a unit. When making any measurement with a self-ranging multimeter, pay specific attention to the unit of measure that is displayed to the right of the measurement. Ignoring the unit of measure will cause you to take an incorrect reading.

## MEASURING VOLTAGE

Voltage is a measurement of electrical pressure; the unit of measure is volts. Voltage measurements are among the most common measurements that an auto technician will perform. To obtain voltage readings, the meter must be connected parallel in the circuit. The meter should be set to read a voltage range that is slightly higher than what is anticipated within the circuit. Connect the red lead to the voltage terminal (V) on the meter. The black lead will be connected to the common (COM) terminal on the meter. Typically, the probe end of the black lead is connected to a known ground, and the red lead will be used to probe connectors and circuits for voltage. The circuit must then be powered up. The amount of voltage on the display is the amount of voltage available **(Figure 8)**.

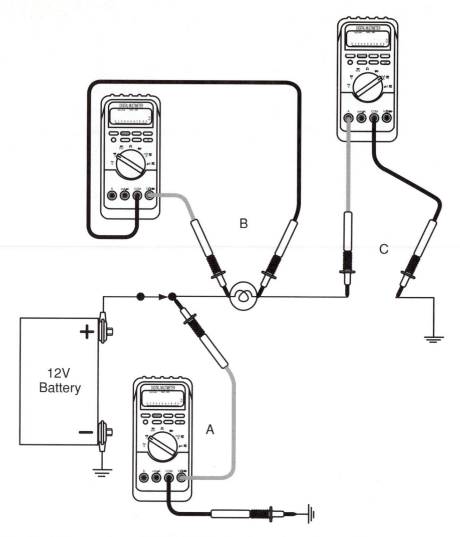

**Figure 8.**   (A) Testing for voltage using a DMM. (B) Testing for voltage drop. (C) Measuring current.

## VOLTAGE DROP

Voltage drop is a measurement of the amount of electrical pressure consumed within a circuit. To measure voltage drop, the meter must be connected in parallel. The meter should be set to the desired voltage scale; again select a range that is slightly higher than the expected reading. The red lead is placed on one side or end of a component or wire that is being tested, and the black lead is placed on the other side. The circuit would then be turned on to operate. The reading obtained on the meter would be the amount of voltage that was used within the component. Voltage drop can be measured across one component (a wire or cable), one connection, or an entire circuit. Using this test can also measure the integrity of a ground circuit. The voltage drop test can be considered an active resistance test. Because the amount of resistance in a circuit directly affects the amount of voltage drop, this test will show the results of resistance in a loaded circuit. This is valuable because the characteristics of an operating circuit can greatly differ from an inoperative one. (Refer to Figure 8).

> **You Should Know** *Reversing the leads of the voltmeter when performing a test will lead to negative readings on the display. The negative readings will be identical to the positive readings. This does not indicate a negative voltage in the circuit itself, but only the interpretation provided by the meter because of the connections.*

## OHMMETER

The ohmmeter is used to measure the resistance present within a circuit. The resistance is displayed on the meter in a unit called ohms. To measure resistance, the ohmmeter will apply a known voltage through the circuit. By knowing the amount of voltage that was applied, and the amount of voltage returned to the meter, the ohmmeter can calculate resistance.

To measure resistance, the meter should be set to the proper ohms scale, indicated by the Greek symbol omega ($\Omega$). The black lead is attached to the COM terminal, and the red lead is connected to the terminal, again identified by the omega symbol. On many meters, this is combined with the voltage terminal. The ohmmeter is connected in parallel with the circuit, and the power has to be turned off. If possible, it is advisable to disconnect the circuit or component being tested. By doing so, you eliminate the possibility of obtaining false readings from other circuit connections that may have been unidentified. The reading on the scale will identify the amount of resistance within the circuit. Typical ohmmeter testing is shown in **Figure 9**.

## AMPERAGE

Amperage is the amount of current that is flowing through a selected circuit. There are generally two amperage terminals on most meters. In high-quality meters, these terminals are protected by internal fuses. The maximum fused loads are usually indicated below the terminals. One terminal is used for accurate low-amperage circuits and can generally withstand up to a 2-amp load. The second terminal is used to measure higher loads, usually up to 10 amps. To protect the meter, it is imperative to select the proper scale and terminal. If you are unsure of the amount of current that is going to be measured, always start with the higher of the two scales. If the amperage is lower than the fuse in the lower setting, move the test lead to the lower setting to obtain a more accurate meter reading.

To measure amperage, select the proper range on the meter. The black meter lead should be connected to the COM terminal and the red should be connected to the proper amperage terminal, usually identified by "A" or "mA." To measure amperage, the circuit being measured must be opened and the meter placed in series. Because the meter is measuring the number of electrons flowing in a circuit, it can be thought of as a flow meter. Once the meter has been connected, the circuit then has to be active. The reading on the display is the amount of current that is flowing in that circuit. (Refer to Figure 8).

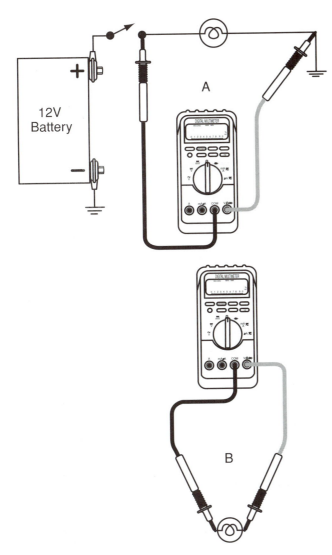

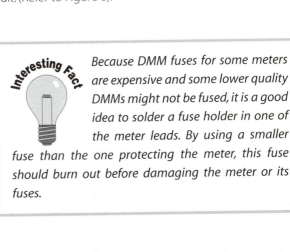

*Because DMM fuses for some meters are expensive and some lower quality DMMs might not be fused, it is a good idea to solder a fuse holder in one of the meter leads. By using a smaller fuse than the one protecting the meter, this fuse should burn out before damaging the meter or its fuses.*

**Figure 9.**  Resistance measurement should only be taken when the circuit is not active (A), or when the component is completely removed from the circuit (B).

## DUTY CYCLE OR PULSE WIDTH MODULATION

The meter setup for measuring duty cycle or PWM is the same as that for measuring voltage. The meter should be placed parallel to the circuit. Select the proper scale and connect the black lead to the COM terminal and a good ground. The red terminal should be connected to the proper terminal on the meter and the wire that is supplying the signal. In most cases, the engine will have to be started for this type of signal to be active. The display will show the frequency of the signal being measured. Refer to **Figure 10** for meter connections.

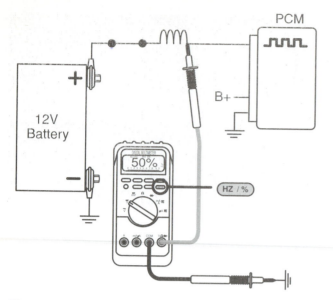

**Figure 10.**   Measuring duty cycle with a DMM.

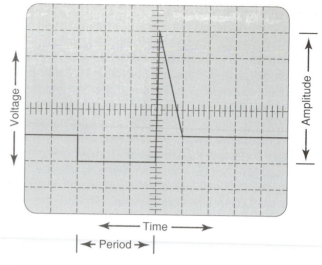

**Figure 12.**   The shape of a signal is a combination of amplitude and period.

## OSCILLOSCOPE

The oscilloscope provides a visual display of the unique characteristics of a signal. Rather than viewing numbers on a digital display, which tells little about the waveform itself, the "scope" allows the operator to "see" specific details. The information provided by an oscilloscope compared with that provided by a DVOM can be likened to comparing a black-and-white picture to a color photograph **(Figure 11)**.

This information may include how much noise is present on a signal or how often the signal changes state, as is the case of fuel injectors. With a voltmeter, we see only the average voltage of a signal. This is displayed only as a

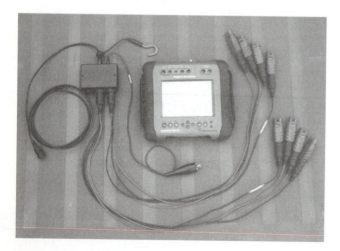

**Figure 11.**   A digital oscilloscope.

number on a screen, leaving the technician without some valuable signal information. Using an oscilloscope, we are able to view a picture of the voltage signal. This is called a trace, which is similar to what you might have drawn with an Etch A Sketch as a child. Three critical viewable signal elements are:

- *Amplitude.* The amplitude is strength of the signal measured in volts. The signal trace will rise vertically on the screen as the voltage increases. This is what gives the signal it height. The amplitude will measure every minor change in the signal. These changes in amplitude are what give a signal its specific shape.

- *Period.* Period is the amount of time in which we look at the signal. This is what gives the signal its width. As time elapses, the voltage trace reaches across the screen.

- *Shape.* Shape is a combination of the amplitude and the period. As the amplitude of the trace moves vertically, the trace is moving from left to right on the screen. This forms a distinct pattern of the signal that is left on the screen **(Figure 12)**.

Any electrical signal can be measured with an oscilloscope, and every signal that we measure will have a distinct pattern. It is through the study of these patterns that we define the finite characteristics of a signal.

Test probes are used to properly connect the signal to the oscilloscope circuits. Most oscilloscopes come equipped with one or more general-purpose probes that can be used for basic measurements. You should not substitute ordinary test leads because they could affect measurement accuracy. Information relating to test probes and leads can be found in the operating manual supplied with the instrument.

## BREAKOUT BOX

A breakout box is a specialized piece of equipment that allows the technician to check voltage and resistance at specific points within a computer or module circuit. The box is equipped with two connectors, a female connector to make connection with the vehicle harness, and a male connector to connect to the computer or module. Each circuit within the connectors is represented by a terminal on the face of the box. The terminals are the test points that the technician uses to take measurements **(Figure 13)**. By connecting the breakout box in series with the computer or module, circuits can be tested live without the need to back probe or compromise the circuit.

**Figure 13.** A breakout box in use.

# Summary

- Back probing prevents damage to fragile electrical terminals and is accomplished by sliding a pin between the wire and the connector to touch the backside of a terminal.
- Discarded electrical connectors are useful for making specialized test harnesses. The correct mating terminals are useful for building jumper wires and front-probing terminals without damage.
- The test lamp can be used to test for voltage and ground and can be substituted for a load component.

- The logic probe is able to locate the power, ground, and pulsed sources without changing connections. The logic probe draws a minimal amount of current.
- The use of a DMM will not alter circuit operation. The DMM has many different functions and can fill the need for several meters in one tool.
- The oscilloscope provides a visual display of the signal characteristics and the quality of a circuit.

# Review Questions

1. Technician A says that back probing prevents damage to terminal mating surfaces. Technician B says that when back probing a terminal, a pin should be inserted between the wire and the weather seal. Who is correct?
   A. Technician A
   B. Technician B
   C. Both A and B
   D. Neither A nor B

2. Technician A says that any type of terminals may be used when front probing terminals. Technician B says that old component connectors might be useful when testing computer circuits. Who is correct?
   A. Technician A
   B. Technician B
   C. Both A and B
   D. Neither A nor B

3. Technician A says that when a test lamp is properly connected, it can be used to test for power or ground. Technician B says that to test for the presence of voltage, the clamp should be connected to a power source. Who is correct?
   A. Technician A
   B. Technician B
   C. Both A and B
   D. Neither A nor B

4. All of the following statements about logic probes are true *except:*
   A. The logic probe has two connections.
   B. The logic probe can detect power.
   C. The logic probe can detect a pulsed signal.
   D. The logic probe can determine an exact amount of voltage.

5. Technician A says that when checking amperage with the DMM, you should start by inserting the lead into the meter terminal with the highest ampere rating. Technician B says that an analog meter can actually change how a circuit is intended to operate. Who is correct?
   A. Technician A
   B. Technician B
   C. Both A and B
   D. Neither A nor B

6. Which statement about DMM measurements is correct?
   A. To measure voltage, the DMM should be connected in series.
   B. A voltage drop is a measurement of the amount of potential consumed by an electrical circuit.
   C. An ohmmeter should be used with the circuit turned on.
   D. The ammeter applies a specific amount of voltage to a circuit.

7. Using an oscilloscope, we can see an actual picture of a voltage signal.
   A. True
   B. False

8. The voltage signal that is displayed on an oscilloscope is called a _____.

9. Shape is a combination of _____ and _____.

# Chapter 19

# Scan Tools

## Introduction

As an engine performance technician, you will find that one of your most valuable assets is the scan tool, sometimes called a scanner. The scan tool is essentially a hand-held computer that is connected to the DLC to interface the vehicle computer and electronic systems. The scanner allows the technician to view DTCs, clear DTCs, view freeze-frame data, view live data, and perform bidirectional tests.

## DESCRIPTION AND OPERATION

The scan tool is a portable hand-held device that is meant to be taken to the vehicle and can be used during a test drive, as opposed to large console-type diagnostic equipment that was commonplace only a few years ago **(Figure 1)**. The scanner connects to the vehicle through a cable connection to the DLC connector. OBDII-equipped vehicles require only one connection, whereas pre-OBDII vehicles might require a power connection as well.

Using the DLC, the scanner can communicate with the PCM and many other electronic systems. Depending on the vehicle, all of the electronic systems may be connected to one another using a serial data network, or individual systems may have separate connections at the DLC.

Modern scan tools are menu driven and therefore are easy to navigate, even for the inexperienced. Menu driven describes the method in which information is input into the tool. This format provides the technician with lists of information from which to choose, making navigation relatively simple **(Figure 2)**.

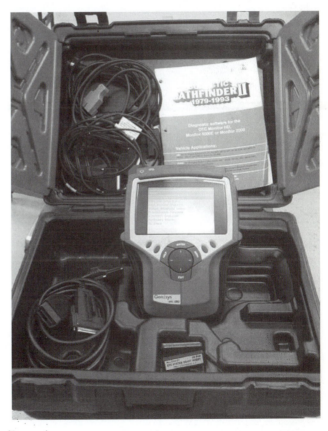

**Figure 1.** A typical scan tool and related accessories.

Technicians may choose from factory-specific scan tools or aftermarket scan tools **(Figure 3)**. The factory-specific tool is designed to interact with the systems of one specific manufacturer. Because these tools are intended for use in

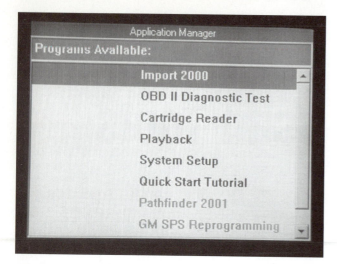

**Figure 2.** Most scan tools are menu driven. The technician must pick the proper selection.

**Figure 3.** OTC Genisys aftermarket scan tool, General Motors Tech 2, and Ford New Generation Star Tester.

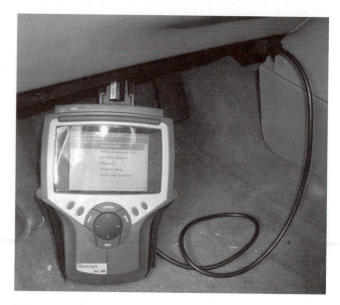

**Figure 4.** Scan tool connection to the DLC.

new vehicle dealerships, these tools are designed with many advanced features that allow the technician a great deal of access to various systems and bidirectional controls.

Aftermarket scanners are intended to work with a variety of vehicles built by various manufacturers. Because of the range and scope of the design and the restriction to specific system information, the aftermarket scanners typically will not have the same variety of functions as the factory tool. Often one company might build and distribute both aftermarket and manufacturer-specific scanners, but due to the contractual obligations, certain features might not be made available on the aftermarket tool.

With the implementation of OBDII, much of the technology relating to fuel, ignition, and emission systems that was shrouded in secrecy is now available to the tool

manufacturers. This allows the tool manufacturer to build equipment that will work with all OBDII systems. In fact, many of the factory-specific tools can be used on generic OBDII systems with a software change. Although OBDII has opened the door to information access, much nonemission-related information and functionality are still available only using the manufacturer-specific equipment.

## Connection

On vehicles equipped with OBDII, the DLC is located under the left side of the dash as shown in **Figure 4**. OBDII-equipped vehicles will use a standard DLC connector. This connector will provide both data and power connections. If the vehicle is not OBDII compliant, a special adapter that matches the DLC and separate power connections will be required.

After connections are made and the scanner is turned on, the first screen that you are introduced to is the main menu. At the menu, you will be given choices of several tasks to perform. Many of these will be tool setup menus used to set things such as screen contrast, date, and time. (refer to Figure 2). To continue with vehicle access, select the option that best describes what you want to do. If you are unfamiliar with the available options of the scanner, select all of the different options to see what those items do. This is how you can best learn the capabilities of your scanner.

## Vehicle Selection

The next menu that appears should have something to do with vehicle selection. In the following menus, you will select all of the options that apply to your vehicle. Some scan tools will have the ability to read the VIN and

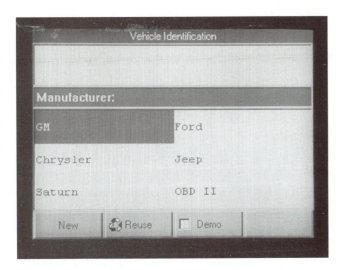

**Figure 5.** Use this menu to input specific vehicle information.

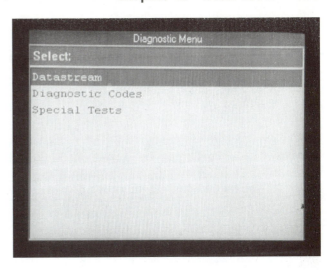

**Figure 7.** This menu allows the technician to select which function to perform.

automatically make vehicle selections. If this is the case, all you must do is verify the data. If your scanner does not do this for you, you will have to select the manufacturer, the year (tenth VIN digit), the make of vehicle, and the engine ID number (eighth VIN digit) **(Figure 5)**.

## Navigation

Now that a vehicle has been properly selected, you will be asked what system you desire to access. This menu will vary a great deal among vehicles due to the differences in available options. Typical selections from this menu might be powertrain (PCM), body (BCM, A/C, and audio), or chassis (ABS, traction control, or electronic suspension). Select the system that you want to access. A systems menu is shown in **Figure 6**.

Once a system has been selected, a list of available options will be displayed for the type of access **(Figure 7)**. Do you want to look at data, DTCs, or perform special functions? Select the desired option from the list. Each of the selections at this menu will take you to yet another menu of options that deal with your selection.

## DTC Menu

The DTC menu will give you options for accessing DTC information. Included in this will typically be options for selecting freeze-frame and failure-record information **(Figure 8)**. From here, select the information which you would like to view. The DTC selection will access all stored DTCs and their descriptions. Once vehicle repairs have been completed, return to this area to clear DTCs

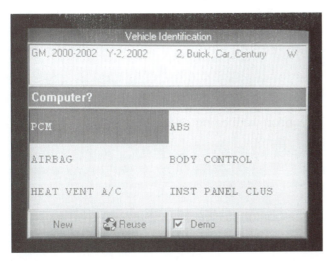

**Figure 6.** This menu is used to select the specific vehicle system.

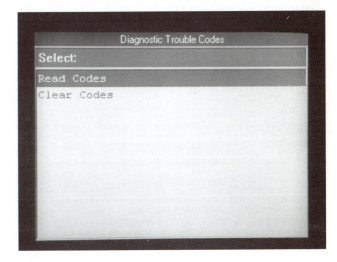

**Figure 8.** Various DTC information choices.

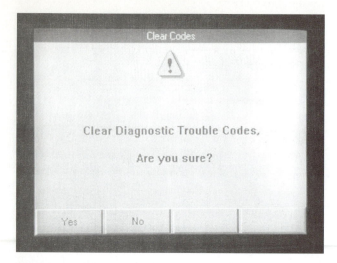

**Figure 9.** Select the clear codes option to clear DTCs, freeze-frame information, and/or failure-record information.

and related freeze-frame or failure-record information **(Figure 9)**.

## Data

You will also have the option of selecting vehicle data. From the selections in this menu, you gain access to live vehicle sensor and actuator feedback data **(Figure 10)**. Due to the vast amount of information that is available from the OBDII-compliant vehicle, data will often be divided up among several lists. These lists are given specific names that accurately describe what it is that they represent. Data lists are comprised of information that is related in some way. This gives technicians easy access to the information that they will likely need.

Depending on the scanner used, the data can be viewed in different forms. Almost all popular scanners will give you the opportunity to customize the way you view data by allowing you to select parameters and cluster them together for easy viewing. Other options may include graphical representations of data parameters, such as graphs or gauge representations.

## Special Functions

From the special functions menu you are able to select specific **bidirectional controls**. Bidirectional control allows you to control the operation of specific actuators using the controls on the scan tool. The number of bidirectional controls depends on what is available from the vehicle manufacturer and what access is made available to the toolmaker **(Figure 11)**. It is commonplace for the manufacturer-specific scanners to have more bidirectional control features than most aftermarket scanners.

Listed below are some of the most common bidirectional functions. This is not an exhaustive list and is only meant to give you an idea of what is available. Common bidirectional controls are:

- *EGR valve control.* EGR valve control allows the technician to open and close electronically operated valves at will. This is useful for testing valve operation, EGR flow ability, and electrical circuit integrity.
- *IAC valve control.* IAC valve control allows the technician to open or close the IAC valve and control the idle speed. This is useful for testing the operation of the valve and its electrical circuits.
- *Cooling fan operation.* Cooling fan operation allows the technician to manually operate the cooling fans. This is used to verify all control circuits relating to the fan, including the relays.

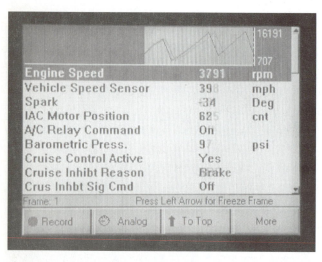

**Figure 10.** Live vehicle data is used by the technician for diagnosis.

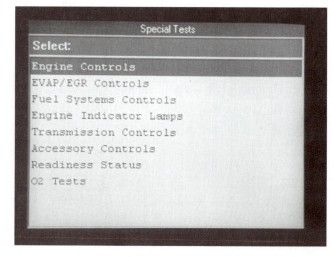

**Figure 11.** Many bidirectional special functions are available.

- *MIL operation.* The technician can test MIL circuitry using this manual override.
- *Fuel injector balance.* Many vehicles equipped with multiport fuel injection may allow control of individual fuel injectors. This may include activation of a specific injector with the engine off or the deactivation of specific injectors with the engine running. Either test can be used to help diagnose injector-related failures.

> **You Should Know** *The specific bidirectional functions available to the technician will greatly vary among different vehicle makes and models as well as different scan tool models.*

## PCM Programming

The PCM programming feature is used to change the operating program that is stored in the EEPROM of the PCM. Depending on the type of scan tool that you are using, the scanner might act as a transport device to transfer information used for programming. In this method, the scanner is used to retrieve information from the vehicle and transport it to a standalone PC that stores all of the possible PCM programs for a particular vehicle. The correct program is then downloaded to the scanner and is transported back to the vehicle, where the new program replaces the old one. Some scan tools might be equipped with memory cards that store available PCM calibrations. This equipment can direct program the PCM without moving data from one machine to another.

## SCAN TOOL ACCESSORIES AND ALTERNATIVES

Modern scan tools have a variety of accessories available. These may be extra leads that allow the scanner to be used as a voltmeter or add-on modules that change the entire function of the scanner. A typical add-on module would be an oscilloscope. This is just a small sample of the capabilities of the modern scanner. The functions available to you will depend wholly on which scanner you have.

For those who find a full-function scan tool to be impractical, there are personal computer (PC)–based alternatives. A variety of manufacturers build software and connection cables to convert any PC or laptop into a scan tool. As with other models, functions will depend solely on the software available.

> **You Should Know** *The PCM calibrations are replaced as complete programs; individual parameters cannot be modified. Calibrations are upgraded by the manufacturer to address specific programming issues that affect component operation or monitoring. The calibrations that are provided by the manufacturers are not designed to significantly increase the "power" of the engine. However, several companies manufacture programming upgrades for the sole purpose of increasing engine power.*

One of the latest and most promising trends is the use of a pocket PC or personal digital assistant (PDA). These devices are finding their way into automotive shops everywhere. With the proper accessories, these devices can be used as scan tools, as well as oscilloscopes and gas analyzers. Software and connection cabling, as well as many other accessories, are available for all popular models. As development for this platform continues, this platform will grow in popularity.

> **You Should Know** *To learn what functions a scan tool can offer you as a technician, select a vehicle and set up a scan tool that will work for the chosen vehicle. Explore each of the different menus and options available. While exploring, think about the situations in which each of these functions might be used in a live diagnostic situation.*

## *Summary*

- The scan tool is a hand-held computer that communicates with the PCM through a connection at the DLC.
- The scan tool can communicate with various electronic systems of the vehicle. Scan tools may be manufacturer specific or may be designed to work on a number of different vehicles.
- Information is accessed through a series of menu selections.
- The scan tool can be used to retrieve data and DTC-related information and perform bidirectional control functions.

# Review Questions

1. Describe what a scan tool is and how it can be useful to the automotive technician.
2. Explain how an aftermarket and manufacturer scan tools are different.
3. Explain what bidirectional control is.
4. The scan tool reads data from a direct connection at the _____.
5. List and describe three functions that can be performed using a scan tool.

# Chapter 20

# Electronic Engine Management Diagnosis

## Introduction

In any specialty area, the diagnosis is the most important part of any repair. As an engine performance technician, you are required to sift through mountains of data, from the customer, from the vehicle, and from your test equipment. You will often be called upon to find the "needle in the haystack." As an auto technician, the amount of time that you spend on a diagnosis can be money that is made or money that is lost. For this reason, it always best to use a logical well-built diagnostic procedure. It might take a little more time to examine all of the variables, but the result will be accurate repairs, customer satisfaction, and a large paycheck for you. Always take pride in your work. In any vehicle diagnosis, the vehicle-specific service manual should be consulted to ensure proper diagnostic procedures.

### DIAGNOSTIC PROCEDURE

A technician's diagnostic procedure is like his or her fingerprint; every technician has one and every one is different. When diagnosing a customer concern, you need to have process or procedure. This provides a logical means of diagnosis to help prevent overlooking the obvious and to prevent wasting time on something that is not contributing to the problem. As your abilities grow and vehicles change, so too will your diagnostic routine. The diagnostic routine should take you through a series of steps that you perform on every vehicle to make sure that you have checked the obvious. The procedure should be logical, starting with the simple and increasing in complexity as you proceed. Refer back to Chapter 3 on diagnosis for a basic diagnostic framework.

## The Intermittent Concern

An intermittent concern is a problem that can surface at any time. This type of problem has no specific pattern or set of conditions in which it occurs and it might last for only a few seconds. Intermittent problems are often temperature, moisture, or vibration sensitive. Therefore, this is the most difficult type of problem faced by a technician. When faced with an intermittent problem, you should follow your diagnostic routine as closely as possible, looking especially for loose or corroded connections. Further adding to the difficulty of diagnosing the intermittent problem is that your test data might be inconclusive and all systems might test normal.

Because intermittent problems are often difficult to duplicate, they are a major contributor to customer dissatisfaction. When you know that you are facing an intermittent problem, you will want to get as much detailed information from the customer as possible. It is a generally accepted practice that if you cannot replicate the problem, you cannot fix it. Most likely your employer will have a policy in place on how intermittent problems are handled. However, you should not feel discouraged if you are unable to repair an intermittent problem. These are difficult for the most seasoned technician.

Intermittent problems are commonly caused by loose or corroded wiring connections, chafed wiring, or faulty connections within electronic circuits or components. Faulty connections within an electronic component are often affected by changes in their operating temperatures. Manipulating associated wiring harnesses might help to induce the fault in the circuit and duplicate the concern. If engine operation does change when manipulating a wiring harness, you need to further inspect suspect wiring **(Figure 1)**.

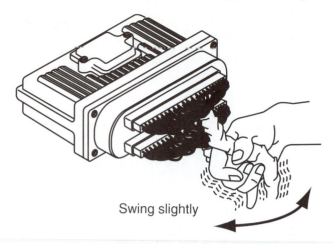

Swing slightly

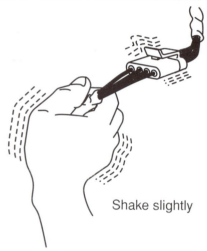

Shake slightly

**Figure 1.** Manipulating system wiring harnesses and connections can often induce an otherwise intermittent fault.

Electronic components can often "glitch" under specific temperature conditions. If you suspect temperature as a fault, there are a couple of ways to induce an artificial environment. If the component can be removed easily, it might be desirable to heat the part using a heat gun on a low setting. A component may be put in a freezer or refrigerator to substitute a cold temperature condition.

If the component cannot be removed, if used carefully the heat gun may still be used. Simulating cold conditions may be done with ice, compressed air, or a manual choke tester. The manual choke tester operates using compressed air and can sometimes get cold enough to produce small particles of ice.

## Hard Faults

Hard faults are the problems that are always present or are easily duplicated. This type of problem will have specific characteristics and indicators and will follow a certain

pattern. Due to the nature of these problems, these are also the easiest to replicate, repair, and verify.

## Pattern Failures

Once you begin working as a technician and experiencing a variety of vehicles, you will begin to see the same makes and models repeatedly. This is especially true if you work in a dealership. Because of this repetition, you might notice that many vehicles might have the exact same symptoms. This is called a pattern failure. Pattern failures usually result from a faulty manufacturing process that results in a part that is out of specifications or wears prematurely. Often if a pattern develops within the manufacturer's warranty period, service bulletins might be available with updated parts and procedures, so it is always a good idea to look for bulletins with this type of problem.

> **You Should Know** *It is a good idea to always check for bulletins and updated service procedures. Service procedures continually change, and not being familiar with the newest procedure can cause you to be responsible for a comeback.*

Technicians tend to like pattern failures because they greatly decrease their diagnosis time. There are, however, a few rules to follow. When faced with a suspected pattern failure, you still need to run your diagnostic procedure, but having a strong idea of what the problem might be will assist you in narrowing the problem down to a system and quickly moving to component diagnosis. Another benefit is that you are likely to remember the steps to a specific test, the data values, and the tools you will need. You might even remember the exact part numbers for replacement parts.

> **You Should Know** *Remember that pattern failures will follow precise patterns and relate to specific parts and part numbers for an exact make, model, and year of vehicle. Furthermore, pattern failures might be directly related to how the vehicle is equipped. When a vehicle is presented to you with concerns that are similar to a known pattern failure but the vehicle does not meet the exact failure criteria, the problem is likely to be something else, and a complete and thorough diagnosis should be done.*

## DIAGNOSTIC TROUBLE CODE DIAGNOSIS

When a monitored system fails a PCM diagnostic test and meets the required criteria, the PCM will record a DTC and possibly a freeze frame and/or failure record that relates to this failure. When this occurs, the MIL might be illuminated for a period of time. When diagnosing a drive-ability concern, you should check for DTCs early in the diagnosis even if the MIL is not illuminated. This procedure is shown in **Figure 2**. If a DTC does exist, start your diagnosis here. At this time in the diagnosis, it is appropriate to view the freeze frame or failure records. This will give you the specifics when the problem last occurred. This information will also be helpful in replicating the problem.

The service information for your specific vehicle will have a list of DTC diagnosis charts. These charts, sometimes called **trouble trees**, are written in a logical manner from the easiest, most likely tests to the more difficult and less likely. However, real-life diagnosis might stray from the chart, but most likely when you find the fault, you will be able to understand why that particular DTC was set **(Figure 3)**. Once you have repaired the problem, you should clear all of the DTC information from the PCM memory before returning the vehicle to the customer.

## DIAGNOSIS WITHOUT DIAGNOSTIC TROUBLE CODES

On many occasions, driveability concerns will not set a DTC in PCM memory. When encountering these types of problems, you must depend on your diagnostic routine and the experience you have acquired. This will also include any information supplied by the customer as well as the information you obtain from your test drive and any test results or vehicle data that are available. Use of the worksheet found in Figure 7 of Chapter 3 might be helpful in gathering specific information from the customer. Here are some common questions to ask yourself when beginning the diagnosis of a problem in which no DTCs have been stored:

1. What system could cause the concern that the customer is describing?
2. When replicating a problem, what system can cause the problem I am experiencing?
3. On what other vehicles have I experienced this same concern, and what did I do to diagnose those vehicles?

The following symptoms are often associated with those problems that do not generate a DTC:

1. Lack of power, poor engine acceleration, and a general feel of a laboring engine.
   a. Low fuel system pressure or volume; will be most noticeable under hard acceleration at higher rpm. Engine may seem to speed up if accelerator is slightly lifted.

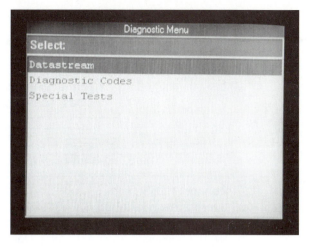

(A)

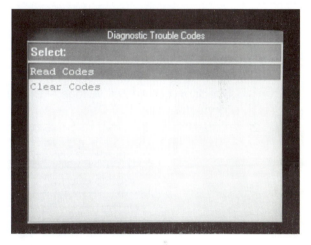

(B)

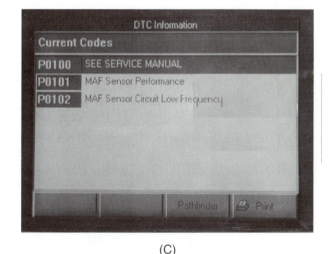

(C)

**Figure 2.** (A) Select your preferred function. (B) The DTC selection menu. (C) DTCs are displayed on the screen.

**DTC 13 HEATED OXYGEN SENSOR (HO₂S) CIRCUIT**

(open / grounded circuit)

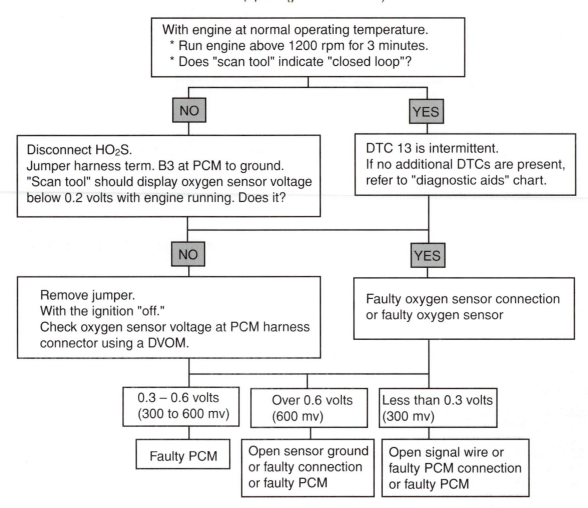

**Figure 3.** A typical diagnostic trouble tree.

b. Restricted exhaust system. Engine may run all right when cold but performance might diminish as the temperature in the exhaust system increases.

c. Severely restricted air filter.

2. Engine miss or chuggle. This problem is intermittent in nature, rarely occurring long enough for the misfire diagnostic to sense it. This problem will usually occur under heavy load, with lean fuel conditions, such as accelerating up a hill with the cruise control on. For complete diagnosis refer to Section 4.

a. Carbon tracking in secondary ignition wiring.

b. Cracked spark plug porcelain.

c. Either internal or external carbon tracking of the ignition coil.

d. Cross-firing from secondary wires.

3. Engine no-start conditions. The engine cranks but will not start and very rarely will set trouble codes. Because the engine has not been able to start, the PCM cannot run any diagnostic monitors. No-start conditions can result from:

a. Ignition system failures; refer to Section 4.

b. Fuel system failures; refer to Section 5.

c. PCM sensors and/or actuators.

4. Poor fuel mileage.

a. In most cases, poor fuel mileage concerns with no other perceived problems are a direct result of the conditions that the vehicle is operated in.

b. In some cases, a sensor value might shift slightly, causing a small loss in mileage. Common sensors that can cause this are the HO₂S and the CTS.

5. Engine hesitation. This problem results as soon as the vehicle is accelerated from a stop. Common problems include:

a. Low fuel pressure.

b. A TPS with a small fault in one segment of the potentiometer.

c. Dirty or faulty MAF. As the MAF sensor is exposed to the rush of incoming air, a layer of debris will build on the resistor and act as an insulator, causing erroneous readings.

## COMPONENT DIAGNOSIS

In the course of any engine performance diagnosis, it will be necessary to perform diagnosis on individual components. You might be directed to do so by a DTC trouble tree or this might be part of your intermittent diagnosis procedure. We will not cover the diagnosis of each specific sensor or actuator but rather each type of component, for example, a potentiometer or switch.

## POTENTIOMETER TESTING

Many sensors used in the engine management system are potentiometers designed to fit specific applications. All potentiometers can be checked in the same manner. In order to test a potentiometer, it is desirable to measure the voltage drop. This can be accomplished by testing the sensor while still connected to the vehicle wiring harness. This also gives you the opportunity to test the vehicle wiring at the same time.

With the ignition in the run position and the engine off, back probe the reference wire and measure the voltage. Five volts should be present. If the proper voltage is present, test the sensor ground by back probing the ground terminal using the meter negative lead. Connect the positive lead to the battery. You should read battery voltage. If this reading is within specifications, back probe the signal wire and observe the meter reading as the sensor is slowly actuated. Connections are shown in **Figure 4**. This may be done by opening the throttle or moving the vane of a vane-type MAF. The voltage drop should change proportionately to the amount and rate of actuation. Oftentimes, the voltage might momentarily drop, indicating a faulty sensor. Because these sensors might have very minor glitches, a DMM with very slow response time might not catch the signal. For this

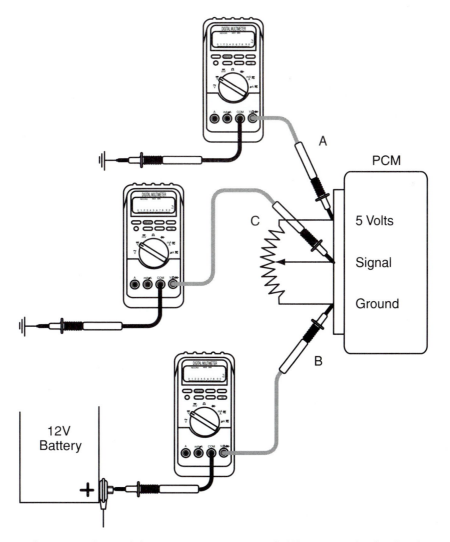

**Figure 4.**   (A) Testing reference voltage. (B) Testing circuit ground. (C) Testing the feedback signal.

reason, an oscilloscope is the preferred tool for this test. The scan tool will also display voltage for many sensors; however, relatively slow data refresh rates might fail to catch many glitches.

A potentiometer can also be tested by measuring resistance; however, this might not be as accurate. When resistance testing is performed, the sensor must be disconnected from the harness, and the terminals must be front probed. Use of specific terminals or jumpers is required to avoid damaging the terminals. The test is performed in the same manner by actuating the device while watching for a smooth linear movement as the resistance changes.

## THERMISTOR

When testing a thermistor, it should be checked throughout the entire operating range in which the sensor is used, in the case of an automobile from as low as 10°F (−12°C) and up to 250°F (122°C). This will depend on your specific environment. If the temperature-resistance graph for the thermistor is available, a basic accuracy check can be made by testing resistance or sensor voltage drop while manipulating the temperature in a controlled situation.

A thermistor can be tested in a live circuit by measuring its voltage drop. Because the thermistor provides a voltage drop for the PCM, the voltage across the device will be proportional to its resistance **(Figure 5)**. The thermistor can

be tested by back probing the sensor connector and measuring the voltage drop across its terminals. This voltage is then compared with the specified values at a given temperature. Temperature-to-resistance and/or temperature-to-voltage charts are typically available in most service manuals.

> **You Should Know** *It should be noted that a thermistor might exhibit normal resistance values at certain temperatures, yet develop resistance values outside of normal tolerances at the upper or lower range of temperature measurements. This type of failure causes the temperature-resistance graph to shift or skew. A skewed characteristic curve can cause inaccurate temperature measurements above and below the normal operating temperature of the engine, resulting in poor engine performance outside of the normal operating area. An engine that exhibits poor cold performance—perhaps during the warm-up cycle—yet operates normally after normal engine temperature has been reached might have a defective engine CTS.*

A CTS can be tested further by removing it from the engine and placing it in a container of water. Attach an ohmmeter across the terminals. Place a thermometer in the water and slowly heat the container, as shown in **Figure 6**. As the temperature begins to rise, observe the change of the resistance and the temperature in which it occurs. The resistance should fall smoothly as the temperature rises. This gives the technician a controlled environment in which to test the sensor.

## TESTING MANIFOLD ABSOLUTE PRESSURE SENSORS

The MAP sensor can be tested by using a vacuum source such as a hand-held vacuum pump equipped with a vacuum gauge. If the pump does not have a built-in vacuum gauge, use a vacuum "T" to connect one, as shown in **Figure 7**. Using a manufacturer's service manual, vacuum data, and the corresponding electrical output signal information, the sensor is easily tested for proper operation. By applying a specified amount of vacuum as indicated in the chart, the corresponding voltage should be the same as indicated. This test can often be accomplished with the sensor still on the vehicle and with the key in and turned to the run position with the engine off. The voltage results can be obtained by viewing the voltage data from a scan tool. Alternately back probing the MAP **barometric pressure**

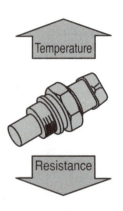

| °F | °C | Ohms | Voltage drop across sensor |
|-----|-----|---------|----------------------------|
| 212 | 100 | 180 | 0.75 |
| 190 | 90 | 240 | 0.97 |
| 160 | 70 | 475 | 1.59 |
| 100 | 38 | 1,400 | 2.97 |
| 40 | 4 | 7,500 | 4.35 |
| 32 | 0 | 10,000 | 4.53 |
| 18 | −8 | 14,600 | 4.68 |
| −40 | −40 | 100,700 | 4.95 |

**Figure 5.** A typical thermistor temperature-to-resistance chart.

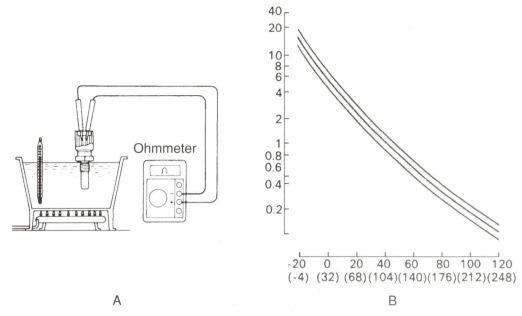

A

B

**Figure 6.**   (A) Testing an ECT sensor off the car. (B) Typical ECT specifications.

**Figure 7.**   A technician uses a vacuum pump to apply vacuum to a MAP sensor while monitoring MAP sensor voltage with a scan tool.

sensor **(BARO)** connector and viewing directly from a DMM or an oscilloscope, the sensor voltage or corresponding voltage trace can be viewed directly.

> **You Should Know** *MAP and BARO sensors can appear to operate normally, yet the calibration of the sensor can be out of range, leading to serious engine problems affecting fuel delivery. Suspect a faulty pressure sensor if oxygen sensor readings indicate an overly rich or lean condition.*

## AIR FLOW SENSORS

The MAF sensors can be tested for basic operation using a shop manual and oscilloscope. Alternately, a scan tool can display MAF data that can be correlated to manufacturer specifications. As stated above, many times a MAF sensor might cause a slight hesitation upon acceleration. When this occurs, the MAF data show a glitch when viewed from the scan tool. When this condition occurs, the sensor should be removed and inspected for debris. A sensor with excessive build-up debris is shown in **Figure 8**. Once you have determined that the MAF is at fault, it may be carefully cleaned with aerosol electronics cleaner. Vane-type meters

**Figure 8.**   A dirty airflow sensor element can cause false readings.

may be tested using the procedure outlined for measuring potentiometers.

> ▽ **You Should Know** *Most manufacturers do not recommend cleaning a MAF sensor. If you have determined that a dirty MAF sensor is the cause of your concern, the customer should be presented with the option of cleaning the sensor or replacing it. If a sensor is cleaned, the resistor might be damaged and a subsequent failure can occur. However, if no damage occurs, the sensor can go on to provide years of reliable service.*

## HEATED OXYGEN SENSOR DIAGNOSIS

Heater circuit failures of the HO$_2$S are commonplace and will more than likely result in setting a DTC. The heater circuit is easily tested by verifying both a power and ground circuit from the sensor harness with the key on. This can also be performed with a test light connected across the heater circuit terminals, as shown in **Figure 9**. The lamp should illuminate when the ignition is turned on. If it does not, a circuit problem might exist. However, if the lamp does illuminate, the sensor is likely at fault.

When operating in closed loop, the sensor should constantly change above and below 450 mV, indicating that the fuel mixture is being controlled. These are called **cross counts**. If the sensor has a fixed voltage below 450 mV, a lean condition may exist, such as a severe vacuum leak. Should the voltage become fixed above 450 mV, the engine might be experiencing a rich condition, such as a

leaking fuel injector or high fuel pressure. The HO$_2$S should be able to cycle between a lean circuit voltage of under 200 mV and a rich circuit voltage of over 800 mV. When the throttle is suddenly opened from idle to WOT and back to idle, the sensor should first show a rich voltage and, as the accelerator is closed, then indicate a lean voltage. This is sometimes called a snap test.

In some cases, an O$_2$S might fail and constantly produce a very low voltage, indicating a lean fuel condition. If the sensor is constantly lean, the fuel mixture will be enriched to the point that black smoke would be emitted from the tailpipe. In this case, the PCM's ability to detect the sensor should be checked. If the PCM is able to detect a good sensor reading, the sensor in the vehicle is bad. This condition will most likely be encountered only in older vehicles utilizing only one O$_2$S.

> ▽ **You Should Know** *When observing an HO$_2$S waveform with an oscilloscope, any situation that affects engine operation, such as ignition misfire or a sticking injector, will show up in the sensor waveform.*

Oxygen sensor data can be viewed on a scan tool or by back probing the connector and attaching a DMM or oscilloscope. Because a scan tool only displays a data sample, rather than continuous real time data, it might not be the tool to use when diagnosing sensors with values that change rapidly. Because accurate operation of the HO$_2$S largely depends on its ability to rapidly respond to an air/fuel ratio change, many technicians prefer to use the oscilloscope

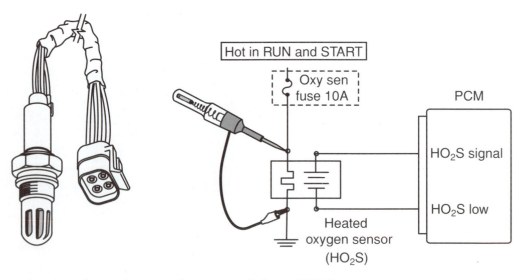

**Figure 9.**  A test lamp can be used to test a heater circuit for an HO$_2$S.

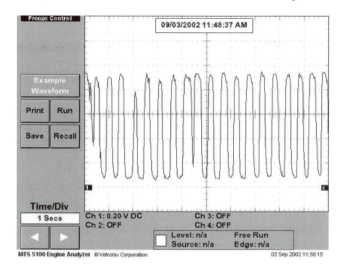

**Figure 10.** This is a good HO$_2$S waveform.

for testing. Because the oscilloscope shows both time and voltage, the quality of the scope pattern can be analyzed for both range and transition time. A good sensor in a proper running engine will have a great deal of activity and should produce crisp rich-lean transitions, as shown in **Figure 10**. A lazy signal will have a jagged appearance, and the transition slopes will be noticeably more gradual. Activity will slow down and will tend to stay either rich or lean too long. A comparison of waveforms is shown in **Figure 11**.

The exception to testing HO$_2$S is the postcatalyst sensor. Its readings should be relatively steady. If these readings are not steady, a faulty catalyst is indicated.

## RELAYS

Relays can be tested in a variety of ways. An ohmmeter can be used to check coil resistance; consult the manufacturer's shop manual for the specified coil resistance **(Figure 12)**. The resistance you measure should be within 5 percent of the manufacturer's specification. Relays that have coil resistances significantly greater or less than specified should be discarded.

A coil that measures infinite resistance is usually considered "open." Relays that have an open coil will not operate and should be replaced. Testing the coil is only part of the relay testing process. Although coils do fail—rendering the relay inoperative—problems with the contacts are far more common. Contacts frequently burn or corrode, causing excessive resistance. The excessive resistance prevents the load from operating at all or causes it to operate with reduced efficiency.

Resistance testing of the contacts is done in a similar manner to testing a switch. The coil of the relay must be activated using jumper wires. An ohmmeter is used to test resistance across the contacts of the relay **(Figure 13)**. If the relay contains more than one set of contacts, check each set separately. There should be less than 0.3 of an ohm resistance across the contacts.

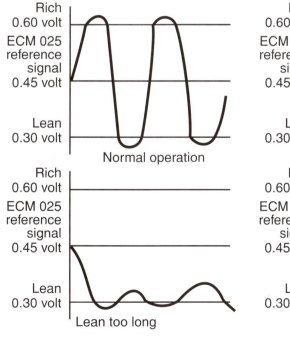

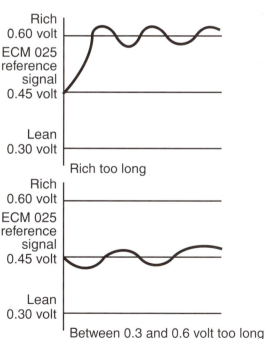

**Figure 11.** A comparison of HO$_2$S waveforms.

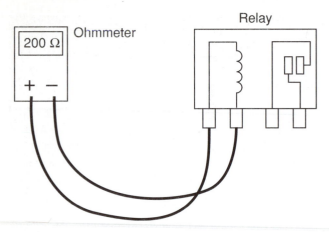

**Figure 12.** Testing the resistance of a relay coil.

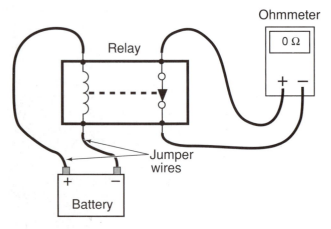

**Figure 13.** Testing the resistance of the closed relay contacts.

Although ohmmeter testing of the contacts is possible, load testing the contacts is preferred. Intermittent or inoperative contacts will be easily located using this method. To verify that the contacts are in good operating order, a load of the same approximate value should be used to simulate actual operating conditions. Connect the relay, power supply, and substitute load. Using the actual load itself is preferred but often impossible. Instead, use an electrical load that draws a similar amount of current to the actual load, such as a light bulb or array of light bulbs. If a load matching the device's current characteristics is not available, a test light can be used as a load. Although not the best technique, it will suffice when nothing else is available. Load testing the contacts with the small load provided by the test light provides a better functionality test compared with a simple resistance check of the contacts alone. Always confirm the voltage drop across the operating contacts, even if you see or hear the load operating. This procedure is shown in **Figure 14**. This is important because it provides information about possible intermittent operation and excessive resistance. Voltage drops greater than approximately 0.25V usually indicate excessive resistance; in this case, replace the relay.

The quick test for checking a relay on the vehicle is to merely switch the relay with one from another circuit. In most cases, there are only one or two types of relays used on a vehicle, so relays with the exact same part numbers are readily available, located in the relay center. Simply switch relays with another circuit. If the component in question begins operating, the relay is at fault. You should double check your findings in this step to ensure accurate results.

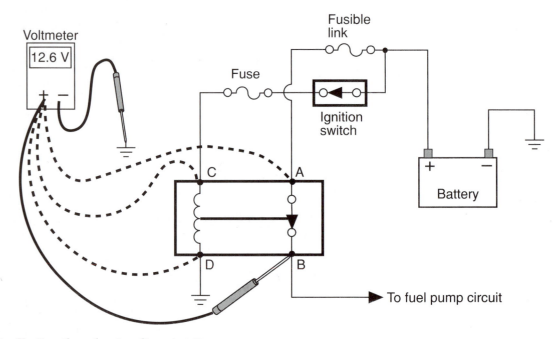

**Figure 14.** Testing the relay in a live circuit.

## TESTING THE POWERTRAIN CONTROL MODULE

The automotive PCM has evolved into a very reliable component and seldom fails. However, because there is somewhat of a mystery surrounding the PCM, it is often misdiagnosed as the root cause when no other problem is found. PCM problems can be either intermittent or hard faults. PCM diagnosis will generally fall into six categories:

1. Reprogramming issues
2. DTC-related problems
3. Multiple DTCs
4. No DTCs
5. Failure of output control
6. Stalling

For the PCM to operate properly, it must have good power and ground connections. Because PCM grounds are often located on the engine, they are often prone to corrosion. Making sure that PCM grounds are clean and tight is the first order of business when diagnosing a PCM problem. Ground connections are often located on the intake manifold area or at the transmission bell housing. If the grounds are in good condition and the concern still exists, further diagnosis is needed.

Because most PCMs have switched to an EEPROM memory for operating parameters, updating the PCM is a common procedure. Generally, it is a good idea when an unusual concern exists to make sure that the PCM has the latest EEPROM calibration installed. A calibration number is available by viewing the scan tool data. In some cases, a service bulletin might be available that recommends an upgrade. In other cases, however, you will have to view update descriptions to determine what operating changes have been made. In any case, you should install the newest available calibration to prevent any future EEPROM concerns. A brief reprogramming sequence is indicated in **Figure 15**, **Figure 16**, and **Figure 17**.

A PCM can cause a DTC to be set. This can be either a system DTC or a specific PCM fault DTC, such as communication error or quad driver failure. In either case, if a code exists, check the trouble tree for the specific DTC. In most cases, the trouble tree will isolate a faulty PCM. In this situation, lightly tapping the PCM may cause the DTC to reset.

The third case of PCM failure is multiple unrelated DTCs being set. As a benchmark, four or more unrelated codes set at about the same time would be unusual. Many times, this is the trademark of a PCM failure. In this case, the codes will have been set at the same time and might be accompanied by an engine stall, MIL illumination, or poor running. Again, a light tap to the PCM case may induce the failure. Before condemning the PCM, grounds should be checked, and DTC diagnoses should be performed to confirm the existence or absence of a reference voltage and the PCM's ability to process information. If these tests are inconclusive, replacement of the PCM might be in order.

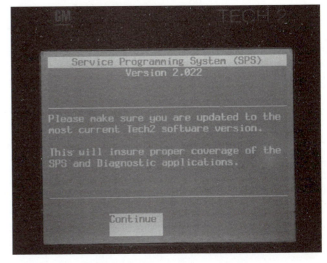

**Figure 15.** Before reprogramming can begin, information must be retrieved from the vehicle's PCM.

**Figure 16.** The scan tool is then connected to a standalone PC, and new program information is downloaded.

In the fourth scenario, the engine might be experiencing noticeable driveability concerns without illumination of the MIL. Because the PCM is in control of the system, whatever information it calculates is assumed to be correct. Often data retrieved with a scan tool can be identified as incorrect. For example, a cold engine might display over 300°F on the scan tool. In a normal situation, the PCM would be able to diagnose this as a fault, but if the PCM is at fault, it reads this as normal.

In this case, the problem can be easily found and identified. A quick check of the sensor circuit to verify that the sensor is operating properly and that the wiring leading to the PCM is in good condition might indicate that the PCM needs to be replaced. Again, grounds must be checked and

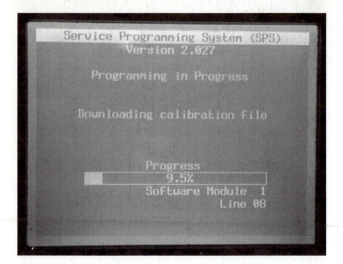

**Figure 17.** The new program information can then be loaded into the vehicle's PCM.

verified, and a light tap to the PCM case might make the parameter in question be corrected.

Scenario number five involves an inoperative output circuit. In this case, PCM commands will fail to be transferred to the output component. In this situation, the technician must verify the operation of the actuator, verify the condition of all of the wiring, and make sure that all of the conditions for operation of that actuator have been met before replacing the PCM.

In some intermittent cases, the PCM might cause the engine to stall. Because the PCM controls the spark timing and fuel systems, an intermittent fault in either of these cases, the engine would stall. It might restart immediately or it might have to sit for a little while before restarting. Unfortunately, these types of problems are the most difficult to diagnose. The tap test can prove useful in some cases.

Each manufacturer has its own method of testing PCM integrity. Often it will require grounding the sensor harness and observing the scan tool voltage. Some tests will require you to use a jumper wire to jump across the signal and the ground wires of a potentiometer or between the signal and the reference wire. These tests are designed to see if the PCM will recognize signals at their extreme limits. Because these methods and expected results differ greatly, the service manual should always be consulted when verifying PCM operation. In some cases, you might have to refer to a DTC trouble tree to locate this information.

## *Summary*

- A logical diagnostic approach will ultimately make the technician more money and better satisfy his or her customers. Individual technicians will have their own diagnostic approach.
- Intermittent problems occur at random with no specific pattern. Intermittent problems are often caused by loose connections or corroded terminals.
- When analyzing data for an intermittent concern, it is possible for all data values to be within specifications. It is difficult to repair any problem that you are not able to replicate.
- Hard faults have specific indicators and occur within a given pattern. Pattern failures are very specific and generally apply only to certain vehicles equipped with exact equipment or components.
- The DTCs that apply to your concern are the first place to begin a diagnosis. Trouble trees are written in a logical manner in order to diagnose the problem.
- Many driveability concerns will not have associated DTCs set. When no codes are present, technicians must rely on their diagnostic abilities and experience to find the cause.
- You may perform component diagnosis when directed by a trouble tree or as part of a personal diagnostic approach.

- Many different sensors can be diagnosed using the same procedures. Potentiometers all perform in the same manner; however, appearance can vary according to application.
- Scan tools are unable to catch many small glitches provided by a sensor. Most sensors can be checked while still on the vehicle.
- A thermistor should be checked throughout its entire operating range.
- MAP/BARO sensors are tested by applying a specific amount of vacuum to the sensor and observing the relative change in the voltage signals.
- A MAF sensor might malfunction if the resistor becomes dirty. Cleaning the MAF sensor can damage the sensing element.
- The $HO_2S$ heater circuit failures will usually result with a DTC being set. Heater operation can be verified by connecting a test lamp between the heater terminals.
- Proper sensor element operation depends on voltage level and rate of response. $HO_2S$ should toggle constantly above and below 450 mV.
- Excessively rich or lean fuel mixtures can cause fixed circuit voltage. A good sensor should be able to cycle in range from lower than 200 mV to over 800 mV when the throttle is quickly moved from idle to WOT and

back to idle. Good HO$_2$Ss should produce crisp rich-lean transitions when viewed on an oscilloscope.

- The most effective method for testing a relay is to activate the relay and measure the voltage drop across the terminals when a load is applied.
- Because technicians do not understand the operation of the PCM, it is often misdiagnosed. Good power and ground connections are essential to PCM operation.

- Reprogramming the EEPROM is the most common PCM service procedure. A PCM can be tested by lightly tapping on the case.
- The technician should make sure that all ground connections are clean and tight before condemning a PCM. The technician should verify that all of the parameters have been met for the PCM to make a decision before assuming that the PCM is at fault.

## Review Questions

1. Technician A says that a logical diagnostic approach is needed to make sure that all of the variables have been examined. Technician B says that every technician can have his or her own diagnostic approach. Who is correct?
   A. Technician A
   B. Technician B
   C. Both A and B
   D. Neither A nor B
2. Technician A says that intermittent problems occur at regular intervals. Technician B says that when diagnosing an intermittent concern, the data will indicate which system is causing the concern. Who is correct?
   A. Technician A
   B. Technician B
   C. Both A and B
   D. Neither A nor B
3. Technician A says that intermittent problems require no special attention to specific details. Technician B says that if a problem has no DTC and it cannot be replicated, it can be difficult to diagnose. Who is correct?
   A. Technician A
   B. Technician B
   C. Both A and B
   D. Neither A nor B
4. Technician A says that hard faults appear all of the time or can be duplicated when the vehicle is operated under specific conditions. Technician B says that a pattern failure relating to a TPS failure in a 1998 Ford Crown Victoria will apply to all 1998 Ford vehicles that use a TPS. Who is correct?
   A. Technician A
   B. Technician B
   C. Both A and B
   D. Neither A nor B

5. Technician A says that if a trouble code exists in PCM memory, the DTC should be diagnosed before any further diagnostics are performed. Technician B says that if a trouble code exists in PCM memory that directly relates to a system that could cause the concern, it should be diagnosed first. Who is correct?
   A. Technician A
   B. Technician B
   C. Both A and B
   D. Neither A nor B
6. Technician A says that when a DTC is not present, no diagnostics should be attempted. Technician B says that many systems can cause driveability symptoms that do not set DTCs. Who is correct?
   A. Technician A
   B. Technician B
   C. Both A and B
   D. Neither A nor B
7. All of the statements about component diagnosis are correct *except:*
   A. Component diagnosis should be performed when directed by a trouble tree.
   B. Component diagnosis can be performed as part of an intermittent diagnostic procedure.
   C. Specific component diagnosis should be performed only when directed by a service bulletin.
   D. The scan tool gathers data too slowly to be effective in many diagnostic situations.
8. Technician A says that a thermistor should always be checked in a range from 10°F to 250°F. Technician B says that voltage drop across a thermistor is proportionate to its internal resistance. Who is correct?
   A. Technician A
   B. Technician B
   C. Both A and B
   D. Neither A nor B

9. Technician A says that the voltage change when testing a potentiometer should be proportionate to the amount and the speed at which the potentiometer is operated. Technician B says that the potentiometer-type sensor should be disconnected and the terminals front-probed when testing voltage drop. Who is correct?
   A. Technician A
   B. Technician B
   C. Both A and B
   D. Neither A nor B

10. Technician A says that a failure in the heater circuit of an $HO_2S$ will usually result in no DTC being set. Technician B says that the $HO_2S$ heater circuit does not need a ground connection to operate. Who is correct?
   A. Technician A
   B. Technician B
   C. Both A and B
   D. Neither A nor B

11. Technician A says that a dirty MAF sensing element can cause a vehicle hesitation. Technician B says that debris on the sensing element acts as an insulation to the heating element. Who is correct?
   A. Technician A
   B. Technician B
   C. Both A and B
   D. Neither A nor B

12. Technician A says that an $HO_2S$ with values fixed below 450 mV indicates that the sensor is faulty. Technician B says that when viewing an oscilloscope pattern for an $HO_2S$, the pattern should show very gradual transitions from rich to lean. Who is correct?
   A. Technician A
   B. Technician B
   C. Both A and B
   D. Neither A nor B

13. Technician A says that an $HO_2S$ signal that constantly toggles above and below 450 mV indicates that good fuel control is present. Technician B says that a good sensor should be able to cycle below 200mV and over 800mV when a snap test is performed. Who is correct?
   A. Technician A
   B. Technician B
   C. Both A and B
   D. Neither A nor B

14. Technician A says that the relay contacts that measure over 0.25V of drop should be replaced. Technician B says that when testing for voltage drop across the relay contacts, a load must be placed across the contact points. Who is correct?
   A. Technician A
   B. Technician B
   C. Both A and B
   D. Neither A nor B

15. Technician A says that tap testing the PCM will indicate if the EEPROM needs to be upgraded. Technician B says that faulty PCM ground connections can lead to multiple unrelated DTCs to be stored. Who is correct?
   A. Technician A
   B. Technician B
   C. Both A and B
   D. Neither A nor B

# Section 4

## Ignition Systems

**Interesting Fact**

*Early ignition systems used a set of mechanical switch contacts to open and close the ignition coil primary circuit. Modern electronic ignition systems use sophisticated computer-controlled electronic switching to accomplish primary current switching. The use of electronic controls affords superb timing accuracy, providing maximum power and fuel economy.*

## SECTION OBJECTIVES

After you have read, studied, and practiced the contents of this section, you should be able to:

- Explain how the ignition process converts potential energy stored in the fuel into useful kinetic energy.
- List the components of the primary and secondary electrical system and explain the purpose of each.
- Explain what is meant by induction.
- Identify and explain the function of a spark plug.
- Discuss how the low-voltage primary signal is increased to a value sufficient to fire the spark plug.
- Identify and discuss the purpose of ignition triggering devices.
- Identify and discuss the purpose of an ignition module.
- Explain the term "dwell."
- Discuss what is meant by "ignition timing."
- Explain why ignition timing is advanced as engine speed increases.
- List several spark plug stress factors.
- Discuss spark plug heat range.
- Explain the difference between distributor (DI) and distributorless (EI) ignition systems.
- Discuss the function of each component of a DI ignition system.
- Explain how an optical distributor works.
- Discuss the function of each component of an EI distributorless ignition system.
- Explain how a "waste-spark" ignition system operates.
- Discuss ignition system primary diagnosis and service.
- Discuss secondary ignition system diagnosis and service.

# Chapter 21

# Ignition System Theory and Operation

## Introduction

Internal combustion engines can be classified as using either **spark ignition (SI)** or **compression ignition (CI)**. The latter is used in diesel engines as a means to ignite the air/fuel charge by raising the temperature of the fuel to the flash point. However, in the case of gasoline-powered engines, the ignition system must supply a high-voltage pulse to the **spark plug**, causing an electrical arc to form on the **electrodes**. The spark plug is an ignition device that contains an air gap. The air gap allows current to flow when the voltage across the gap is of sufficient magnitude. The spark plug is located within the combustion chamber. The heat created by the spark causes the air/fuel charge within the combustion chamber to ignite **(Figure 1)**.

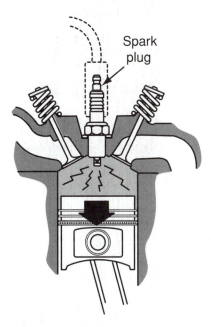

**Figure 1.** Heat created by the ignition spark causes the air/fuel charge within the combustion chamber to ignite.

> **Interesting Fact**
>
> A "hot" spark means there is an adequate flow of electrons traveling across the spark plug electrodes. Sharp electrodes help the electrons jump the air gap, minimizing ignition misfires. Engine misfire wastes fuel, robs the engine of power, and increases tailpipe emissions.

The ignition process converts the potential energy stored in the fuel into useful kinetic energy that develops horsepower. Sufficient spark energy must be available to prevent cylinder misfire, which results in decreased power and increased tailpipe emissions.

The ignition system must increase the low voltage provided by the storage battery and alternator to a value high enough to fire the spark plugs at the precise time. This high-voltage signal is typically around 10,000V but can range from approximately 6kV to over 40kV.

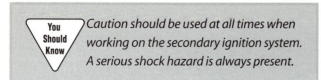

*Caution should be used at all times when working on the secondary ignition system. A serious shock hazard is always present.*

## COMPONENTS

The ignition system consists of a variety of components, some of which are dependent upon the specific ignition system design. In the most basic form, the ignition system requires at least one **ignition coil**, a coil triggering device, and one or more spark plugs. The number of plugs varies according to the number of cylinders and whether the engine uses more than one plug per cylinder to ignite the air-fuel charge. The ignition coil develops the high-voltage signal that fires the spark plug.

Early coil triggering systems used a set of switch contacts located within the **ignition distributor** to trigger the coil. The ignition distributor is a high-voltage switch that directs the spark energy to the appropriate spark plug at the proper time. As technology progressed, magnetic and optical sensors replaced mechanical switch contacts, with magnetic sensors being the most popular triggering device. Modern distributorless ignition systems also depend on magnetic sensors for coil triggering. Both magnetic and optical sensors require some form of signal amplification to trigger the ignition coil because the sensors operate at low current levels.

An **ignition module**, or **ignition control module (ICM)**, provides the proper interface needed to trigger the ignition coil. It contains active devices, including transistors, which can safely handle the electrical load provided by the ignition coil.

In most applications, a crankshaft position (CKP) sensor provides position and engine speed information to the ignition module and powertrain control module (PCM). In certain applications, the ignition module is integrated into the PCM. Position information is important because each spark plug must fire at a precise point in the engine's cycle. Cylinder number 1 is used as a reference cylinder. Based upon the firing order of the engine, the remaining spark plugs are fired sequentially by triggering the ignition coil through the ignition module.

## PRIMARY AND SECONDARY ELECTRICAL SYSTEMS

The automotive ignition system is divided into a primary and a secondary circuit. The primary circuit contains all of the low-voltage components. The secondary circuit is often called the high-voltage side of the ignition system.

The typical primary ignition system consists of the following:

- *Storage battery and alternator.* The storage battery and alternator are the power sources that supply the low-voltage (14-volt) primary electrical system.
- *Primary wiring.* Primary wiring connects low-voltage electrical components that are part of the primary electrical system.
- *Ignition switch.* The ignition switch controls how current is fed to the various electrical components within the primary electrical system.
- *Ignition coil primary winding.* The ignition coil primary winding is an electromagnetic energy-storage device that induces a high-voltage signal into the secondary winding.
- *Ignition control module.* The ignition control module contains a transistor that opens and closes the ignition coil primary circuit in response to crankshaft position and speed.
- *Crankshaft position sensor.* The crankshaft position (CKP) sensor produces a varying signal that is used by the ICM to control primary coil current.
- *Camshaft position sensor.* The camshaft position (CMP) sensor produces a reference signal that is used by the PCM to indicate the location of cylinder number 1.

The typical secondary ignition system consists of the following:

- *Spark plug.* The spark plug provides an air gap in the combustion chamber where an arc can be developed to initiate the combustion process. It also removes heat from the combustion chamber.
- *Ignition coil secondary winding.* The ignition coil secondary winding develops a high-voltage signal used to jump the air gap of the spark plug. Under normal operating conditions with a properly functioning ignition system, the magnitude of this voltage is approximately 10,000V.
- *Distributor cap.* The distributor cap acts as a high-voltage switch to distribute the secondary high-voltage signal to the appropriate spark plug in accordance with the firing order.
- *Distributor rotor.* The distributor rotor is located within the distributor. This rotating component completes the secondary electrical circuit between the center terminal of the distributor cap and the corresponding cap tower.
- *Ignition cables.* Ignition cables are heavily insulated wires that carry the high-voltage ignition signals to each spark plug.
- *Coil wire.* The coil wire connects the high-voltage tower of the coil to the center tower of the distributor.

A typical ignition system is shown in **Figure 2**.

## IGNITION COIL

The ignition coil performs two important functions:

- It changes the low voltage available from the vehicle's primary electrical system (14V) into a high-voltage

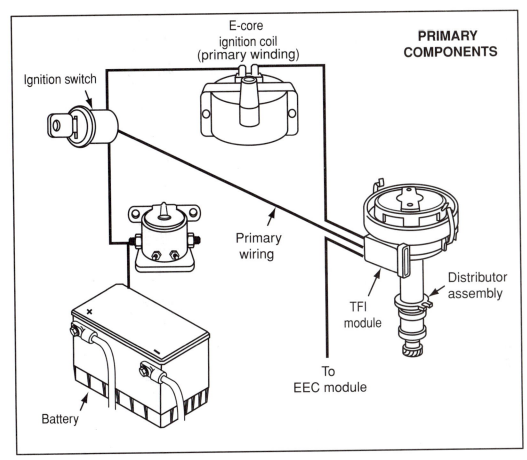

PRIMARY COMPONENTS

E-core
ignition coil
(primary winding)

Ignition switch

Primary
wiring

Distributor
assembly

TFI
module

To
EEC module

Battery

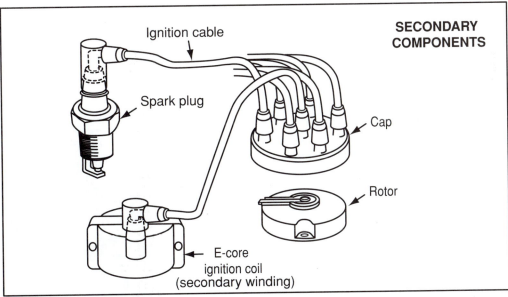

SECONDARY COMPONENTS

Ignition cable

Spark plug

Cap

Rotor

E-core
ignition coil
(secondary winding)

**Figure 2.**  A typical ignition system showing primary and secondary components.

signal capable of initiating an arc at the spark plug.

• It stores energy within a magnetic field. This energy is released at the proper time in the engine's cycle to fire the spark plug.

The coil is often referred to as the dividing line between the primary and secondary electrical system. Several types of ignition coils are shown in **Figure 3**.

An ignition coil consists of a primary and a secondary winding. Together, the windings form a **transformer**

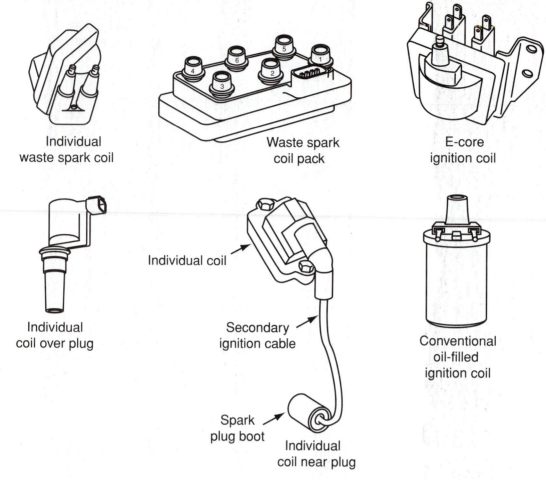

**Figure 3.**   Various styles of ignition coils.

**(Figure 4)**. The primary winding is usually labeled positive and negative, although this depends upon the application **(Figure 5)**.

The ignition coil operates using the principle of **induction (Figure 6)**. A magnetic field is used to induce a voltage from one winding into the other.

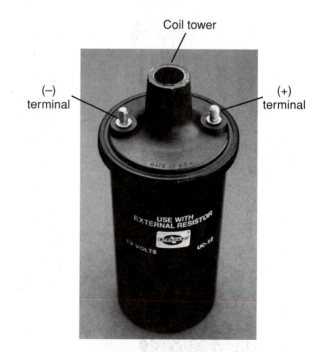

**Figure 5.**   An ignition coil primary winding is labeled positive (+) and negative (−).

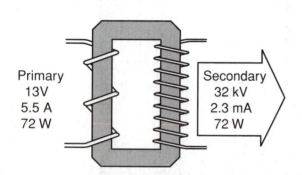

Primary
13V
5.5 A
72 W

Secondary
32 kV
2.3 mA
72 W

**Figure 4.**   Primary and secondary windings form a step-up transformer.

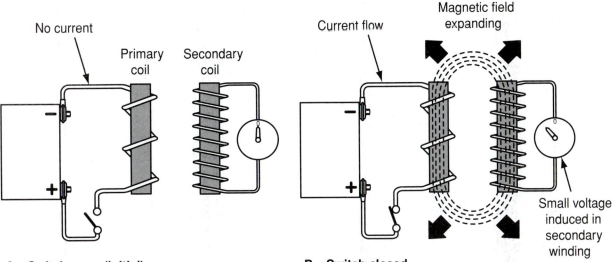

**A  Switch open (initial)**
- No current flows in primary winding.
- No magnetism is produced.
- Zero voltage across secondary winding.

**B  Switch closed**
- Current flows in primary winding.
- Magnetic field expands.
- A small voltage is induced in secondary winding. (This voltage is not sufficient to overcome spark plug gap resistance.)

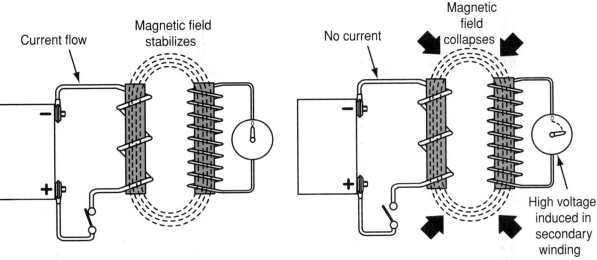

**C  Field stabilized**
- After a short period of time, the magnetic field reaches maximum strength and stabilizes.
- Zero voltage across secondary winding.

**D  Switch opened**
- Magnetic field collapses quickly.
- High voltage is induced into secondary winding causing electrons to jump the gap.
- This results in an arc being formed across spark plug electrodes.

**Figure 6.**  An ignition coil operates on the principle of induction.

## Turning Up the Voltage

An ignition coil is a step-up transformer. A step-up transformer can increase the voltage across the secondary winding to several thousand volts—enough to jump the air gap of a spark plug. This voltage is usually about 10kV but varies depending upon many factors, including spark plug electrode spacing, cylinder compression, and air/fuel ratio. Most ignition coils are capable of developing up to 100kV (100,000V) across the secondary winding.

In the case of a step-up transformer, there are more turns of wire in the secondary winding compared with those in the primary. The primary winding of the typical ignition coil is wound with about 200 turns of number 20

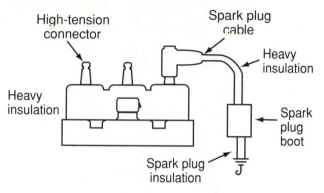

**Figure 7.**   All secondary ignition components must be heavily insulated.

copper wire. This is in contrast to the secondary winding that consists of about 20,000 turns of a very small diameter (number 38) copper wire. As a rule of thumb, most ignition coils have a secondary to primary turns ratio of about 100:1. More secondary turns result in a higher induced voltage being developed. The **secondary high-voltage** system refers to components connected to the ignition coil secondary winding. Because extremely high voltages are involved, all secondary ignition components must be heavily insulated **(Figure 7)**.

### Coil Polarity

The primary winding of the coil is connected so that the polarity of the secondary winding that connects to the spark plug center electrode is negative relative to the side, or ground, electrode. The center electrode is always hotter than the side electrode because of its location. The side electrode is part of the spark plug shell, which provides better heat transfer to the cylinder head, so it is always cooler than the center electrode **(Figure 8)**.

Placing a negative polarity on the hot center electrode allows electrons to jump more easily across the gap to the side electrode. This occurs because the high-temperature center electrode increases the energy levels of the electrons, making it easier for them to leave the surface of the electrode.

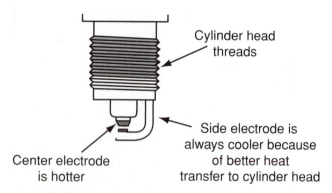

**Figure 8.**   Electrons easily jump from the hot center electrode to the cooler side electrode.

If the polarity of the primary winding is reversed, the polarity of the secondary winding will also be reversed. This polarity reversal increases the voltage needed to fire the spark plug by as much as 30 percent. This occurs because the cooler side electrode does not provide the electrons with the same excitation level as the hotter center electrode. More voltage is needed to allow the electrons to jump the gap.

In the case of **waste-spark** ignition systems, allowances must be made for increased secondary voltage requirements caused by reverse-fired spark plugs. Waste-spark ignition systems fire two spark plugs simultaneously, one in the forward direction and the other in the reverse direction. On each firing event, one spark plug always fires with a positive polarity on the center electrode. This is the reverse-firing plug. The remaining plug fires normally in the forward direction with a negative polarity on the center electrode. Information related to waste-spark ignition systems is presented in Chapter 23.

### The Ignition Coil as an Energy Storage Component

The ignition coil is an energy storage component, similar to a battery. Unlike a battery, however, it stores its energy in a magnetic field instead of using chemical means.

In the case of an automotive electrical system, the low-voltage primary system operates on DC. In order to use a transformer in such an application, it is necessary to use **pulsating DC** to step up the low-voltage DC signal. The pulsating signal provides the changing magnetic field needed to allow induction between the primary and secondary windings of the transformer. A triggering circuit that uses a transistor or other suitable electronic switch supplies this pulsating DC signal **(Figure 9)**. The triggering circuit must fire the spark plug at the precise point in the engine's cycle.

## TYPES OF IGNITION COILS

A variety of ignition coil designs are used, depending upon the application. The traditional oil-filled coil and E-core designs are used in distributor-based systems, whereas coil packs and compact coil designs are used in distributorless ignition systems.

### Traditional Oil-Filled Ignition Coil

In this design, the center of the coil contains a soft iron laminated core. This type of coil is often referred to as an **autotransformer** because the primary and secondary windings are not electrically separated. Instead, several turns of the windings are shared by both the primary and secondary on a common core **(Figure 10)**. The lower (primary) portion of the coil is wound with heavier-gauge wire compared with that of the upper (secondary) winding. The

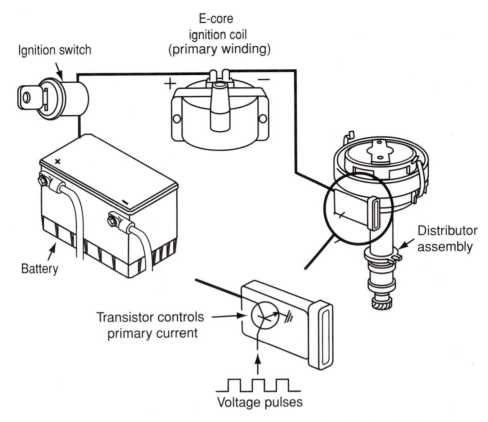

Ignition switch

E-core
ignition coil
(primary winding)

Battery

Distributor
assembly

Transistor controls
primary current

Voltage pulses

**Figure 9.** A coil primary triggering circuit using a transistor to supply pulsating DC signal to coil (−) terminal.

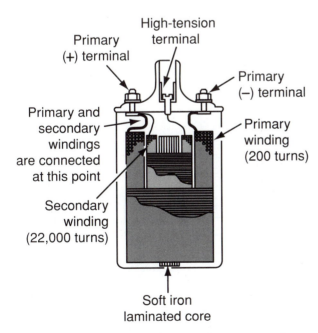

High-tension
terminal

Primary
(+) terminal

Primary
(−) terminal

Primary and
secondary
windings
are connected
at this point

Primary
winding
(200 turns)

Secondary
winding
(22,000 turns)

Soft iron
laminated core

**Figure 10.** An autotransformer shares several turns of the windings between the primary and secondary.

case is filled with a special transformer oil to provide cooling and insulation of the windings. This type of coil is seldom used on production vehicles anymore.

## E-Core Ignition Coil

The E-core ignition coil provides a closed magnetic path between the primary and secondary windings. The E-core design gets its name from how it is made. The laminated iron core is shaped like the letter "E" **(Figure 11)**.

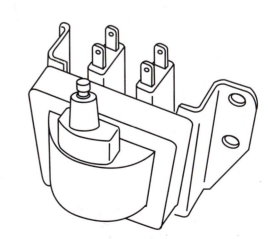

**Figure 11.** An E-core ignition coil.

Because of the closed magnetic field, the energy transfer between the primary and secondary windings of the E-core coil is increased. The E-core ignition coil is also more compact and does not use oil as a cooling medium. Epoxy is used as an insulating material to protect the windings against moisture and vibration.

## Multiple Coil Packs

Coil packs are used with ignition systems that use waste-spark distributorless ignition systems. Coil packs are so named because they contain more than one coil in a single unit **(Figure 12)**.

## Coil-On-Plug

In coil-on-plug (COP) ignition systems, the ignition coil is mounted directly over the spark plug **(Figure 13)**. Each spark plug uses its own coil to fire the spark plug.

A variation of the COP system is the coil-near-plug (CNP) design. In this system, the ignition coils are mounted close

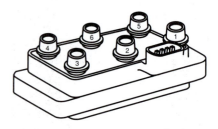

**Figure 12.**   Coil packs contain more than one ignition coil.

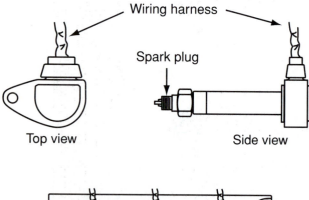
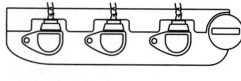

**Figure 13.**   COP ignition systems use an ignition coil that is mounted directly over the spark plug.

to their respective spark plugs. Short ignition cables connect the spark plugs to the individual coils.

## COIL TRIGGERING CIRCUITS

In the case of the ignition coil, one side of the primary winding is connected to the primary electrical system. This connection is made to the positive battery terminal, resulting in the application of 14V to the positive (+) coil primary terminal. In order to develop a magnetic field within the coil primary winding, a means of providing a return path to the battery is needed. This return path is supplied by the switching device. A set of **breaker points** completes the circuit **(Figure 14)**. Breaker points are a type of mechanical switch used to trigger the coil primary at the precise instant. This system is sometimes referred to as contact point ignition.

If the points are closed, current will begin to flow in the primary winding, causing the magnetic field to increase. A voltage is induced in the secondary winding but is of such low amplitude that the spark plug does not fire. Once the magnetic field builds to a maximum, no further increase in primary current occurs. At this point, the coil is **saturated**. Once the saturation point is reached, no additional increase in secondary voltage can occur when the field collapses because primary current is at its maximum.

The process of increasing the magnetic field within the primary winding is sometimes called "coil charging." It is so named because the coil temporarily stores energy within its magnetic field for release at a slightly later time. This process is similar to the storage battery receiving a charge from the alternator. Unlike the battery, which depends on a chemical reaction to store energy, coil charging occurs from the magnetic field created by current flowing within the primary winding.

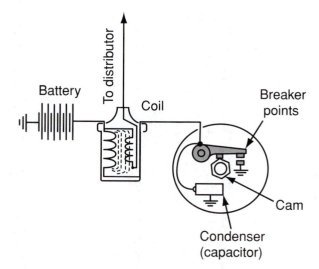

**Figure 14.**   Ignition coil is triggered using breaker points.

**Dwell** is a measure of the total time the switching device is on. Dwell time includes the time needed for the coil to reach saturation and any additional time that occurs before the current flow is interrupted. Dwell can be measured in milliseconds or degrees of camshaft rotation. It is sometimes called **dwell angle**.

Dwell time must be of sufficient duration to allow the magnetic field strength to build to its maximum value. This provides maximum coil secondary output voltage and current. Keeping primary current flowing through the ignition coil beyond the time needed to reach saturation can cause the coil to overheat. This can lead to premature coil failure. Most late-model ignition systems use a **variable-dwell**

circuit for this reason. Variable-dwell operation allows the dwell time to vary based upon engine speed yet ensures adequate time is provided for the coil to reach saturation.

Ordinary contact point ignition systems are incapable of variable-dwell operation. The ignition control module (ICM) is used in all modern ignition systems and has been standard equipment on vehicles for over 25 years. It provides many advantages over standard breaker point ignition systems.

> **You Should Know** *It is the collapse of the magnetic field, not the activation of primary current, that initiates the spark plug arc.*

## IGNITION CONTROL MODULE

Ignition control modules use solid-state electronics to trigger the ignition coil. Using sophisticated circuitry, the ignition module limits coil current to about 5 or 6 amperes, depending upon the ignition system. At the proper time, determined by various engine operating conditions, the switching device within the module opens. Opening the transistor interrupts coil primary current. This causes the magnetic field to collapse. Because the magnetic field tends

> **Interesting Fact**
> Early ignition systems used a ballast resistor or resistance wire to limit coil primary current. This protected the coil from excessive current. During engine crank, when battery voltage is lowest, the ignition switch bypassed the resistor, providing maximum secondary output voltage. This helped with engine starting, especially in cold temperatures (**Figure 15**).

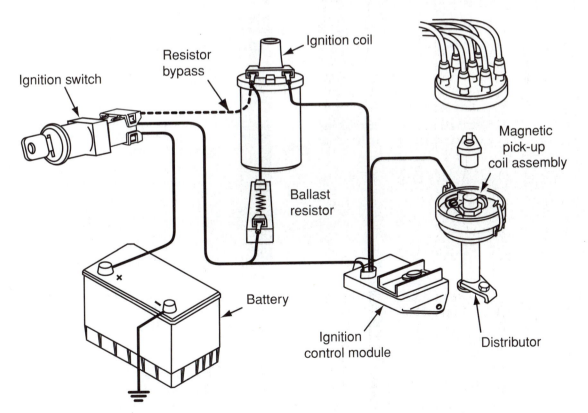

**Figure 15.** Ballast resistor by-pass circuit.

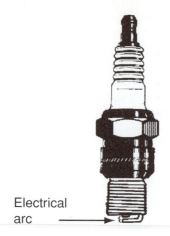

Electrical
arc

**Figure 16.** Increasing secondary voltage builds on spark plug electrodes until the air gap ionizes, igniting the fuel mixture.

to sustain the original primary current, a voltage is induced in the coil's secondary winding at this instant. The increasing secondary voltage builds on the spark plug electrodes until the air gap **ionizes**, igniting the air/fuel mixture **(Figure 16)**.

The ICM applies and interrupts the primary ignition coil current at the precise moment. The ICM uses information supplied by the CKP sensor, the CMP sensor, and the PCM to determine when the coil primary should be activated. This allows a magnetic field to build within the coil. At the precise instant, the ICM interrupts primary current to the ignition coil, causing the magnetic field to collapse, thus initiating a spark event across the spark plug air gap **(Figure 17)**.

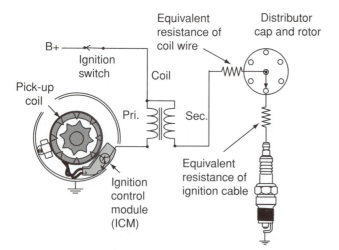

**Figure 17.** The ICM interrupts coil primary current. This causes the magnetic field to collapse, initiating a spark across the spark plug electrodes.

# IGNITION TIMING

Ignition timing refers to the exact point in the engine's cycle where a spark event is targeted. The spark event should occur at approximately top dead center (TDC) of the compression stroke on each cylinder.

> **You Should Know** *Ignition timing varies depending upon engine speed and load and is usually controlled by the PCM. The initial timing, or base timing, is usually set to 10 degrees BTDC.*

The ignition system must have a point of reference in order to establish proper ignition timing. This reference point is called **initial timing** or **base timing**. Base ignition timing is usually set to about 10 degrees before top dead center (BTDC) using cylinder number 1 as the reference cylinder **(Figure 18)**. Consult the shop manual or emissions decal for the exact base timing specification.

If the engine has four cylinders, the spark plugs will fire at least once per complete camshaft revolution. The spark events follow the engine's firing order. For example, if the firing order is 1-3-4-2, cylinder number 1's spark plug will fire, followed by cylinder number 3's, followed by cylinder number 4's, followed by cylinder number 2's, and so on. The firing order is a function of how the engine is designed, including how the camshaft is ground.

## Spark Advance

Under a given set of conditions, maximum spark advance provides optimum engine torque. Because the **burn time** of the fuel is relatively constant under fixed air/fuel ratios, as engine speed increases, the combustion process must begin earlier in the cycle. The process of initiating the combustion event sooner is called spark advance, or timing advance. As engine rpm increases, the pistons are traveling at higher speeds so spark advance must be increased. Conversely, as engine speed decreases, piston speed is reduced, so spark advance must be decreased.

Engine load affects timing because at light loads, the air/fuel charge is leaner. This causes a longer ignition delay, creating the need for more timing advance. As engine load increases, the air/fuel mixture is richer. Combustion occurs rapidly. Spark advance must be advanced less—or even retarded in certain cases—to prevent **engine ping**. Engine ping is often called **spark knock**. It occurs if ignition timing is too far advanced for a given engine speed and load. Low fuel octane can also cause spark knock.

The engine control system calculates the optimum spark advance based upon input data supplied by various sensors. If the engine is equipped with a knock sensor (KS), spark advance can be reduced to bring spark knock

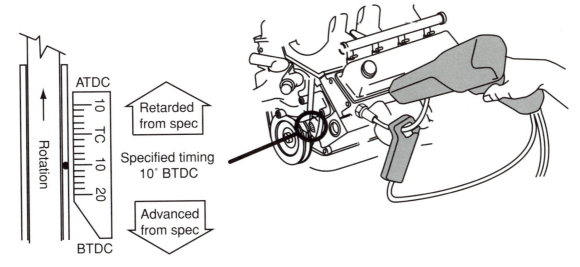

**Figure 18.** Setting base ignition timing.

into control. Some engine control systems can selectively retard ignition timing to a specific cylinder. This allows maximum overall spark advance under a variety of engine operating conditions.

## SPARK PLUG FUNCTION

The spark plug has two important functions:
1. To act as an ignition device.
2. To remove heat from the combustion chamber.

The spark plug uses a high-voltage electrical signal to create an arc, which ignites the compressed air/fuel charge. The arc is developed across the spark plug electrodes **(Figure 19)**.

The spark plug must reliably perform in cold-weather starting. It must also ignite the air/fuel mixture under a variety of engine operating conditions including hard acceleration. Even after prolonged engine operation at maximum power,

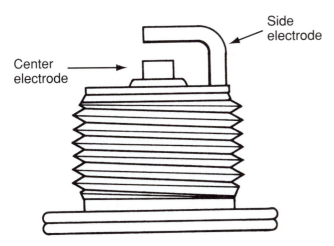

**Figure 19.** Spark plug electrodes.

the spark plug must maintain a proper seal at the cylinder head and be free of overheating.

As a heat exchanger, the spark plug transfers unwanted thermal energy away from the combustion chamber. It is important to understand that a spark plug does not create heat. Rather, it dissipates heat using thermal conduction. This characteristic is called **heat range** and has no relationship to the spark plug's electrical characteristics.

## SPARK PLUG CONSTRUCTION

Spark plugs must be constructed of materials that can withstand the high operating temperatures and pressures encountered during engine operation. The various components of the spark plug have specific purposes essential for long life and high performance.

### High-Voltage Terminal and Insulator

Glass, metal, and ceramic materials are used in the construction of a spark plug. A steel **high-voltage terminal** is melted into the **insulator** that electrically connects the terminal to the center electrode. This terminal is often called the terminal stud. In certain applications, the stud is removable to accommodate the ignition cable. In most applications, the secondary ignition cable is connected to this high-voltage terminal using a friction-fit (push-on) connector that is a part of the spark plug **boot**. The boot insulates the high-voltage signal at the terminal stud.

The spark plug insulator is constructed of a special ceramic material that can withstand extremely high temperatures. This material must maintain excellent electrical insulation characteristics over the entire operating temperature range. The surface of the insulator is glazed. This minimizes leakage currents that might otherwise flow, because moisture and dirt do not adhere well to a smooth, glassy surface. The insulator must provide mechanical strength and resistance to chemicals.

## Shell

The **shell** of the spark plug is made of steel. It must secure the spark plug into the cylinder head using threads. The spark plug shell is plated for corrosion resistance. This is important to prevent the threads from seizing within the cylinder head. Depending on the application, a sealing gasket may be fitted.

> **You Should Know** *The use of an antiseize compound can reduce the possibility of spark plug threads seizing within the cylinder head. A small amount of this compound should be applied to the threads of each spark plug before installation.*

## Electrodes

The **ground (side) electrode** is welded to the spark plug shell. In most designs, it has a rectangular cross section. The terms "front" and "side" electrodes are used to distinguish between how the ground electrode is positioned relative to the center electrode (**Figure 20**).

Different spark plug designs may feature the use of multiple ground electrodes or no side electrode at all. In the latter case, the entire rim of the shell forms the ground electrode (**Figure 21**).

The center electrode of a conventional spark plug is molded into the insulator. A special seal is used to ensure a gas-tight seal of combustion gases. The cylindrical center electrode protrudes from the nose of the insulator (**Figure 22**).

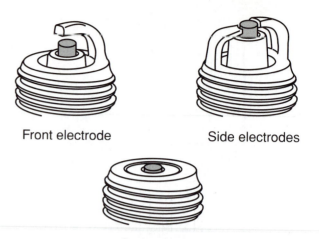

Front electrode    Side electrodes

Surface-gap
(without ground electrode)

**Figure 21.** Electrode styles.

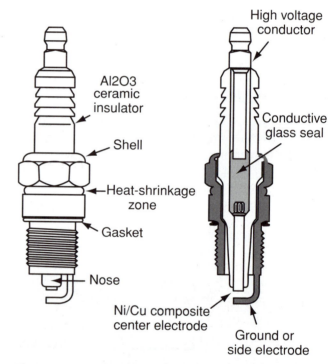

High voltage conductor

Al2O3 ceramic insulator

Shell

Heat-shrinkage zone

Gasket

Nose

Conductive glass seal

Ni/Cu composite center electrode

Ground or side electrode

**Figure 22.** Spark plug construction. Note how the center electrode protrudes from the insulator nose.

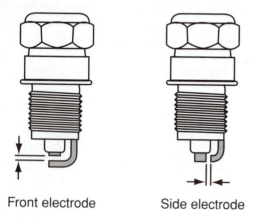

Front electrode    Side electrode

**Figure 20.** Comparison between front and side electrodes.

In the case of spark plugs fabricated using precious metals such as platinum, the diameter of the center electrode is often smaller than that of electrodes using a copper core with a nickel-alloy jacket.

Commonly called the spark plug air gap, or simply "plug gap," this dimension is the shortest distance between the center electrode and the ground electrode. Air gaps range from about 0.025 inch to 0.100 inch, depending upon the application. The smaller the gap, the lower the ignition voltage requirement.

In the case of short air gaps, although the ignition voltage requirement is lower, less transfer of spark energy to the air/fuel charge occurs. The chance for ignition misfire increases. In contrast, wider spark plug air gaps require significantly higher voltages to cause an arc to form across the gap. Although a wide gap can transfer more energy into the air/fuel charge, it is not without a penalty. The accompanying reduction in ignition voltage reserves can be problematic in terms of ignition misfires.

The shape of spark plug electrodes also plays an important part in reducing the possibility of misfire. Sharp electrodes provide the highest reliability in terms of ignition events. As the electrodes wear, rounding of the electrodes occurs, which increases the possibility of misfire.

## Spark Plug Stress Factors

The spark plug must be capable of withstanding the harsh conditions encountered within the combustion chamber. These include electrical, mechanical, chemical, and thermal stresses.

Because voltages in excess of 30kV can be present, the insulator must be able to withstand these voltages without arcing over—both externally and internally—to the combustion chamber.

In terms of mechanical considerations, the spark plug must be able to maintain a gas-tight seal under the high pressures (approximately 100 bar) encountered within the combustion chamber. Additionally, the shell and insulator must withstand the forces encountered when installing and removing the plug.

High-temperature chemical reactions can take place within the combustion chamber. Deposits caused by carbon, fuel, oil, engine coolant, and additives can become electrically conductive at certain temperatures, causing ignition breakdown. This occurs because the insulator is unable to provide sufficient resistance to current flow. Ignition breakdown can cause cylinder misfire, which results in power loss, excessive fuel consumption, and increased tailpipe emissions.

The spark plug is subjected to severe temperature extremes. This occurs because the plug absorbs heat from the hot combustion gases resulting from the combustion event, yet is subjected to the cold air/fuel charge pulled in during the intake stroke. Therefore, the plug must be able to survive under conditions of extreme thermal shock. As another concern, the external (terminal) part of the spark plug should remain at the lowest temperature possible to prevent premature boot failure.

Depending upon the type of electrode materials used, spark plugs can perform satisfactorily for up to 100,000 miles, provided no abnormal combustion conditions occur. Misfire and the use of leaded gasoline can severely reduce spark plug life.

## ELECTRODE MATERIALS

The choice of electrode materials used in the manufacture of the spark plug is important to obtain a long service life and prevent ignition misfires. Depending upon the required service life, special materials are often used to minimize electrode wear.

### Standard Spark Plugs

Standard, as opposed to long-life, spark plugs have long used a nickel-based alloy for the center electrode. Design enhancements to improve corrosion resistance include the use of a composite, or compound material. The composite material features a copper core with a nickel-based alloy jacket. The copper core provides better heat transfer to the insulator.

In the case of spark plug electrodes, pure materials offer better thermal characteristics than do alloys. Unfortunately, pure metals such as nickel are less resistant to chemical substances found in combustion gases, as well as solid deposits.

To obtain long spark plug life, the jacket material consists of mostly nickel, which is alloyed with elements such as chromium and manganese. Silicon is often added because both it and manganese offer increased chemical resistance against sulfur dioxide, an aggressive compound found in gasoline and motor oil.

Similar compounds are used in the construction of the side, or ground, electrode. In addition to being resistant to erosion and deposits, the side electrode must also be flexible enough to allow for proper spark plug electrode gap adjustments. Typically, the side electrode is fabricated from copper. It is then plated using a nickel alloy similar to that used for the center electrode.

### Long-Life Spark Plugs

Long-life spark plugs use precious metals, either in pure form or as an alloy to enhance electrode life. One precious metal that withstands the extremely hostile combustion chamber environment is silver (Ag). This element exhibits the highest thermal and electrical conductivity of any metal. It also offers excellent resistance to chemical erosion. Because of silver's excellent heat-transfer abilities, when it is used as a spark plug center electrode, the diameter of the electrode can be reduced **(Figure 23)**.

Platinum and platinum alloys are commonly used in the construction of long-life spark plugs because of their extreme resistance to corrosion and oxidation. Variations of platinum-plated spark plugs include single-platinum and double-platinum designs. The single-platinum spark plug has only one electrode platinum-plated. Usually, this is the center electrode, but for original equipment manufacturer (OEM) applications, the side electrode can be platinum-plated for use with waste-spark ignition systems. In this

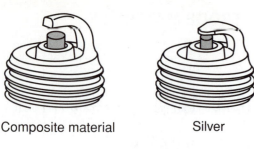

Composite material        Silver

Platinum

**Figure 23.** A comparison of different electrode materials. Note the smaller electrode diameter of silver and platinum types compared with a conventional design using composite material.

application, the forward-firing plug uses a platinum-plated center electrode, whereas the side electrode of the reverse-firing plug is platinum-plated. This process is done as a cost-saving measure to conserve on the use of platinum, a precious metal costing several hundred dollars per ounce. The appropriately plated plug must be inserted into the proper cylinder, based upon the coil's firing polarity. Interchanging the forward- and reverse-firing plugs will cause extremely accelerated electrode wear.

Double-platinum spark plugs are more costly to produce, compared with single-platinum plugs. However, there is no concern regarding improper installation in waste-spark ignition systems because both electrodes are platinum-plated and will wear normally.

Spark plugs using platinum-plated electrodes are more costly than are standard spark plugs, costing a few dollars per plug. Silver spark plugs are the most expensive, costing several dollars per plug. Some spark plug manufacturers are using other precious metals such as iridium to increase spark plug life.

Special materials and coatings are used in the fabrication of long-life spark plugs, in addition to the electrodes themselves. Easy spark plug removal is a must, even at extended service intervals approaching 100,000 miles. Coatings and materials that resist corrosion and thread seizing within the spark plug hole are important.

## SPARK PLUG OPERATING TEMPERATURE AND HEAT RANGE

To minimize spark plug misfires, it is important to control the formation of deposits on the spark plug. Carbon deposits are produced if incomplete combustion occurs, as in the case of a cold engine start. These deposits can form on the insulator nose of the spark plug, which produces a conductive path. The conductive path causes the current to flow along the insulator, reducing the ignition energy available for the arc.

The amount of deposit formation on the insulator nose is highly temperature dependent. The hottest part of the spark plug is the insulator tip. At tip temperatures below 500°C, the possibility of deposit formation increases, whereas higher temperatures cause the carbon accumulation to burn off. Tip temperatures above 900°C can cause the air-fuel mixture to autoignite, causing severe internal engine damage. It is desirable to keep spark plug operating temperatures above 500°C, which ensures self-cleaning of the spark plug, yet below the point where autoignition can occur (**Figure 24**).

The heat range is a measure of the spark plug's capacity to dissipate heat to the cylinder head. The heat range is determined by the ability of the center insulator to absorb and conduct heat away from the combustion chamber. Several factors influence this process:

- Length and construction of ceramic insulator nose.
- Type of materials used in the construction of the center electrode and ceramic insulator.
- Gas volume around the insulator nose.

The distance from the firing tip of the insulator to the point where the insulator meets the metal shell is called the insulator nose length. A hotter spark plug has a longer insulator nose, which conducts less heat away from the combustion chamber. This forces the heat to travel farther into the cylinder head, increasing internal spark plug temperature (**Figure 25**).

In the case of a colder spark plug, the insulator nose is shorter. This permits faster heat transfer to the cylinder head, reducing internal spark plug temperature (**Figure 26**). Colder spark plugs reduce the chance of preignition and detonation. As a rule, a one-step decrease in spark plug heat range removes approximately 100°C from the combustion chamber.

Spark plug heat range has no relationship to spark intensity. A high heat range spark plug does not provide a hotter spark.

## Resistor Spark Plugs

Most OEM applications use resistor-type spark plugs to minimize electromagnetic radiation. This radiation is called **electromagnetic interference (EMI)**. Electromagnetic interference can interfere with electronic control systems, causing them to malfunction. EMI can scramble information from important sensors, such as those found in engine management systems. EMI can come from a variety of sources, including the charging and ignition systems. Whenever a signal is generated by either mechanical or electrical means, the possibility of generating EMI exists.

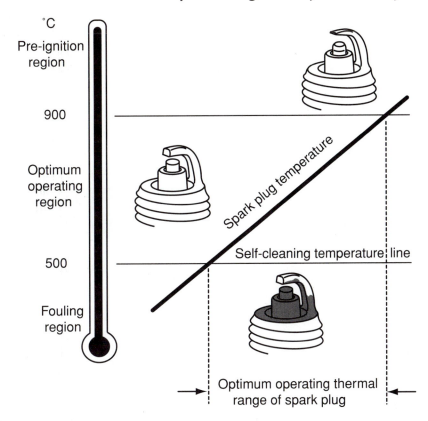

**Figure 24.**  Spark plug operating temperature.

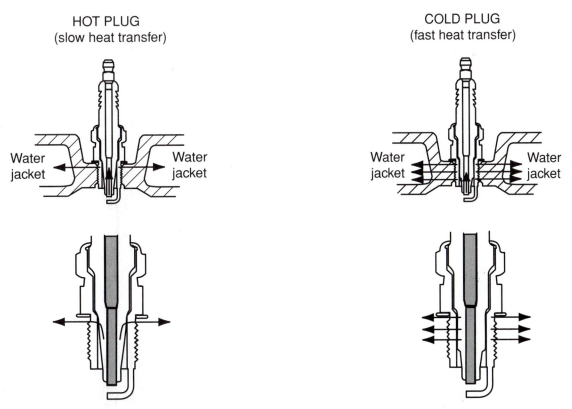

**Figure 25.**  A hot-type spark plug has a longer insulator nose, which conducts less heat away from the combustion chamber.

**Figure 26.**  A cold-type spark plug has a shorter insulator nose. This provides faster heat transfer to the cylinder head.

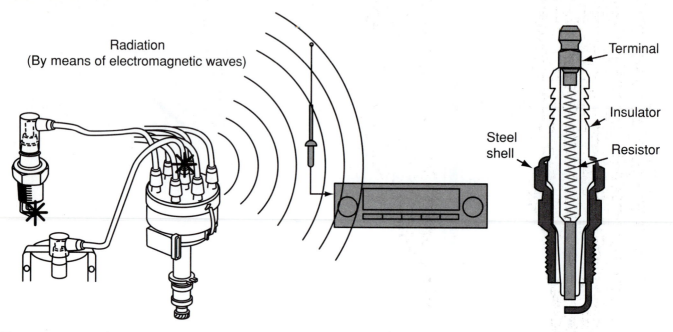

**Figure 27.** Use of a resistor spark plug to reduce electromagnetic interference.

Also, EMI causes **radio frequency interference (RFI)**. The interference is caused by variations in amplitude caused by fast-rising high-voltage pulses developed in the secondary ignition system. RFI can cause interference to the vehicle's own radio receiver, as well as to nearby receivers. The interference is most prominent when listening to AM (amplitude modulation) broadcasts. Little noise is present when listening to FM (frequency modulation) broadcasts. This is because the signal's content is contained in frequency variations and not amplitude variations, as in the case of AM radio reception.

Incorporating a resistor within the spark plug reduces the magnitude of the secondary ignition pulse, thereby reducing interference **(Figure 27)**. The value of resistance varies but is usually between 5,000 ohms and 15,000 ohms. Resistor-type spark plugs can be used wherever nonresistor plugs are used. The use of resistor plugs does not affect engine performance in a properly functioning ignition system.

The process of reducing EMI is called **interference suppression**. Suppressing interference can be accomplished using a variety of methods. Suppression techniques are most effective when applied to the source of EMI, rather than trying to reduce the offending signals once they are radiated. Resistor spark plugs are most effective at reducing EMI when used in conjunction with electronic suppression ignition cables.

## Secondary High-Voltage Wiring

Depending upon the ignition system design, a coil wire and secondary ignition wires may be used. The core of secondary ignition wires is usually fabricated from carbon to reduce EMI to the radio and other electronic devices. Solid-core cables act as antennas, increasing undesirable electromagnetic radiation, and are no longer used on production vehicles. The use of solid-core ignition cables is reserved for certain race applications only. The resistance value of an ignition cable varies, depending on its length.

The insulation of secondary ignition wires must be capable of withstanding the high electrical stress imposed by the high operating voltages. In applications where high temperatures will be encountered, special silicone insulation is used. Silicone provides excellent electrical insulation qualities. It also offers superior heat-resistant properties, making it ideal for use near exhaust system components.

Spark plug boots are usually made from silicone because the temperature of spark plug insulators can reach extremely high values. The boot covers the spark plug insulator, forming an electrical connection between the ignition wire and the plug terminal. The boot is necessary to contain the spark. A cross section of a secondary ignition wire is shown in **Figure 28**.

A coil wire is used with distributor-based ignition systems. It connects the ignition coil to the ignition distributor **(Figure 29)**.

## TYPES OF IGNITION SYSTEMS

Ignition systems are classified into two main categories:
- Distributor-based
- Distributorless

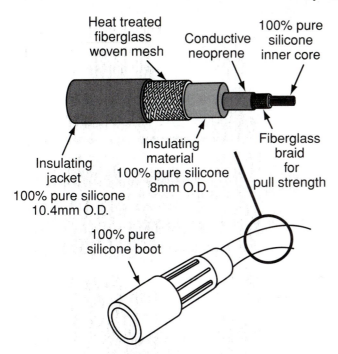

**Figure 28.** A secondary ignition cable showing spark plug boot.

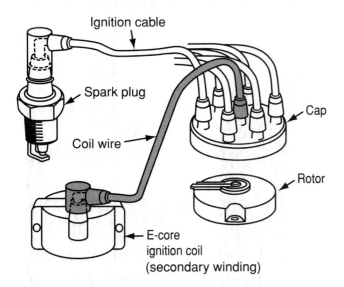

**Figure 29.** The coil wire connects the ignition coil to the distributor.

## Distributor-Based Ignition

Ignition systems are classified by how the distribution of spark takes place. For example, distributor-based ignition systems use a **distributor** and **rotor** to selectively distribute the spark energy to the appropriate spark plug at the proper time. A spark plug wire carries the high-voltage pulse to each spark plug. This type of ignition system is

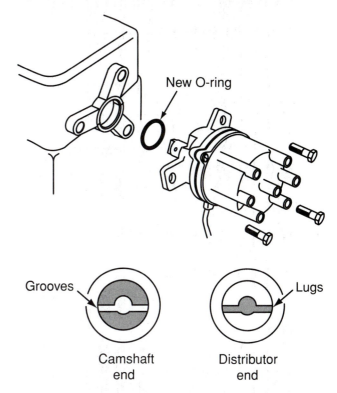

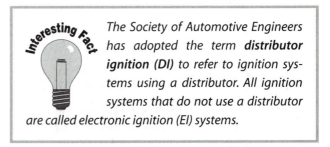

**Figure 30.** The ignition distributor is driven off the camshaft through a gear or timing belt.

often referred to as conventional coil ignition. The ignition distributor is usually driven off the camshaft through a gear or timing belt **(Figure 30)**.

> **Interesting Fact**
> The Society of Automotive Engineers has adopted the term **distributor ignition (DI)** to refer to ignition systems using a distributor. All ignition systems that do not use a distributor are called electronic ignition (EI) systems.

The distributor contains a CMP sensor, rotor, and cap **(Figure 31)**. Sometimes the ignition control module is attached to the distributor **(Figure 32)**.

## Distributorless Ignition Systems

Newer ignition system designs have eliminated the distributor altogether. This improves system reliability and reduces maintenance while reducing manufacturing costs. EI systems using distributorless technologies employ coil packs that fire two cylinders simultaneously. This is commonly

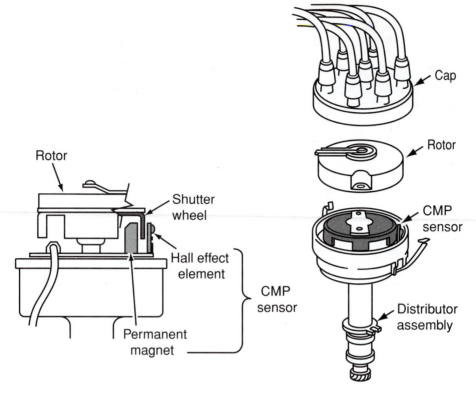

**Figure 31.** The distributor contains a CMP sensor, rotor, and cap.

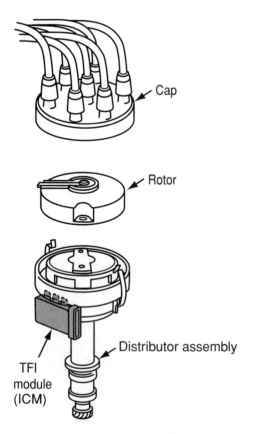

**Figure 32.** A conventional distributor-based ignition system with an ICM mounted to the
distributor housing.

called a waste-spark ignition system because one spark event occurs on the exhaust stroke.

Engines using waste-spark ignition systems fire two plugs simultaneously. In this system, one spark plug fires at the specific point on the compression stroke, while the companion cylinder's spark plug fires on the exhaust stroke. The system gets its name from the fact that the spark fired on the exhaust stroke is essentially wasted because it contributes no power.

Another distributorless design uses coil-on-plug (COP). COP systems use an individual ignition coil for each cylinder. The COP design offers a distinct advantage because it needs no ignition wires. Coil-on-plug ignition systems do not use the waste-spark theory. Instead, the spark plugs are fitted with individual ignition coils that are fired at the appropriate time, based upon information from the ignition module.

Variations of the COP design are coil-near-plug (CNP) systems. The main difference between COP and CNP systems is the location of the coil relative to the spark plug **(Figure 33)**.

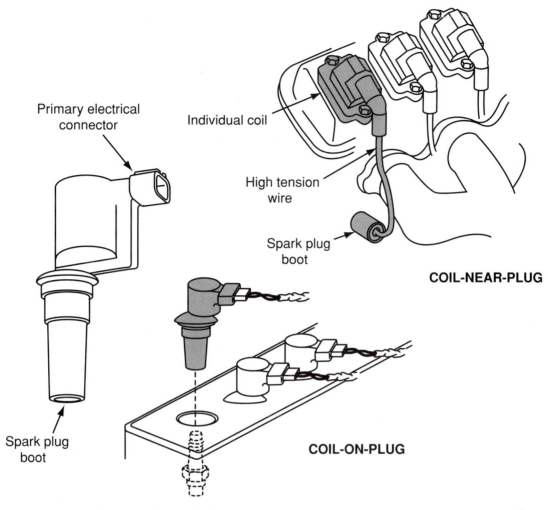

**Figure 33.** A comparison between COP and CNP ignition systems.

# *Summary*

- The purpose of the ignition system is to ignite the air/fuel charge within the engine cylinders at the proper time.
- All ignition systems use at least one ignition coil.
- The ignition coil contains a primary and a secondary winding.
- A low-voltage signal from the storage battery is boosted to a high-voltage signal in the secondary winding. The low-voltage signal is applied to the primary ignition coil winding.
- Magnetic induction is used to transform the low-voltage primary signal into the high-voltage secondary signal.
- The spark plug contains an air gap, allowing an arc to form when a voltage of sufficient magnitude is applied.

- A spark plug acts as an ignition device within the combustion chamber.
- The coil primary is controlled by a switching device. This device can be a set of breaker points or a solid-state ICM.
- When referring to DI systems, a distributor cap and rotor distribute spark energy to the proper spark plug at the appropriate time.
- Ignition cables carry the high-voltage ignition signal from the distributor cap to the individual spark plugs.
- Dwell is a measure of how long the ignition coil primary is energized.
- Ignition timing refers to the exact point in the engine's cycle where a spark event is to occur.
- The reference point of ignition timing is called "base ignition timing."

■ Ignition timing must be advanced as engine speed increases because the burn time of the fuel is relatively constant.

■ In addition to acting as an ignition device, the spark plug removes heat from the combustion chamber.

■ The heat range of a spark plug is a measure of the spark plug's ability to dissipate heat.

■ Hotter spark plugs have a longer insulator nose compared with that of colder plugs.

■ Long-life spark plugs use silver, platinum, or iridium plating on the electrodes.

■ Resistor spark plugs help reduce electromagnetic interference.

■ The insulation of secondary cables must be capable of withstanding the high electrical stress imposed by the high operating voltages encountered in the ignition system.

■ EI systems do not use a distributor. Distributorless ignition systems operate with much higher reliability than their DI counterparts.

# Review Questions

1. Technician A says that the ignition process converts potential energy stored in the fuel into useful kinetic energy that develops horsepower. Technician B says that the ignition system applies a high-voltage pulse to the spark plug, causing an arc to form on the electrodes. Who is correct?
   - A. Technician A
   - B. Technician B
   - C. Both A and B
   - D. Neither A nor B

2. Technician A says that all ignition systems use a distributor. Technician B says that all ignition systems use at least one ignition coil. Who is correct?
   - A. Technician A
   - B. Technician B
   - C. Both A and B
   - D. Neither A nor B

3. Technician A says that the ignition coil operates on the principles of magnetism. Technician B says that an ignition coil relies on magnetic induction to boost the low-voltage primary signal. Who is correct?
   - A. Technician A
   - B. Technician B
   - C. Both A and B
   - D. Neither A nor B

4. Technician A says that the ignition coil stores energy within the magnetic field. Technician B says that an ignition coil is actually a transformer. Who is correct?
   - A. Technician A
   - B. Technician B
   - C. Both A and B
   - D. Neither A nor B

5. Technician A says that some ignition coils are oil filled. Technician B says that a coil pack contains more than one ignition coil. Who is correct?
   - A. Technician A
   - B. Technician B
   - C. Both A and B
   - D. Neither A nor B

6. Technician A says that dwell refers to the time that the coil secondary delivers the high-voltage signal to the spark plug. Technician B says that dwell refers to the time that the ignition coil primary is turned off. Who is correct?
   - A. Technician A
   - B. Technician B
   - C. Both A and B
   - D. Neither A nor B

7. Technician A says that ignition timing refers to how long fuel burns within the combustion chamber. Technician B says that ignition timing refers to the point in the engine's cycle where the spark plug is fired. Who is correct?
   - A. Technician A
   - B. Technician B
   - C. Both A and B
   - D. Neither A nor B

8. Technician A says that ignition timing must be advanced as engine speed increases. Technician B says that ignition timing should remain fixed regardless of engine speed. Who is correct?
   - A. Technician A
   - B. Technician B
   - C. Both A and B
   - D. Neither A nor B

9. Technician A says that a spark plug removes heat from the combustion chamber. Technician B says that a colder spark plug has a longer insulator nose. Who is correct?
   - A. Technician A
   - B. Technician B
   - C. Both A and B
   - D. Neither A nor B

10. Technician A says that long-life spark plugs should always be left in an engine for at least 100,000 miles. Technician B says that long-life spark plugs should be removed and inspected every 7,500 miles for best performance. Who is correct?
    A. Technician A
    B. Technician B
    C. Both A and B
    D. Neither A nor B

11. Technician A says that platinum spark plugs are considered to be long-life spark plugs. Technician B says that double-platinum–plated spark plugs are plated with two layers of platinum on each electrode. Who is correct?
    A. Technician A
    B. Technician B
    C. Both A and B
    D. Neither A nor B

12. Technician A says that heat range is a measure of how much heat a spark plug can dissipate. Technician B says that a hotter spark plug has a shorter insulator nose. Who is correct?
    A. Technician A
    B. Technician B
    C. Both A and B
    D. Neither A nor B

13. Technician A says that resistor-type spark plugs increase the chances of cylinder misfire. Technician B says that resistor spark plugs reduce electromagnetic interference. Who is correct?
    A. Technician A
    B. Technician B
    C. Both A and B
    D. Neither A nor B

14. Technician A says that OEM ignition cables are constructed of solid copper cores. Technician B says that the core of secondary ignition wires is fabricated from carbon to reduce electromagnetic interference. Who is correct?
    A. Technician A
    B. Technician B
    C. Both A and B
    D. Neither A nor B

15. Technician A says that spark plug boots are made of silicone to withstand high temperatures. Technician B says that ordinary pliers should be used to remove spark plug boots from the spark plugs. Who is correct?
    A. Technician A
    B. Technician B
    C. Both A and B
    D. Neither A nor B

# Chapter 22

# Distributor Ignition Systems

## Introduction

Distributor ignition (DI) systems contain a distributor, **distributor cap**, and rotor. The ignition distributor performs the function of a mechanical switch that distributes secondary voltage to each spark plug at the appropriate time. This is accomplished using a distributor cap and rotor to direct spark energy from the ignition coil to each spark plug ignition cable or wire **(Figure 1)**. In this case, a single ignition coil is used to fire more than one spark plug.

The ignition distributor also contains an ignition coil triggering device. Early DI systems used breaker points to control the coil directly. Later systems use a magnetic or optical sensor to trigger an ignition module, which controls coil operation. In this case, the sensor operates in conjunction with an ICM and the PCM. Some systems integrate the ignition control function into the PCM.

### COMPONENTS OF DISTRIBUTOR IGNITION

The components of the DI system include:

- Distributor
- Distributor cap and rotor
- Ignition coil
- Coil triggering components (breaker points, CMP sensor, ignition module, PCM, etc.)
- Coil primary current-limiting device
- Secondary ignition wires, sometimes called ignition cables (including the coil wire)
- Spark plugs

## Distributor Cap and Rotor

The distributor cap and rotor are constructed of plastic materials, offering excellent electrical insulation characteristics. This is important because the components must be able to withstand voltages as high as 50,000V without breaking down.

Metal terminals are molded or otherwise secured into the cap at each spark plug cable tower location. The terminals extend inside the cap to form a small air gap with the rotor head as it passes each terminal **(Figure 2)**. An air gap is necessary to prevent wear between the spark plug cable terminals and the end contact of the rotor.

The rotor is fastened to the top of the distributor shaft. The rotor completes the electrical current path between the coil wire tower contact and the spark plug wire tower terminal as the distributor shaft rotates. The coil and spark plug wires are fitted with terminals that snap into the tower terminals of the cap.

The high-voltage signal from the ignition coil is connected to the coil tower of the distributor cap by the coil wire. The coil wire tower is similar to the spark plug wire towers except that the contact point inside of the cap uses an electrically conductive carbon button. The carbon button provides lubrication for the center contact of the rotor. In some applications, the carbon button is spring-loaded to maintain a good electrical contact with the rotor.

## Firing Order

To ensure that the proper **firing order** of the engine occurs, each spark plug wire must be connected to the appropriate spark plug wire tower. The firing order is the

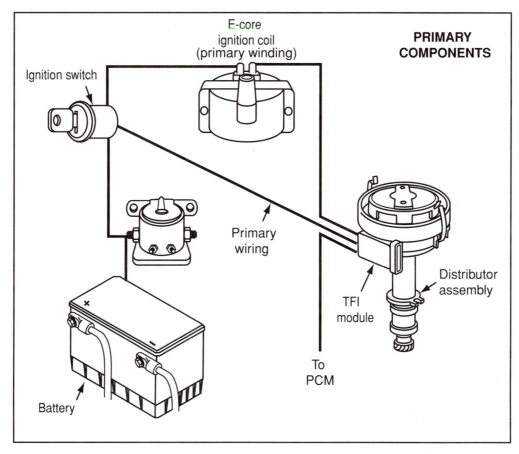

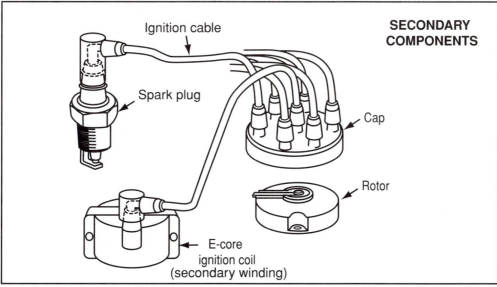

**Figure 1.** A single ignition coil is used to fire each spark plug.

sequence in which each spark plug is fired during an entire revolution of the camshaft. This sequence is based upon the camshaft profile.

As the rotor advances in position, the spark is delivered to the corresponding tower, firing the next plug in succession

**(Figure 3)**. Each cylinder must fire in the correct order when its respective piston reaches a point near TDC on the compression stroke.

Consult a shop manual for the proper firing order because it varies with different engine designs. Even engines

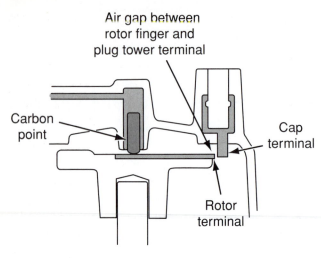

**Figure 2.** Terminals inside the cap form a small air gap with the rotor head.

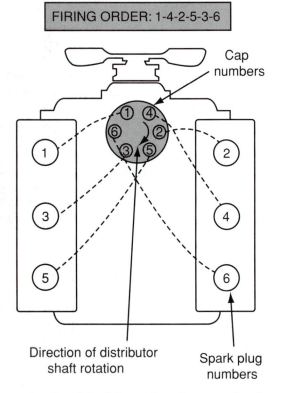

**Figure 3.** Spark is delivered to the next plug in succession according to the firing order.

from the same manufacturer and engine family can have a completely different firing order.

## Distributor Shaft Rotation

Depending on the engine design, the distributor shaft can rotate either clockwise or counterclockwise. Although this has no effect on engine performance, it can have a

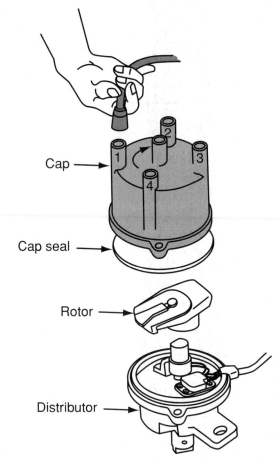

**Figure 4.** Ignition cables must be connected to the proper towers according to the direction of rotor rotation.

tremendous effect on troubleshooting if the spark plug wires are installed incorrectly. It is important to determine the direction of rotation so the ignition wires can be connected to the proper towers, following the engine's firing order **(Figure 4)**.

### Rotor Gap

Because of the additional air gap created between the rotor end contact and the spark plug tower terminal, the resistance of the secondary circuit is increased. The added resistance reduces secondary current, which helps to suppress RFI.

Unfortunately, the additional circuit resistance increases the voltage requirements of the ignition system needed to fire the spark plugs. The ignition coil must have adequate capacity to reliably jump both the rotor and the spark plug gaps.

## BREAKER POINT IGNITION

Early DI systems used a set of breaker points to control coil operation. Breaker points are so named because they break, or interrupt, coil current at the precise instant to cause the coil's magnetic field to collapse **(Figure 5)**. This causes a high-voltage pulse to be developed in the coil secondary, creating the spark that initiates a combustion event.

The ignition breaker points are located under the distributor cap, along with the **capacitor** (condenser) and distributor cam. The cam is splined to the distributor shaft, which causes the breaker points to open and close as the shaft rotates.

*A technician replaced the ignition cables on a vehicle only to discover that it would not start afterward. The number 1 ignition cable was properly connected between the number 1 spark plug and the corresponding distributor tower. The firing order was confirmed to be correct, but the car still would not start. Only after another technician examined the car was the problem determined. Although the first technician followed the correct firing order when installing the ignition cables, he assumed the distributor rotated clockwise. In fact, the distributor shaft rotated counterclockwise, causing a no-start condition.*

When the points close, the coil primary current increases, causing the magnetic field to build. Eventually, the primary current saturates and the field is at its maximum. The time the points are closed is called the dwell angle or dwell. It is measured in degrees of camshaft rotation and typically amounts to less than 5 msec. Keeping the points closed longer than the time needed to saturate the coil's magnetic field serves no purpose and causes heat to build in the coil windings.

*A condenser is another term used to describe a capacitor. Years ago radio capacitors were called condensers, and the name carried over into the automotive field.*

When the cam forces the points open, the field collapses, causing a high-voltage signal to be induced into the coil's secondary winding. The energy is delivered to the spark plug, initiating a combustion event.

The capacitor prevents arcing at the breaker points when they open. Arcing can significantly reduce the amount of high-voltage developed in the secondary winding, causing a weak spark. Arcing also causes the contact surfaces to deteriorate, reducing ignition efficiency.

A ballast resistor, or resistance wire, limits ignition coil primary current. This prevents overheating of the coil. During engine cranking, full voltage is applied to the coil through the by-pass circuit of the ignition switch, providing

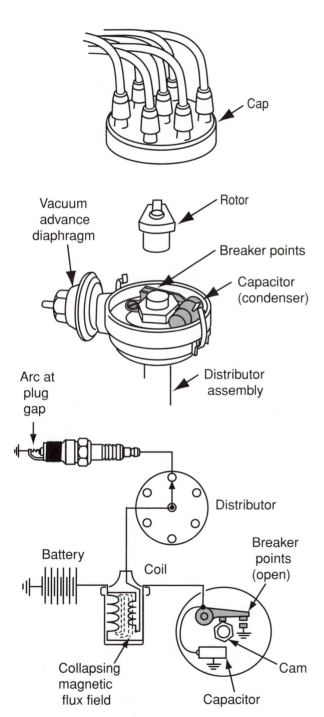

**Figure 5.** Coil operation controlled by breaker points.

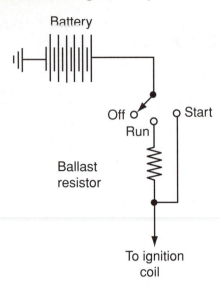

Battery

Off  Start
Run

Ballast
resistor

To ignition
coil

**Figure 6.** During cranking, the ballast resistor is bypassed, providing maximum spark energy.

maximum spark energy to start the engine **(Figure 6)**. This is especially important when starting a cold engine because the engine's ignition requirements are increased.

Once the engine is running, the ignition switch opens the by-pass contacts. The primary resistance becomes a part of the primary circuit, limiting coil current. The engine's ignition requirements are less once the engine is running.

Because the ignition breaker points are a mechanical switch, they are subject to electrical and mechanical wear. Misalignment, low spring tension and cam-lobe follower wear contribute to poor ignition system performance, especially under heavy engine loads.

A radio suppression capacitor is mounted close to the ignition coil. Its purpose is to reduce electromagnetic interference (EMI) generated when the primary current is switched on and off. The capacitor is connected between the ignition coil primary (+) terminal and the engine ground **(Figure 7)**.

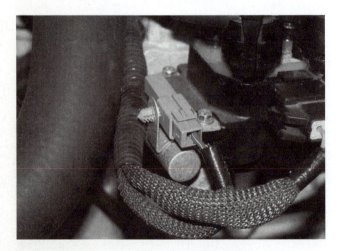

**Figure 7.** Radio suppression capacitor.

## BASE IGNITION TIMING

Base ignition timing is usually adjustable on DI systems. This is accomplished by loosening the distributor hold-down bolt and rotating the distributor housing. An instrument called a **timing light** illuminates the timing marks in relation to the crankshaft. Information related to timing adjustments is discussed in Chapter 24.

## MECHANICAL AND VACUUM ADVANCE

Early ignition systems used a combination of a vacuum advance unit and mechanical centrifugal advance weights to control ignition timing **(Figure 8)**. The centrifugal mechanism advances ignition timing as engine rpm increases.

On the other hand, the vacuum advance mechanism uses engine load to determine the amount of spark advance. As manifold vacuum drops, indicating a heavy engine load, spark timing is returned closer to the base timing setting. At light engine loads, ignition timing is advanced.

 *Interesting Fact* *Drag racers often disable the vacuum advance mechanism and use only the centrifugal mechanism, which consists of springs and weights. This allows the ignition advance curve to be fine-tuned for maximum performance.*

A significant amount of interaction occurs between the vacuum and mechanical advance systems. The characteristics of each system must be carefully designed to complement one another for maximum performance and best fuel economy.

## ELECTRONIC IGNITION USING A DISTRIBUTOR

Electronic ignition (EI) systems eliminate the weaknesses found in ordinary breaker point ignition systems. Modern ignition systems rely on the PCM to control spark advance. This method is much more precise compared with centrifugal and vacuum advance systems. Electronically controlled spark timing can easily compensate for variations in engine operation as well as ambient factors such as intake air temperature.

A variety of engine sensors, including a camshaft position (CMP) sensor, provide the powertrain control module (PCM) with information relating to specific engine operating conditions **(Figure 9)**. In some applications, a knock sensor (KS) is used to retard ignition timing in the event spark knock is detected. The use of a KS permits maximum

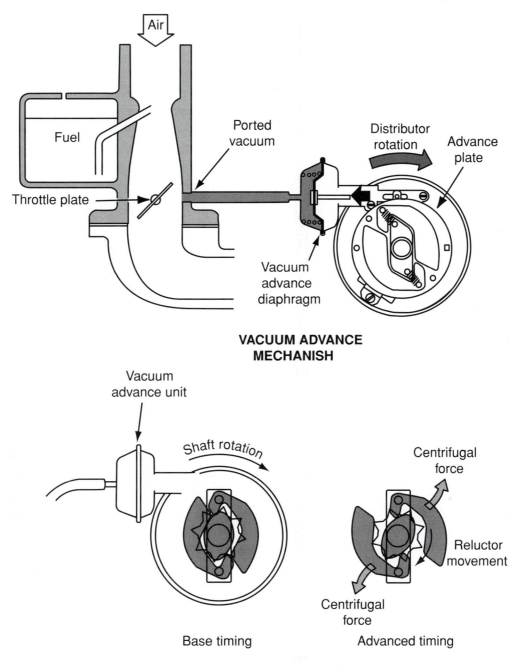

**Figure 8.** Typical mechanical-vacuum advance mechanism.

spark advance under a variety of engine loads, affording optimum power and fuel economy.

Instead of a set of breaker points, the electronic ignition distributor contains a pickup coil, Hall effect switch, or optical pickup assembly. The sensor is commonly called a **position indicator pulse (PIP) sensor**. A PIP sensor generates a signal that represents, or profiles, engine crankshaft position and speed. Some manufacturers refer to this signal as "profile ignition pickup." In the case of DI systems, the PIP sensor measures camshaft position, not crankshaft position. For this reason, the PIP sensor is often referred to as a CMP sensor. Because the camshaft rotates at exactly one-half crankshaft speed, the CMP sensor can easily develop a signal that precisely represents crankshaft position and speed **(Figure 10)**.

The PIP sensor signals the ignition control module (ICM), which triggers the ignition coil. The ICM contains solid-state components that amplify and shape the signal generated by the sensor.

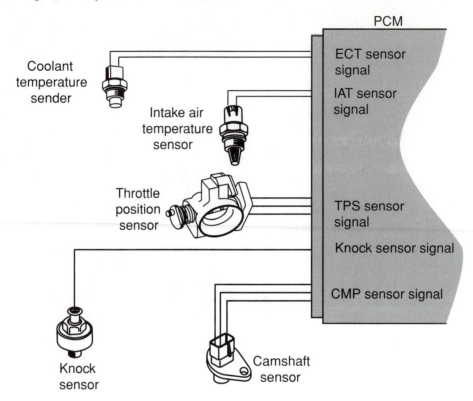

**Figure 9.** Sensors provide the PCM with information about specific operating conditions.

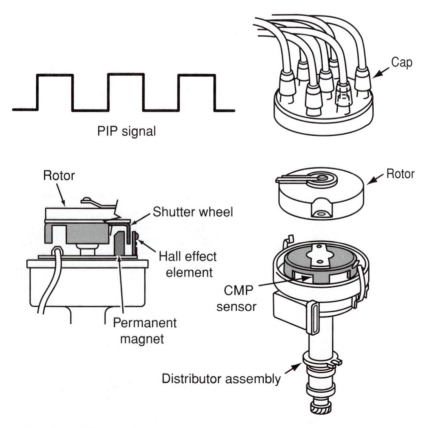

**Figure 10.** CMP sensor develops PIP signal that represents crankshaft speed and position.

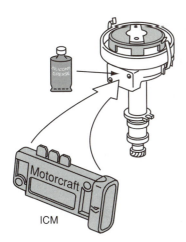

**Figure 11.** Silicone heat sink compound is applied on the mounting flange of the ICM and the mounting surface.

One important advantage of electronic ignition is the ability to minimize ignition coil heating by using a technique called variable dwell. Variable dwell limits how long primary current flows in the ignition coil, based upon engine speed. At low engine speeds, the ICM limits coil charging time, minimizing coil heating.

Depending upon the application, the ICM can be mounted on the distributor or remotely mounted. Because the ICM contains components that dissipate a significant amount of heat, special silicone thermal grease must be used when mounting the module. This thermal compound helps transfer heat between the module flange and the mounting surface **(Figure 11)**.

## ELECTRONIC SPARK ADVANCE

The PCM causes the ICM to interrupt ignition coil primary current at the proper time, firing the coil. The controlling signal sent to the ICM is usually called **SPOUT**, the acronym for "spark output." It represents the spark advance signal developed by the PCM, and it is always based upon the PIP signal.

Variations in engine operating conditions such as engine load and speed require different amounts of spark advance. By monitoring the various engine operating parameters, the PCM is able to determine optimum ignition timing for the particular conditions. The PCM adjusts the SPOUT signal to provide the proper amount of spark advance **(Figure 12)**.

If the SPOUT signal is missing as a result of a circuit fault, the ICM will fire the coil at base timing. Driving the vehicle with no spark advance will result in significantly

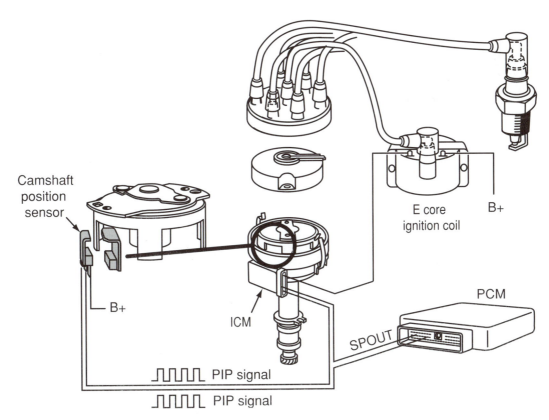

**Figure 12.** Spark advance is controlled by the SPOUT signal.

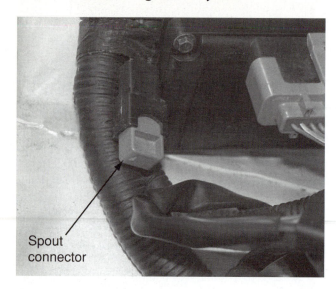

**Figure 13.** Timing is adjusted by unplugging the SPOUT connector.

increased engine operating temperatures. A substantial reduction in engine output power will also occur.

In some applications, the SPOUT signal to the ICM can be interrupted for setting base timing. This is often accomplished by removing the timing advance (SPOUT) connector **(Figure 13)**.

> **You Should Know** *The PIP signal is an input to the PCM, whereas the SPOUT signal is an output to the ICM. The PIP signal is also sent to the ICM for a base timing reference. Both are considered to be digital (on/off) signals.*

## TYPES OF DISTRIBUTOR IGNITION SYSTEMS

Automotive manufacturers have used a variety of DI systems over the years. Most rely on either magnetic or optical sensors to signal the ICM and PCM. Magnetic sensors are by far the most popular signaling devices.

### Hall Effect Ignition System

A typical Hall effect DI system is shown in **Figure 14**. The Hall effect switch is located within the distributor. The PIP signal developed by the sensor signals the ICM and PCM with engine position and speed data. The Hall effect switch is sometimes called a "vane switch" or "stator," because it is the stationary part of the sensor assembly. Spark advance is electronically controlled by the PCM through the SPOUT signal without the use of any mechanical advance mechanisms.

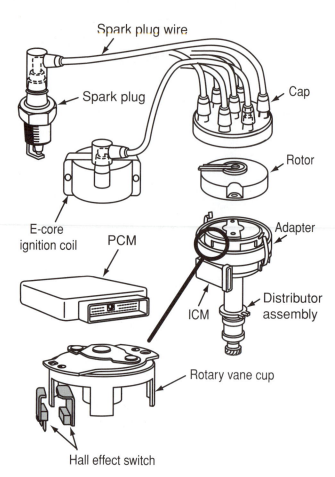

**Figure 14.** A typical Hall effect DI system.

## System Specifics

The Hall effect switch contains the sensor element on one side and a magnet on the other. A set of vanes is attached to the distributor shaft. The number of vanes matches the number of engine cylinders. As the shaft rotates, the vanes pass between the sensor and the magnet **(Figure 15)**. The vanes are made of a ferrous material.

When the vane is within the Hall effect switch opening, the magnetic flux path is blocked by the vane. This causes the sensor assembly to develop an output of approximately 12V. At this point, PIP is high **(Figure 16)**.

As the shaft rotates, the vane moves out of the sensor's opening, placing a "window" between the magnet and sensor. This causes the magnetic field to pass from the magnet to the Hall effect sensor. At the point where the vane moves past the sensor opening, the edge of the window causes the PIP signal to switch states. PIP is now low, or about 0V **(Figure 17)**.

Various components within the sensor assembly, including a **Schmidt trigger** circuit, assist in shaping the output signal from the Hall effect sensor. This produces a square wave signal with sharp edges, ensuring accurate spark timing.

It should be noted that the output signal from the Hall effect sensor element is inverted by the signal processing

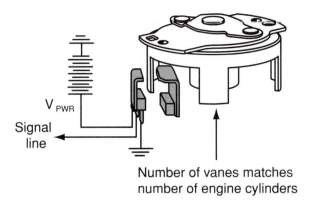

**Figure 15.** A Hall effect vane switch.

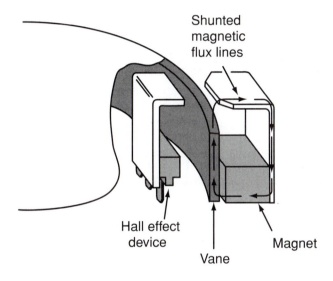

PIP is "high" when vane blocks magnetic flux

**Figure 16.** The vane is in the Hall effect switch opening.

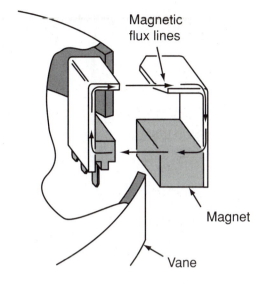

PIP is "low" when vane moves out of sensor opening

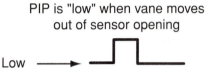

**Figure 17.** The vane is out of the Hall effect switch opening.

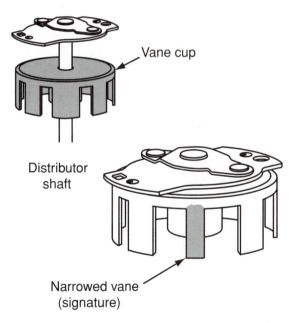

**Figure 18.** A narrowed vane represents cylinder number 1.

circuits within the sensor assembly. Whenever the Hall effect element is exposed to a magnetic field, the sensor develops a high output state. The sensor produces a low output state when no magnetic field is present. Even though the sensor assembly output is inverted, it is of no consequence to the ICM or PCM. Circuits within these modules are designed to compensate for the inverted signal.

To ensure correct fuel delivery timing, the vane that represents cylinder number 1 is often made narrower than the other vanes **(Figure 18)**. By using a narrow vane, a different "signature" is sent to the PCM. Not all the pulses will be alike. When the PCM recognizes the signal representing the number 1 cylinder, a reference point is established for timing the firing of the fuel injectors. This is necessary with sequentially fired injection systems.

## Circuit Operation

The signal developed by the Hall effect sensor signals the power switching transistor inside of the ICM to switch the coil primary on and off. The coil is fired when the ICM

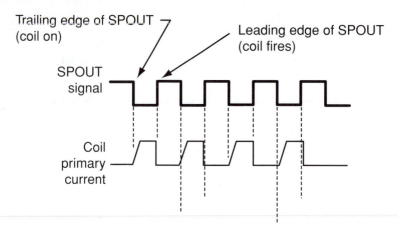

**Figure 19.** The rising edge of the SPOUT signal turns the coil primary "off," firing the coil. The trailing (falling) edge is often used to turn the coil primary "on."

interrupts coil primary current. The ICM uses the rising edge of the SPOUT signal to trigger the coil. The trailing edge of the SPOUT signal is often used to turn the coil primary on **(Figure 19)**. Because the SPOUT signal controls ignition dwell variably, coil heating is minimized.

## Ignition Module Connections

The typical ICM of a Hall effect DI system requires several connections to control ignition system operation. These connections include:

- PIP signal (CMP data from vane switch)
- PIP "pass through" signal to PCM (used by PCM to establish engine position and speed data)
- CMP sensor power (to power Hall effect switch assembly)
- CMP sensor ground (to complete power circuit of CMP sensor; provides PIP signal ground to ICM)
- ICM power (usually connected through the ignition switch)
- ICM ground (for ICM circuitry and coil driver circuit)
- Coil driver (controls coil [−] terminal through ICM transistor)
- SPOUT signal (spark advance signal from PCM)
- **Ignition diagnostic monitor (IDM)** signal to PCM for diagnostics purposes (used as tachometer signal)

## Induction-Type Distributor Ignition System

Induction-type pulse generators operate in a manner similar to the Hall effect system previously described. The pulse generator consists of a stator and trigger wheel **(Figure 20)**. The trigger wheel is often called a reluctor wheel or timer core.

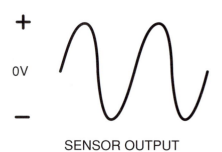

SENSOR OUTPUT

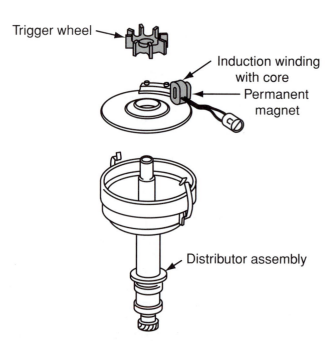

**Figure 20.** An induction-type pulse generator using a trigger wheel. As the teeth move across the coils, the magnetic flux changes.

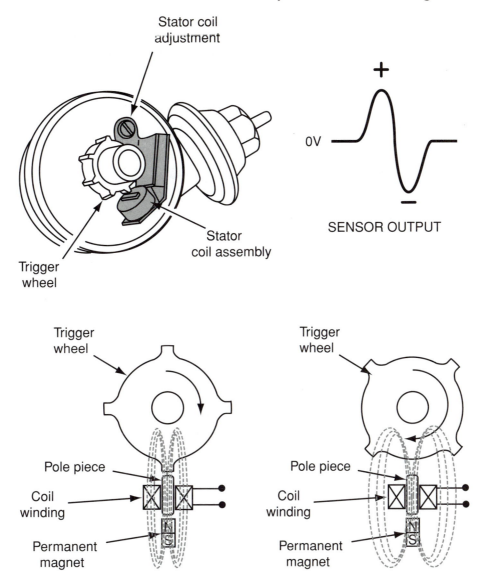

**Figure 21.** The continually changing magnetic field causes an AC voltage to be developed at the output of the stator coil.

## System Specifics

The trigger wheel is mounted to the distributor shaft, rotating at one-half crankshaft speed and is constructed of soft magnetic steel. The number of teeth on the trigger wheel corresponds to the number of engine cylinders. The stator contains a permanent magnet with a coil wound around a soft magnetic core. The magnetic core enhances the inductive properties of the coil.

There is a small air gap between the stator and trigger wheel, which changes as the trigger wheel rotates. Because of the change in air gap distance, the magnetic flux also changes. The changing magnetic flux induces an AC voltage into each coil. The induced voltage will be highest when the trigger wheel teeth approach their corresponding coils. Increasing engine speed causes the output voltage from the stator to increase accordingly. As the teeth move away from the coils, the magnetic flux decreases because the air gap is increasing. At this point, the polarity of the stator reverses, causing the output signal to move in a negative direction. The continually changing magnetic field causes an AC voltage to be developed at the output of the stator coils **(Figure 21)**.

## Circuit Operation

The inductive pickup coil sends position and speed information to the ICM in the form of an AC signal **(Figure 22)**.

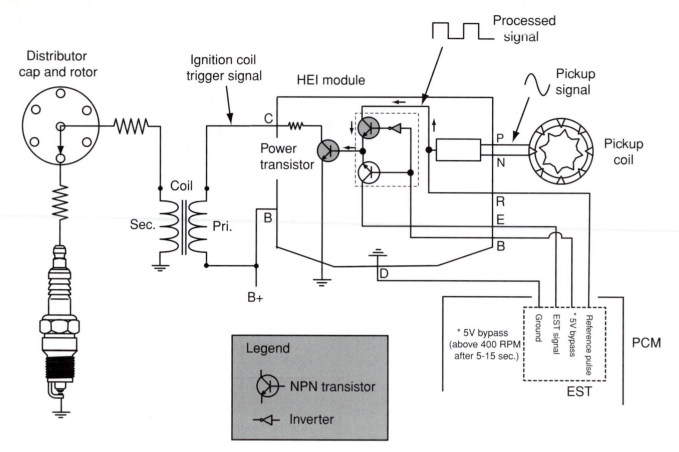

**Figure 22.**  Low-rpm signal from pickup coil disables spark advance.

This AC signal is processed and sent to the power switching transistor inside of the ICM to switch the coil primary on and off. During engine cranking, no spark advance is required. A low-rpm signal disables electronic spark advance. This mode is often referred to as module mode because spark timing is determined by the ICM, rather than by the PCM.

When the switching transistor is in the on state, current flows in the coil primary, causing the magnetic field to build. Using position and speed information provided by the pickup coil, the power switching transistor inside the ICM switches off at the precise instant. This causes the interruption of primary coil current, allowing the magnetic field to collapse. The collapsing field causes a high-voltage signal to be developed within the secondary winding, firing the spark plug of the appropriate cylinder.

The ICM activates PCM-controlled spark advance once engine rpm reaches a preset speed (typically around 400 rpm). The PCM applies a 5-volt signal to the by-pass line of the ICM, enabling electronic spark advance **(Figure 23)**. Sometimes, electronic control of spark advance is called the **electronic spark timing (EST)** mode.

A switching circuit within the ICM electrically disconnects the pickup coil signal from the ICM circuits and substitutes the EST signal from the PCM. The PCM uses a variety of sensor signals including engine coolant temperature (ECT), intake air temperature (IAT), mass airflow (MAF), barometric pressure (BARO), and engine speed to calculate optimum spark advance.

## Optical Ignition System

Optically triggered DI systems use a light-emitting diode (LED) and a photodiode to provide precise signaling to the ICM. The LED is usually of the infrared variety and emits virtually no visible light. The photodiode operates as a detector, responding to light pulses generated by a shutter plate as it rotates on the distributor shaft.

An optically triggered DI system is shown in **Figure 24**. The PCM contains an integrated ICM, with the exception of the power switching transistor. This transistor controls ignition coil primary current and is located between the distributor and ignition coil.

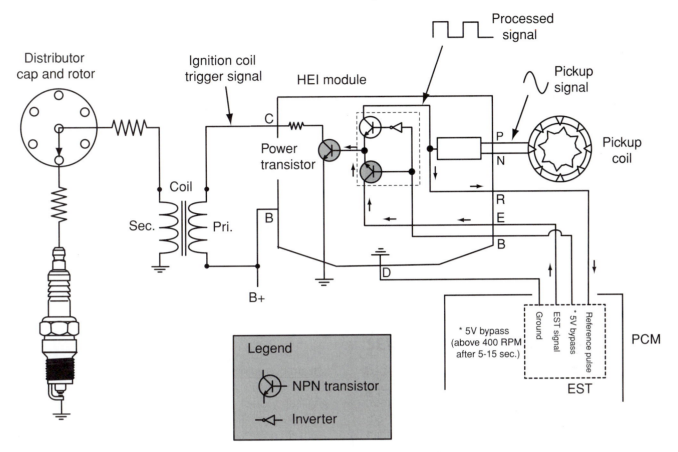

**Figure 23.** The PCM applies a 5-volt signal to activate electronic spark advance.

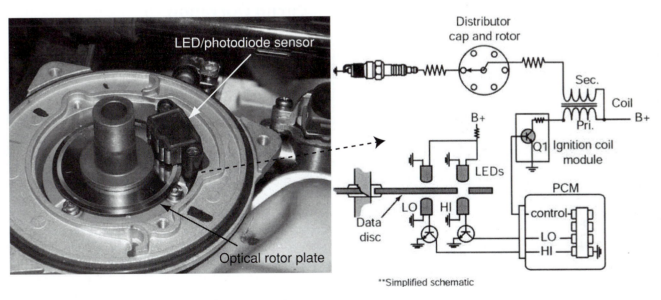

**Figure 24.** Optically triggered distribution ignition system.

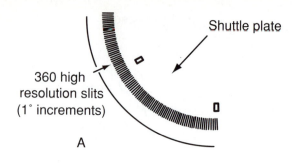

360 high
resolution slits
(1° increments)

Shuttle plate

A

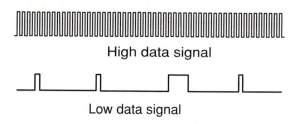

High data signal

Low data signal

Pulse shaping produces a digital ON - OFF signal

**Figure 26.**   The output signal produced by an optical sensor assembly.

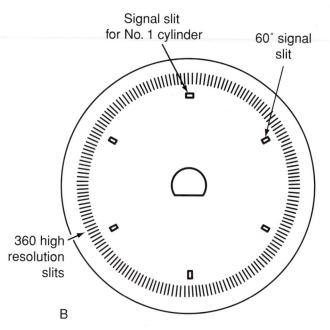

Signal slit
for No. 1 cylinder

60° signal
slit

360 high
resolution
slits

B

**Figure 25.**   (A) Optical rotor plate is used to indicate camshaft position. (B) Additional slits identify cylinder number 1 and crankshaft position at 120-degree intervals (V6 engine).

## System Specifics

The shutter plate, sometimes called a rotor plate, contains a number of rectangular windows or slits stamped into the plate **(Figure 25)**. In the example, 360 slits located along the outer edge of the plate correspond to degrees of crankshaft rotation. This provides 1-degree increment signals representing engine crankshaft speed.

Additional slits, including a uniquely shaped slot indicating cylinder number 1, are provided to identify the crankshaft position **(Figure 26)**. Considering that the application is a V6 engine, slits positioned every 60 degrees indicate

camshaft position at 120-degree intervals, corresponding to cylinder firing.

Light emitted from the LED passes though the slits of the shutter plate and is detected by the photodiode on the opposite side of the shutter plate. A digital, on-off signal is generated as the plate rotates.

To ensure reliable engine speed data, two emitter-detector pairs are used, with one mounted slightly offset from the other. Light is allowed to pass through the shutter slit at one emitter-detector pair location but is blocked by the rotor plate at the other. This allows a flip-flop signal to be generated. Signal conditioning circuits help shape the pulses much like the Schmidt trigger circuit shapes the signal from a Hall effect sensor.

Another emitter-detector pair is used to establish crankshaft position. The width of the slit in the rotor plate is wider to identify the location of cylinder number 1 at TDC. Circuits within the PCM are designed to recognize the wider slit.

## Circuit Operation

The signal generated by the emitter-detector pair signals the power switching transistor inside of the ICM to switch the coil primary on and off. At the proper time determined by information supplied to the PCM by the CMP sensor, the PCM sends a signal to the power transistor. This energizes the ignition coil primary winding, causing the magnetic field to build to its maximum value **(Figure 27)**.

Based upon the various engine sensors, including the CMP sensor, the PCM determines optimum spark advance. At the precise time, the signal to the power switching transistor is switched off. This interrupts coil primary current, causing the magnetic field to collapse. As the field collapses, it induces a high-voltage signal into the secondary winding, firing the spark plug of the appropriate cylinder **(Figure 28)**.

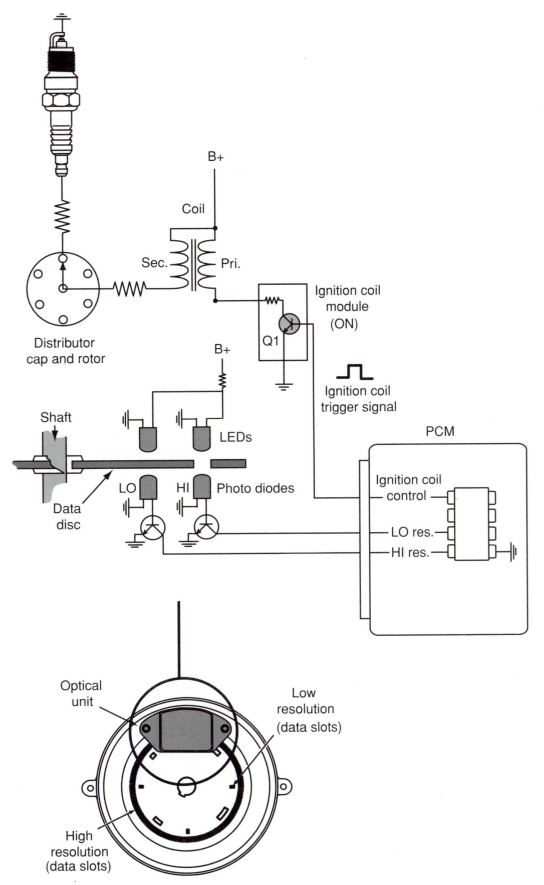

**Figure 27.** Transistor Q1 is on, energizing coil primary winding.

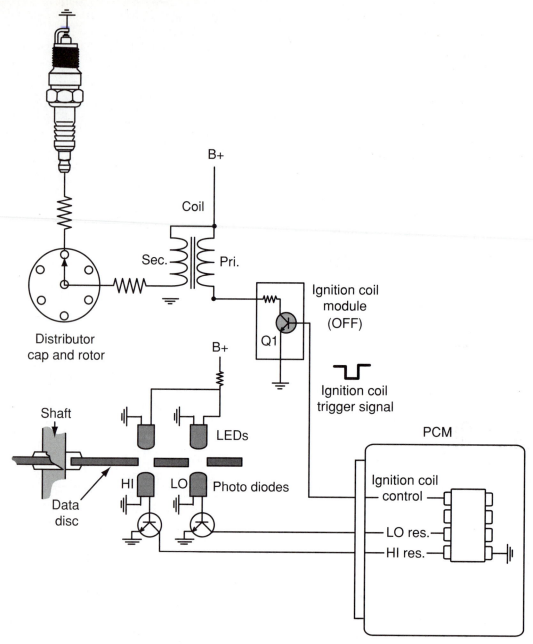

**Figure 28.** Transistor Q1 turns off, causing the magnetic field to collapse. This causes the spark plug to fire.

# Summary

- DI systems contain a distributor, distributor cap, and rotor to direct the secondary high-voltage to each spark plug.
- The distributor contains a coil triggering device that supplies a signal to trigger the ignition coil directly or through an ignition module.
- The firing order of the engine specifies the sequence in which spark plugs are to be successively fired.
- Ignition wires (sometimes called ignition cables) supply spark energy from each distributor tower to the corresponding spark plug according to the firing order of the engine.
- Early ignition systems used a set of breaker points to interrupt ignition coil primary current. A high-voltage signal is induced into the secondary winding when the breaker points open.

- Modern DI systems use magnetic or optical sensors to provide camshaft position and speed information to the ICM and PCM.
- Electronic spark advance is much more accurate than mechanical methods of spark control.
- Hall effect switches contain a magnet on one side and a solid-state sensor element on the other. A set of vanes corresponding to the number of engine cylinders provides a camshaft position signal to the ICM, which controls ignition coil primary current.

- Induction-type pulse generators contain a trigger wheel and pickup assembly. As the trigger wheel rotates past the pickup, the induced voltage changes. This signal is amplified and applied to the ICM circuitry, which controls ignition coil primary current.
- Optical distributors use LEDs and photodiodes in conjunction with a slotted trigger wheel to indicate camshaft position. The light pulses generated as the wheel rotates produce a digital on-off signal that is used by the ICM to control ignition coil primary current.

## Review Questions

1. Technician A says that the purpose of an ignition distributor is to distribute the high-voltage current pulse to each spark plug according to the firing order of the engine. Technician B says that the rotor completes the electrical path between the coil cable tower contact and the spark plug cable tower. Who is correct?
   A. Technician A
   B. Technician B
   C. Both A and B
   D. Neither A nor B

2. Technician A says that the firing order is based upon which way the distributor shaft rotates. Technician B says that the firing order is based upon the camshaft profile. Who is correct?
   A. Technician A
   B. Technician B
   C. Both A and B
   D. Neither A nor B

3. Technician A says that the ignition coil fires when the breaker points close. Technician B says that the ignition coil fires when the breaker points open. Who is correct?
   A. Technician A
   B. Technician B
   C. Both A and B
   D. Neither A nor B

4. Technician A says that dwell refers to the time that the breaker points are closed. Technician B says that dwell refers to the time that the ignition coil primary is energized. Who is correct?
   A. Technician A
   B. Technician B
   C. Both A and B
   D. Neither A nor B

5. Technician A says that base ignition timing is adjustable on DI systems. Technician B says that electronic spark advance is more accurate than are mechanical mechanisms. Who is correct?
   A. Technician A
   B. Technician B
   C. Both A and B
   D. Neither A nor B

6. Technician A says that the CMP sensor is used to supply a signal to the PCM and ICM circuits. Technician B says that the CMP sensor informs the ICM if spark knock occurs. Who is correct?
   A. Technician A
   B. Technician B
   C. Both A and B
   D. Neither A nor B

7. Technician A says that SPOUT is the acronym for "spark output." Technician B says that the SPOUT signal tells the ICM how far to advance ignition timing. Who is correct?
   A. Technician A
   B. Technician B
   C. Both A and B
   D. Neither A nor B

8. Technician A says that a Hall effect switch contains an LED and a shutter wheel. Technician B says that a Hall effect switch uses a solid-state sensor that responds to magnetic fields. Who is correct?
   A. Technician A
   B. Technician B
   C. Both A and B
   D. Neither A nor B

9. Technician A says that the signal sent to the PCM from the Hall effect switch is an AC signal. Technician B says that the Hall effect switch supplies a digital on-off signal to the PCM. Who is correct?
   A. Technician A
   B. Technician B
   C. Both A and B
   D. Neither A nor B

10. Referring to an optical distributor, Technician A says that another name for a rotor plate is a shutter plate. Technician B says that the sensor within the optical distributor produces an AC signal. Who is correct?
    A. Technician A
    B. Technician B
    C. Both A and B
    D. Neither A nor B

# 23 Chapter

# Distributorless Ignition Systems

## Introduction

Distributorless ignition systems offer accurate spark timing without the need for an ignition distributor. Eliminating the distributor improves system reliability because deterioration of the distributor cap and rotor is eliminated.

The Society of Automotive Engineers (SAE) refers to this distributorless ignition system as EI (electronic ignition). Some manufacturers break this system down into low data rate and high data rate versions **(Figure 1)**. This classification is based upon the speed at which data is supplied to the powertrain control module (PCM). High data rate systems are better able to respond to changing engine conditions, offering superior control of spark advance.

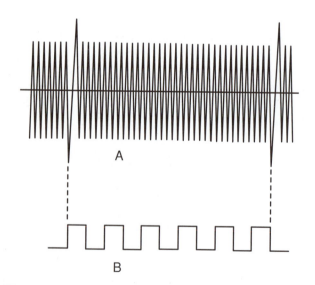

**Figure 1.** EI. (A) High data rate. (B) Low data rate.

EI systems use the waste-spark theory of firing two spark plugs simultaneously—one on compression, the other on exhaust. In effect, the spark plug that fires on the exhaust stroke is wasted.

Other variations of distributorless ignition systems include coil-on-plug (COP) and coil-near-plug (CNP) systems. These systems use individual ignition coils for each spark plug. Because a separate ignition coil is used for each engine cylinder, the waste spark is eliminated. All EI systems use variable dwell coil control to minimize coil heating.

### ADVANTAGES OF USING DISTRIBUTORLESS IGNITION SYSTEMS

When compared to a conventional DI system, distributorless ignition systems offer several important advantages:
- Having no ignition distributor results in fewer moving parts to break or wear.
- With no cap or rotor, component replacement is eliminated, reducing maintenance.
- Because there is no rotor gap, there is reduced electromagnetic interference (EMI).
- The precision ignition timing requires no adjustments.
- Without a distributor cap, condensation is eliminated, reducing engine misfires and no-starts.
- Having no ignition coil secondary cable reduces circuit complexity and EMI.

### WASTE-SPARK IGNITION SYSTEMS

Waste-spark ignition systems operate by firing two spark plugs simultaneously. The cylinders must be companion cylinders in the engine's firing order. One spark plug

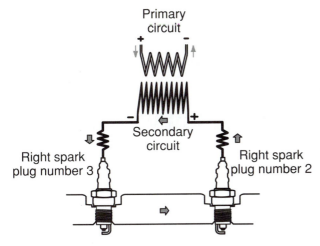

Figure 2. The ignition coil is electrically located between the spark plugs of companion (paired) cylinders.

fires on the compression stroke of the appropriate cylinder, whereas the other fires on the exhaust stroke. The process alternates on the next revolution of the crankshaft. Pairs of companion cylinders are fired according to the engine's firing order.

The spark plugs are connected in series, with the ignition coil configured in the center between the two plugs **(Figure 2)**. Only companion cylinders share a common secondary ignition coil.

Little ignition energy is used by the spark plug fired on the exhaust stroke. The presence of exhaust gases and no compression in the cylinder allows the spark to easily jump the spark plug electrodes. Only about 3,000V is needed to jump the gap. The air gap of this waste-spark cylinder behaves similar to the rotor gap of DI systems. However, little potential for EMI exists because the arc occurs within the cylinder head, where it is shielded. The majority of the ignition coil's energy is used to fire the spark plug on the compression stroke.

*As cylinder number 1 approaches TDC on the compression stroke, the companion cylinder also approaches TDC. However, the companion cylinder is on the exhaust stroke and contributes no power. Although both spark plugs fire simultaneously, only the cylinder on the compression stroke containing an air/fuel charge contributes power to the crankshaft.*

## COMPONENTS OF A WASTE-SPARK IGNITION SYSTEM

The components of a waste-spark ignition system include the following:

- Ignition control module (ICM)—may be integrated within PCM
- Crankshaft position (CKP) sensor
- Camshaft position (CMP) sensor (some systems)
- PCM
- Primary low-voltage wiring and connectors
- One or more ignition coil packs
- Secondary ignition cables
- Spark plugs

## SYSTEM SPECIFICS

A CKP sensor is used to provide engine speed and position information to the ICM. In some systems, no separate ICM is used. Instead, the ignition control functions are incorporated into the PCM. The CKP sensor can be of either the variable reluctance or Hall effect type. In the case of a **variable reluctance sensor (VRS)**, the trigger wheel is often mounted to the crankshaft **(Figure 3)**. Alternately, the trigger wheel can be integrated into the flywheel **(Figure 4)**.

Some means of identifying the position of the crankshaft is necessary for spark timing purposes. This is accomplished by either using a camshaft position (CMP) sensor or adding a position reference indicator to the trigger wheel in the case of a VRS. This position reference indicator is often nothing more than a missing tooth on the wheel **(Figure 5)**. The missing tooth produces a unique signal compared to that of the remaining trigger wheel teeth as it passes the sensor **(Figure 6)**. It is indexed to a specific piston position of a cylinder pair. The cylinder pair always includes cylinder number 1 because it is always regarded as the reference cylinder.

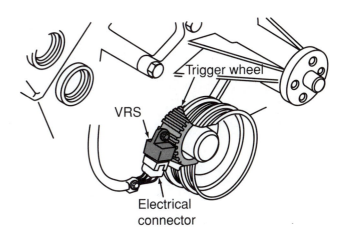

Figure 3. The trigger wheel is often mounted to the crankshaft. A VRS sends crankshaft position and speed data to the ignition control circuits.

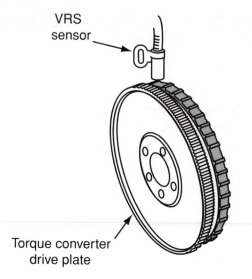

Figure 4. Trigger wheel shown as part of flywheel assembly. The VRS sends crankshaft position and speed data to ignition control circuits.

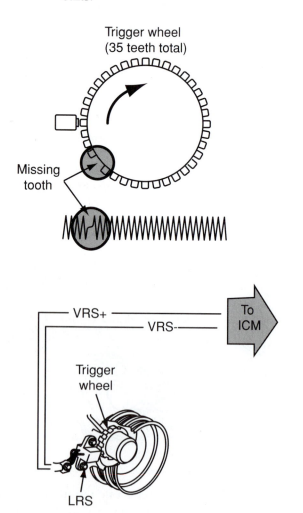

**Figure 5.** Missing tooth on trigger wheel identifies a reference point on the crankshaft, usually cylinder number 1.

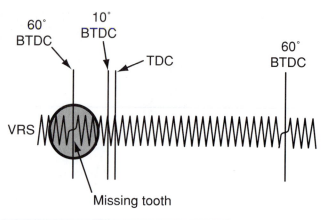

**Figure 6.** A VRS signal showing the unique waveform produced by a missing tooth.

The missing tooth produces a unique signal pattern that allows the ICM to recognize the location of cylinder number 1 **(Figure 7)**. This synchronizes the ICM for proper spark timing. Based upon the firing order of the engine, once the ICM knows the position of the missing tooth, it can determine which cylinder pairs will be approaching top dead center (TDC). This permits firing the proper coil at the appropriate time. The remaining coils are fired in sequence, according to the engine's firing order **(Figure 8)**.

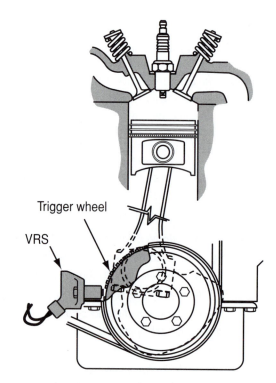

**Figure 7.** A missing tooth allows the ICM to recognize the location of piston number 1 in its cycle.

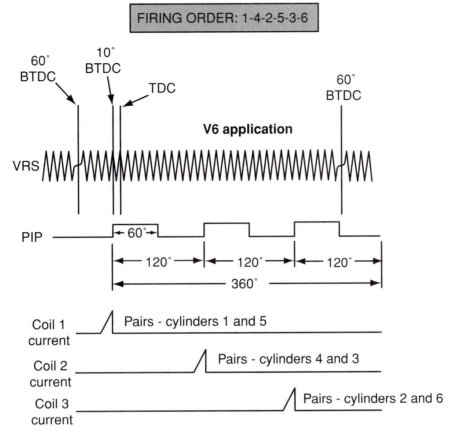

**Figure 8.** Ignition coils are fired in sequence according to firing order once the ignition circuit recognizes the missing tooth.

The missing tooth is usually located 60 degrees before the piston reaches TDC on a V6 application. For a four-cylinder engine, the missing tooth is typically located 90 degrees before top dead center (BTDC), and 50 degrees BTDC in the case of a V8 application. This is illustrated in **Figure 9**.

> **Interesting Fact**
>
> With **sequential electronic fuel injection (SEFI)** systems, fuel timing is important, so a CMP sensor is needed anyway. The engineer may choose to use the CMP sensor for both fuel and ignition timing or to use the CMP sensor for fuel timing alone. In this case, the CKP sensor must produce a "signature" so the ignition system can recognize the number 1 cylinder.

## Ignition Control Module

The ignition control module of a waste-spark ignition system operates in a manner similar to conventional DI. The primary purpose of the ICM is to charge the appropriate coil and interrupt the primary current at the precise instant determined by the PCM. In the event that the SPOUT timing advance signal is lost, the ICM will fire the coils at the base timing specification.

> **You Should Know**
>
> When diagnosing ignition concerns on systems that use integrated ignition control circuits, use extreme caution. A careless slip with a test probe can destroy the ignition circuitry, requiring that the entire PCM be replaced. This can be an expensive lesson to learn.

In the case of engine control systems incorporating the ICM circuits within the PCM, the technician will be unable to access the SPOUT signal at the PCM terminals. The signal path confines the SPOUT signal within the PCM itself. The SPOUT information will be available only when using a scan tool.

## Coil Packs

Depending upon the number of engine cylinders, more than one coil pack might be used. In the case of a

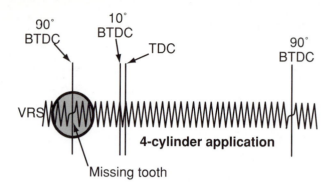

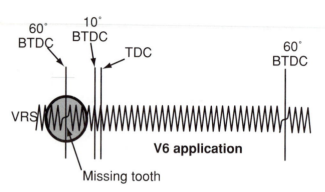

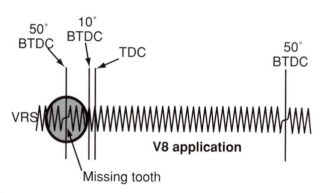

**Figure 9.**   Missing tooth positions for four-cylinder, six-cylinder, and eight-cylinder applications.

four-cylinder engine, a single coil pack containing two ignition coils is common. The V8 engine applications typically use two coil packs consisting of two coils each **(Figure 10)**. In the case of six-cylinder engines, a triple coil pack containing three ignition coils is often used **(Figure 11)**.

In some cases, the coils are individually replaceable **(Figure 12)**. In others, the complete coil pack must be replaced, even if only one coil is defective. This can be costly, especially in the case of a triple coil pack.

One side of the primary winding from each ignition coil is connected to vehicle power. The connection is usually though a fuse and the ignition switch. Each of the coil primary terminals is connected to the appropriate ICM coil

drivers **(Figure 13)**. The ICM uses switching transistors to control primary current in each of the individual coil's primary windings. A "pull to ground" signal causes the coil to charge. At the proper time, the switching transistor releases the ground, causing the coil to fire.

## CIRCUIT OPERATION

The basic operation of the VRS has been described in Chapter 14. Waste-spark ignition systems using this type of CKP sensor use a trigger wheel that contains a number of teeth. High data rate systems use many teeth, whereas low data rate systems might use only a few. High data rate systems provide extremely accurate spark timing, offering maximum fuel economy and power.

In **Figure 14**, the trigger wheel contains 35 teeth. This represents a 36-tooth wheel with a missing tooth. The location of the missing tooth is the crankshaft reference position. Each tooth represents 10 degrees of crankshaft rotation. This system is a high data rate system because only 10 degrees of crankshaft rotation is required before an updated signal is received by the ICM. Small changes in engine speed are easily detected, allowing for rapid adjustments to ignition timing. This results in optimum spark advance at any engine speed.

## Engine Starting

During engine cranking, the ICM looks for the missing tooth to establish the proper coil firing sequence. No coils are energized until the position reference indicator is located. Depending upon the firing order, once the ICM is synchronized, the ignition coil for cylinders 1 and 4 will be charged and subsequently fired on the next crankshaft revolution. The coil that fires the spark plug of cylinder number 1 is always referred to as coil 1.

To make engine starting easier, ignition timing is usually fixed by the ICM at base timing, typically 10 degrees before top dead center (BTDC). Coil 1 will be charged by the ICM for about 3 msec before the piston reaches this position. This provides sufficient time for the coil primary current to easily reach its maximum (saturated) value.

Allowing adequate time for the coil primary to achieve maximum current results in maximum secondary output voltage. This is especially important during engine cranking because it is more difficult to ignite the air/fuel charge under these conditions.

## Engine Running

Once engine speed reaches a minimum value that indicates the engine is running, the PCM takes control of spark advance. A position indicator pulse or profile ignition

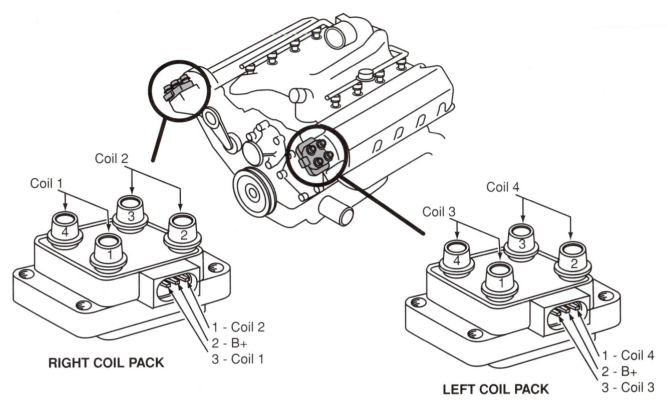

Coil 1
Coil 2

Coil 3
Coil 4

1 - Coil 2
2 - B+
3 - Coil 1

**RIGHT COIL PACK**

1 - Coil 4
2 - B+
3 - Coil 3

**LEFT COIL PACK**

**Figure 10.** A typical V8 application using two coil packs consisting of two coils each.

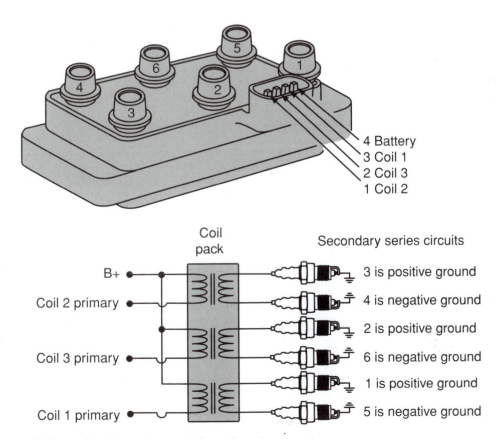

4 Battery
3 Coil 1
2 Coil 3
1 Coil 2

Coil pack

Secondary series circuits

B+

Coil 2 primary

Coil 3 primary

Coil 1 primary

3 is positive ground
4 is negative ground
2 is positive ground
6 is negative ground
1 is positive ground
5 is negative ground

**Figure 11.** A typical V6 application using a triple-coil pack.

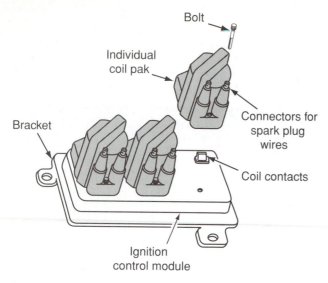

**Figure 12.** Individually replaceable coils.

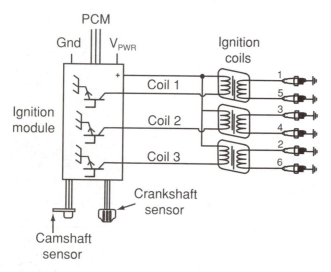

**Figure 13.** The ICM uses switching transistors to control ignition coil primary current.

Missing tooth identifies location of cylinder #1

**Figure 14.** The location of the missing tooth is the crankshaft reference position.

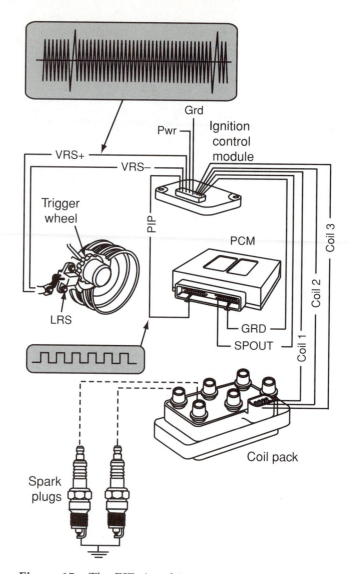

**Figure 15.** The PIP signal is generated by the ICM.

pickup (PIP) signal is generated by the ICM using the signal supplied by the CKP sensor. The PIP signal is a square wave that has a 50 percent duty cycle. The amplitude of the signal is typically about 13V, depending upon battery voltage. The signal is fed to the PCM so that spark advance calculations can be made **(Figure 15)**.

The leading edge of the PIP signal is precisely related to the position of the missing tooth on the CKP sensor. The signal occurs at 10 degrees BTDC, which is the base timing specification. On four-cylinder applications, the signal remains in the high state for 90 degrees. In the case of the V6 engine, the signal stays high for 60 degrees, and for the V8, the PIP signal continues in the high state for 45 degrees **(Figure 16)**.

When the PCM circuits recognize the presence of the PIP signal, the PCM uses data supplied by the various input

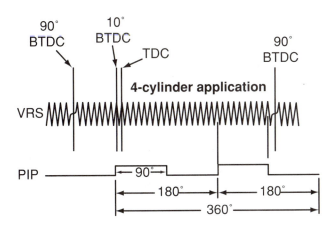

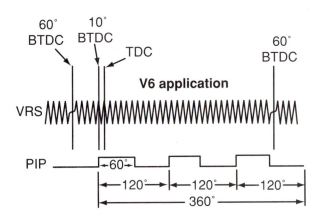

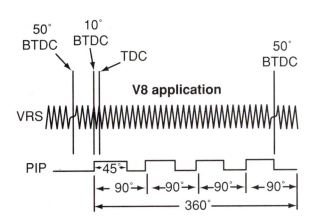

**Figure 16.** The PIP signal remains in the high state for 90 degrees (four-cylinder), 60 degrees (six-cylinder), and 50 degrees (eight-cylinder) applications).

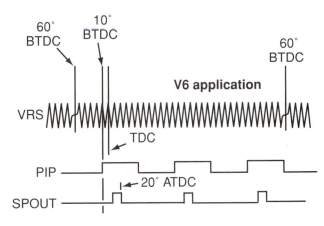

**Figure 17.** The spark advance signal from the PCM is called SPOUT.

**Figure 18.** A spark retard connector. Removing a plug retards ignition timing several degrees to reduce spark knock.

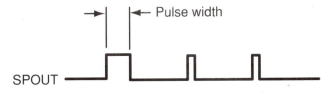

**Figure 19.** A pulse width–modulated SPOUT signal.

sensors and programmed instructions to determine optimum spark advance. The PCM transmits a SPOUT signal to the ICM, targeting the spark events accordingly **(Figure 17)**.

In order to reduce stubborn cases of spark knock, some systems provide a means using a scan tool to retard ignition

timing slightly. Early systems often used a two-pin connector-plug arrangement that retarded spark advance a few degrees when the plug was removed **(Figure 18)**.

Depending upon the application, the SPOUT signal might be a pulse width–modulated signal **(Figure 19)**. The use of such a digital signal allows the transmission of diagnostics data from the ICM to the PCM, if necessary. The PCM can set a diagnostic trouble code (DTC), which can be retrieved in the event a fault occurs. This helps the technician pinpoint the problem.

## INTEGRATED WASTE-SPARK IGNITION SYSTEMS

To reduce the number of electronic modules, improve reliability, and reduce costs, vehicle manufacturers began integrating the ICM into the PCM. This trend began several years ago and has helped reduce interconnect wiring, improving system reliability.

An integrated ignition system is shown in **Figure 20**. For clarity, only connections related to the ignition system are shown. In some cases, certain sensors are used by both the ignition and engine control systems. This includes the CKP sensor and in some cases, the CMP sensor.

Diagnosing integrated ignition systems can be somewhat more difficult because not all signals are available at the PCM terminals. The use of a scan tool is important when troubleshooting integrated ignition systems. Much of the data formerly available at the ICM terminals is available only when viewing scan tool data.

## COIL-ON-PLUG IGNITION SYSTEMS

The coil-on-plug (COP) ignition system represents the next generation of ignition systems. There are no troublesome secondary ignition cables to break down or cause EMI.

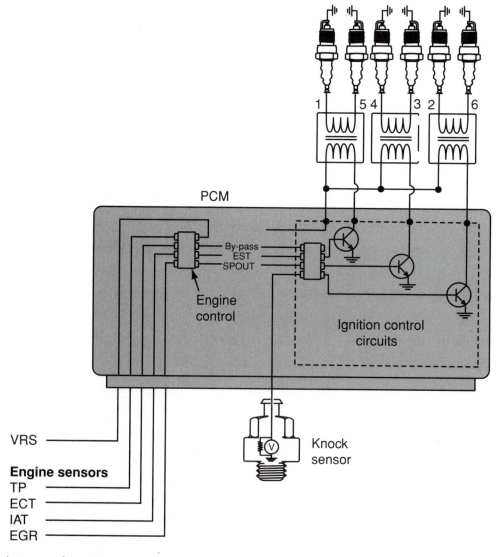

**Figure 20.**   An integrated ignition system.

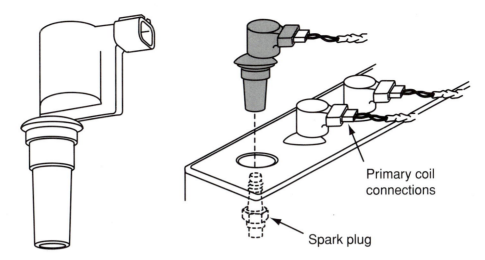

**Figure 21.** A COP ignition system.

The concept of COP ignition is extremely simple. An ignition coil is mounted over each spark plug, offering reliable secondary ignition performance that is less affected by moisture and corrosion. The ignition coil is fired at the proper time by the ignition control circuit. Only low-current primary wiring is used to connect the compact, encapsulated ignition coils **(Figure 21)**.

COP ignition systems operate somewhat differently than waste-spark ignition systems because they use a separate ignition coil for every cylinder. No waste-spark is generated. The spark plug of each cylinder is fired on the compression stroke near TDC. The exact point is based upon spark advance calculations carried out by the PCM.

## SYSTEM SPECIFICS

The ignition control functions are usually integrated within the PCM with coil-on-plug systems. Accordingly, only a handful of connections related to the ignition system are available at the PCM terminals. These include the CKP sensor and coil driver for each coil. The system is often treated as a "black box" because of its level of integration.

As with all ignition systems, a means of locating a point of reference is required. The reference point is usually provided by a CMP sensor that provides information regarding camshaft position.

Rather than firing spark plugs in pairs, as in the case of waste-spark ignition systems, the COP ignition system fires spark plugs individually. At the specific time determined by the PCM, the ignition control circuits energize and fire the appropriate ignition coils of each cylinder, according to the firing order of the engine.

Each coil is a one-piece assembly that is bolted to the cylinder head. A silicone or neoprene boot seals the secondary terminal to the top of each spark plug. The primary electrical connection is made using a two-pin connector that plugs into the ignition coil.

## CIRCUIT OPERATION

The CKP sensor supplies engine speed information to the PCM. A CMP sensor ensures that the correct coil fires at the proper time in accordance with the engine's firing order. A CMP signal is essential because companion cylinders do not fire simultaneously, as is the case with waste-spark ignition systems. Instead, each spark plug fires independently, based upon a command from the PCM to the respective ignition coil.

Most COP ignition systems use Hall effect CKP and CMP sensors. The sensors typically operate from a 5-volt reference source supplied by the PCM. In addition to ignition control, the CKP and CMP sensors are used by the PCM to control fuel delivery and emissions control equipment.

A typical COP ignition system is shown in **Figure 22**. Based upon information received from the CKP and CMP sensors, the ignition control circuits energize the proper coil in the firing sequence. This is accomplished using a switching transistor for each ignition coil primary **(Figure 23)**.

At the precise instant, the switching transistor releases the ground connection to the ignition coil primary. This causes the magnetic field to collapse, inducing a high-voltage signal into the secondary winding, causing the spark plug to fire. The ignition circuits within the PCM determine which coil will be charged next (according to the firing order) and the process repeats.

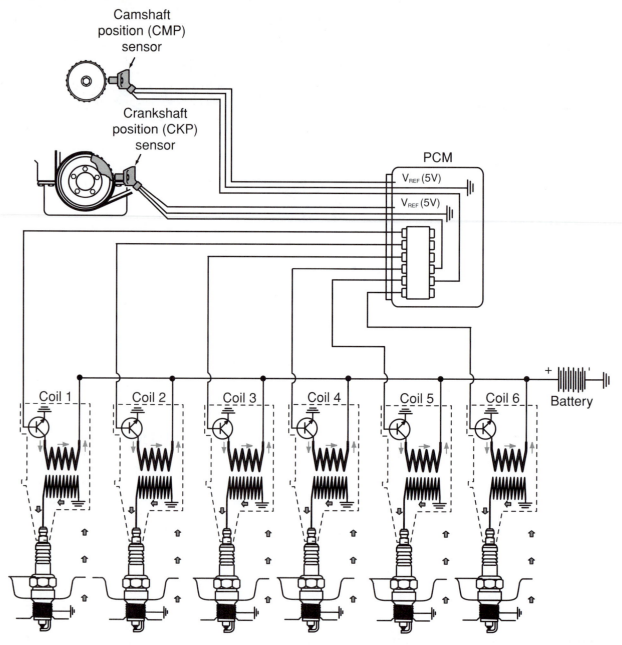

**Figure 22.** A COP ignition system using CMP and CKP Hall effect sensors.

As with other ignition system designs, optimum spark advance is determined by engine operating conditions. The PCM adjusts the point where the ignition coils fire, using information supplied by various input sensors. Variations in engine speed, load, ambient temperature, and driver demand are some of the factors considered in calculating spark advance.

Because each coil is independently controlled, ignition timing can be individually adjusted for each cylinder. This system works particularly well when used in conjunction with a knock sensor (KS). If spark knock is detected in cylinder number 2, for example, the PCM can selectively retard spark advance to this cylinder on the next firing event. The PCM is able to track which cylinder is knocking because it knows what cylinder was fired just before knocking occurred.

One disadvantage of COP ignition systems is the inherent inability to measure secondary ignition waveforms. Because it is usually not practical to access the secondary circuit due to how the individual coils are mounted, all ignition system diagnostics must be carried out using primary signals. More information related to ignition system diagnostics is presented in Chapter 24.

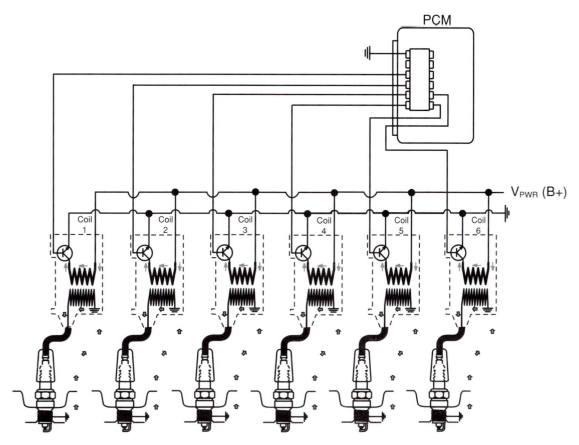

**Figure 23.**  Switching transistors are used to control each coil primary individually.

## COIL-NEAR-PLUG IGNITION SYSTEMS

CNP ignition systems operate virtually identically to COP systems. The primary difference is where the ignition coil is located relative to the spark plug. If the ignition coil cannot be located over the spark plug, the vehicle manufacturer can choose to locate the coils near the spark plugs. A short secondary ignition wire is used to connect each spark plug to the respective ignition coil **(Figure 24)**.

The disadvantage of CNP systems is that they use a secondary ignition cable for each spark plug. This presents a number of disadvantages, including secondary cable breakdown and radiation of EMI. On the other hand, the availability of connecting an inductive pickup to the secondary cable offers distinct troubleshooting advantages.

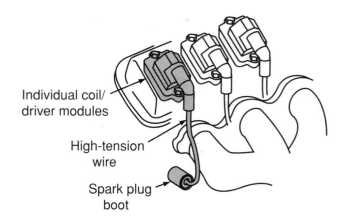

**Figure 24.**  Short secondary cables connect a spark plug to the ignition coil in CNP applications.

## *Summary*

- Distributorless (EI) ignition systems provide precise spark timing without using a distributor.
- EI ignition systems use the waste-spark theory of firing two plugs simultaneously—one on compression, the other on the exhaust stroke.

- Variations of distributorless ignition systems include COP and CNP designs.
- Because there is no rotor gap for the spark to jump, EMI is reduced.

- A CKP sensor using a position reference indicator can be used to signal the ICM, which in turn fires the appropriate ignition coil. The position reference indicator identifies cylinder number 1.
- If a position reference indicator is not used, a CMP sensor must supply a reference signal to the ICM, identifying cylinder number 1.
- The ICM operates in a manner similar to conventional DI systems. It controls the charging and interruption of coil primary current.
- Coil packs containing more than one ignition coil are often used in EI systems.
- Newer ignition system designs use an integrated ignition control circuit, which is located inside of the PCM.

- COP ignition systems use a separate ignition coil for each cylinder.
- No waste spark is generated with COP ignition systems.
- COP ignition systems produce less electromagnetic radiation because there are no secondary ignition wires to act as antennas.
- The control of each ignition coil is similar to conventional (DI) and distributorless (EI) systems.
- CNP ignition systems also use a separate ignition coil for each cylinder.
- CNP systems use a short secondary ignition wire for each cylinder.

# Review Questions

1. Technician A says that EI ignition systems use a distributor. Technician B says that all ignition systems use a distributor. Who is correct?
   A. Technician A
   B. Technician B
   C. Both A and B
   D. Neither A nor B

2. Technician A says that waste-spark ignition systems fire two plugs at once—one on exhaust, the other on intake. Technician B says that waste-spark ignition systems are so named because the spark fired on the compression stroke wastes most of the ignition energy. Who is correct?
   A. Technician A
   B. Technician B
   C. Both A and B
   D. Neither A nor B

3. In discussing waste-spark ignition systems, Technician A says that each ignition coil fires alternate pairs of cylinders. Technician B says that each ignition coil fires a pair of companion cylinders. Who is correct?
   A. Technician A
   B. Technician B
   C. Both A and B
   D. Neither A nor B

4. Technician A says that two coil packs containing one coil each are used in a V8 application using a waste-spark ignition system. Technician B says that two coil packs containing two coils each are used in a V8 application using a waste-spark ignition system. Who is correct?
   A. Technician A
   B. Technician B
   C. Both A and B
   D. Neither A nor B

5. Technician A says that each coil primary (−) terminal is connected to battery voltage through the ignition switch. Technician B says that all coil primary (−) terminals are connected to the vehicle ground connection. Who is correct?
   A. Technician A
   B. Technician B
   C. Both A and B
   D. Neither A nor B

6. Technician A says that integrated waste-spark ignition systems use an external ICM to control coil charging. Technician B says that when discussing integrated waste-spark ignition systems, all ignition control circuitry is contained within the PCM. Who is correct?
   A. Technician A
   B. Technician B
   C. Both A and B
   D. Neither A nor B

7. Technician A says that diagnosing integrated waste spark is sometimes more difficult because not all signals associated with the ignition system are available at the PCM terminals. Technician B says that much of the data formerly available on nonintegrated waste-spark ignition systems is available when using a scan tool. Who is correct?
   A. Technician A
   B. Technician B
   C. Both A and B
   D. Neither A nor B

8. Technician A says that COP ignition systems use one ignition coil for each pair of cylinders. Technician B says that COP systems use a separate coil for each cylinder. Who is correct?
   A. Technician A
   B. Technician B
   C. Both A and B
   D. Neither A nor B

9. Technician A says that spark advance is not used with COP ignition systems. Technician B says that CNP ignition systems use a separate ignition module for each ignition coil. Who is correct?
   A. Technician A
   B. Technician B
   C. Both A and B
   D. Neither A nor B

10. Technician A says that when discussing COP ignition systems, a separate coil driver is used for each ignition coil. Technician B says that coil drivers are used only with conventional DI ignition systems. Who is correct?
    A. Technician A
    B. Technician B
    C. Both A and B
    D. Neither A nor B

# Chapter 24

# Primary Ignition System Diagnosis and Service

## Introduction

Diagnosing ignition system concerns can sometimes be difficult because problems can occur in either the primary or secondary ignition system. In stubborn cases, simultaneous failures in both systems can be responsible for engine performance complaints. An orderly diagnostic approach is needed to effectively carry out the repair.

Because fuel system problems can sometimes be mistaken for ignition problems, it is important to rule out fuel system concerns before proceeding with ignition system diagnosis. For example, a faulty fuel injector can be responsible for a cylinder misfiring, though many technicians often associate misfire with ignition faults.

Although specialized test instruments are often used in diagnosing ignition system concerns, many of the same items used to troubleshoot electrical system faults are used. These instruments include a digital volt-ohm-milliammeter (DVOM), scan tool, and oscilloscope. As with any electrical system diagnosis, access to electrical schematics and component locations is vital to an effective repair.

### CHARTING A COURSE OF ACTION

Although problems in the secondary high-voltage electrical system can be caused by component failures in the secondary circuit itself, many times they are the result of primary ignition system failures. Poor connections, loss of signals from the crankshaft position (CKP) or camshaft position (CMP) sensors, and ignition module failures can cause the secondary high-voltage circuit to develop weak spark or no spark at all. Yet, each area of concern is part of the primary ignition system. For this reason, vehicle manufacturers often suggest starting pinpoint testing in the primary ignition circuit, once a handful of visual checks are performed. These visual checks are designed to quickly rule out obvious secondary component failures such as faulty ignition cables, cracked or worn ignition distributor caps, and broken or excessively worn spark plugs. Pay special attention to all connections and repair any that are found loose, corroded, or damaged. Any visibly worn or defective components must be replaced before proceeding with further diagnosis.

> **You Should Know** *The use of a spark tester is important in verifying the presence and quality of the ignition spark. Simply pulling ignition cables loose on a running engine can produce uncontrolled spark energy, which can damage sensitive ignition and powertrain control modules.*

Depending on the type of ignition system, there might not be a distributor cap to check, so the exact course of action often depends upon the specific system being serviced. The service manual will provide details regarding the various components used in the system, as well as diagnostic procedures that should be followed.

In all cases, a logical approach to diagnosing the system begins with verifying the complaint. Assume anything can seriously affect the time spent on the repair. Always duplicate the problem whenever possible. This action becomes difficult when intermittent concerns are present.

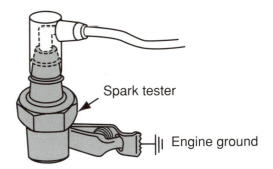

**Figure 1.** Using a spark tester to check for spark.

## CHECKING FOR SPARK

If the vehicle fails to start, it could mean no spark is present. If the vehicle runs poorly, weak spark might be to blame. In either case, a simple check of spark quality using a spark tester is important **(Figure 1)**. The spark tester should be of the adjustable variety and be set to a gap of about ¾ inch. This will load the secondary ignition system sufficiently to confirm the spark intensity is adequate to ignite the air/fuel mixture. Verifying the presence of spark is the first step in diagnosing a no-start complaint.

> **You Should Know** *Always make certain the storage battery has an adequate charge before performing any tests requiring the engine to be cranked. A battery with insufficient charge can cause you to obtain false results, making diagnosis difficult, if not impossible.*

A number of component failures can cause no ignition spark or a weak spark. These include:

● Faulty ignition coil.
● Defective ignition control module (ICM) (or PCM in the case of integrated ignition control circuits).
● Defective CKP or CMP sensors.
● Faulty connections between any of the primary ignition system components, including low or no voltage at the ignition coil primary (+) terminal.
● Ignition cable failures, including excessive resistance or high-voltage breakdown.

Each of the components and connections must be carefully checked to isolate the cause of the ignition failure. However, the order in which the components are tested is important to resolve the problem quickly.

If testing with a spark tester reveals no spark, a simple, quick check should be performed using a test light to verify the coil primary negative (−) terminal is switching.

Connect a 12-volt test light between a known good ground and the negative (−) coil primary terminal. With

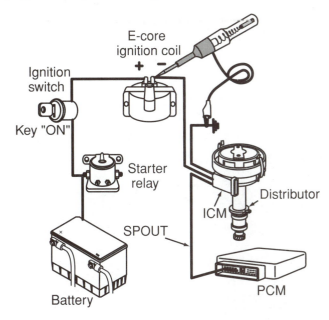

**Figure 2.** Using a test light to check for coil power.

the ignition switch in the key-on position, the test light should illuminate brightly **(Figure 2)**. If it fails to illuminate, proceed to the "Ignition Coil Diagnostics" section and perform a check of ignition coil primary voltage. A problem between the ignition switch and positive (+) ignition coil terminal might exist.

*Note:* A DVOM can be used in place of a test light to measure the voltage at the (−) ignition coil terminal. The DVOM should read battery voltage with the ignition switch in the key-on position.

Next, turn the ignition switch to the engine crank position and observe the test light, which should flash as the ICM triggers the ignition coil primary **(Figure 3)**. If the test light fails to flash, there might be wiring or connection problems between the (−) ignition coil terminal and ignition control circuits. Other possibilities include a defective CKP sensor or CMP sensor. If the light flashes, it indicates the ICM is receiving a signal from the CKP or CMP

> **You Should Know** *Just because the coil primary (−) terminal is switching does not mean that the ICM and all sensors are good. Although it is quite likely these components are functioning normally if the coil primary is switching, unusual engine performance concerns are still possible. Diagnostics should continue until the problem is resolved using pinpoint tests to determine full component functionality.*

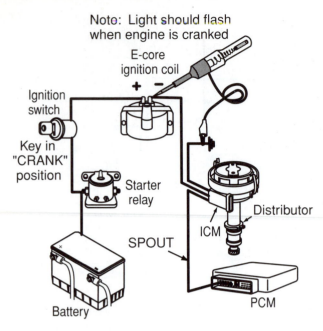

Note: Light should flash
when engine is cranked

E-core
ignition coil

Ignition
switch

Key in
"CRANK"
position

Starter
relay

Distributor

ICM

SPOUT

Battery

PCM

**Figure 3.** Checking coil primary operation using a test light.

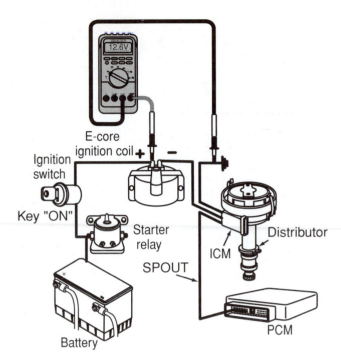

E-core
ignition coil

Ignition
switch

Key "ON"

Starter
relay

Distributor

ICM

SPOUT

Battery

PCM

**Figure 4.** Using a DVOM to check for battery voltage at coil primary.

sensors. It also indicates the switching circuits within the ICM are operating.

*Note:* For multiple coil ignition systems, repeat the test for each coil independently.

## IGNITION COIL DIAGNOSTICS

Before testing ignition coil winding resistance, a check of battery voltage at the ignition coil primary terminal should be performed. Testing for battery voltage at the (+) ignition coil terminal is easily accomplished by connecting a DVOM between the (+) coil terminal and a good ground **(Figure 4)**.

With the ignition switch in the **key-on–engine-off** position, voltage at the (+) ignition coil terminal should be close to battery voltage, or about 12.6V. During engine cranking, the voltage should be greater than 10.5V.

When troubleshooting ignition systems containing multiple ignition coils, a separate check of each (+) ignition coil terminal should be performed. Repeat the test for both key-on and cranking modes.

If the voltage is low, or nonexistent, trace the connections back to the ignition switch using the resistance function of a DVOM. Check for corroded connections that can cause excessive resistance in the circuit.

## Checking the Primary and Secondary Coil Windings

If the previous tests indicate the voltage at the (+) ignition coil primary terminal is normal, the coil itself might

be defective. To check the ignition coil for open windings, a DVOM is used. The DVOM is set to measure resistance.

To measure primary resistance, connect the DVOM test leads across the ignition coil (−) and (+) terminals with the coil disconnected from the vehicle harness wiring **(Figure 5)**. A reading between 0.5 and 4 ohms is typical, but the service manual should be consulted for the particular coil being measured.

Ordinarily, secondary resistance does not affect primary operation. However, it should be measured to verify it is within specifications while checking the primary resistance. This is done to rule out insulation breakdown between the primary and secondary windings. Typical secondary winding resistance falls between 6,000 ohms and 25,000 ohms. Consult the service manual for secondary winding resistance specifications.

An ignition coil that falls outside of the prescribed primary or secondary resistance values should be replaced. Visually inspect the coil carefully. If it is cracked, shows signs of carbon paths, or appears otherwise damaged, replace it.

## Coil Wiring Concerns

Other problems that can prevent the coil from triggering include a faulty ICM or wiring problems between the coil triggering (−) terminal and the ICM. With PCM integrated ignition control, the ICM may be located within the PCM. The use of a schematic diagram and DVOM set to measure

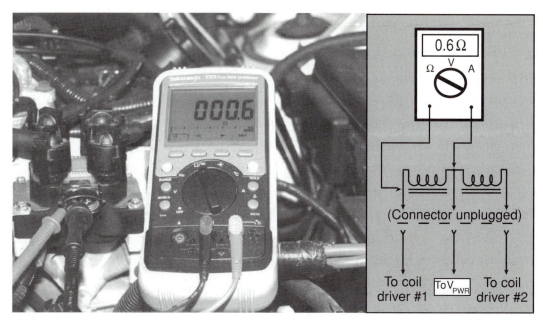

**Figure 5.** Checking coil primary resistance using an ohmmeter.

resistance can easily pinpoint wiring or connection concerns. Before proceeding with this course of action, however, it is appropriate to verify the presence of a crankshaft and camshaft signal at the ICM.

## Checking the Radio Suppression Capacitor

If excessive static is noted when listening to the radio, the suppression capacitor might be open. Radio suppression capacitors almost always fail in the open state, increasing the amount of radio static. A DVOM capable of measuring capacitance is required to check the capacitor **(Figure 6)**.

**Figure 6.** Checking the radio suppression capacitor using the capacitance function of a DVOM.

An open radio suppression capacitor will not usually cause poor ignition system performance. However, if the radio suppression capacitor short-circuited, battery voltage to the ignition coil(s) would be shunted away from the coil, causing a failure in the primary ignition system. The ignition coil primary circuit protection device will open in response to the excess current flow.

## Verifying Crankshaft and Camshaft Signal Quality

Verifying both CKP and CMP sensor signal quality is best accomplished using an oscilloscope. However, in the case of magnetic sensors, a DVOM can be used to verify if any signal is present at all **(Figure 7)**. If a faulty CKP or CMP sensor is suspected, the sensor should be tested and visually inspected for damage.

In the case of optical distributors, it is important to carefully examine all internal components. An optical sensor produces a digital on-off signal, which must be tested using an oscilloscope because of the frequency of operation. A typical pattern generated by an optical distributor is indicated in **Figure 8**.

Bench testing of optical sensors requires specialized equipment and is normally not performed in the service bay. Instead, the sensor is tested as part of the distributor assembly in the vehicle. If the appropriate output signals are not obtained during engine crank and engine run, the distributor should be replaced.

If the sensor is dirty or contaminated with engine oil, follow the manufacturer's service procedures to clean the sensor. Always replace any light shields, as this can impair

NOTE: Crank engine while observing DVOM on AC volts

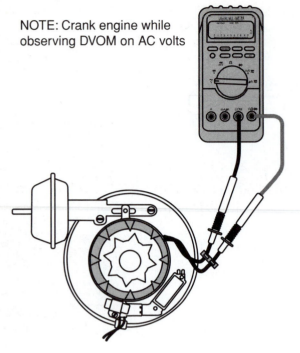

**Figure 7.** Using a DVOM to check a magnetic sensor.

triggering performance because ambient light can affect sensor operation.

If the CKP and CMP sensor signals are normal, proceed to the next section, "Checking Ignition Module Operation."

## CHECKING IGNITION MODULE OPERATION

Even if the quality of the CKP and CMP sensors is found to be acceptable, the possibility of poor connections to the ICM or PCM cannot be overlooked. A thorough check of sensor wiring and connections should be performed, including continuity tests of all signal wiring between the sensors and their respective modules.

If the signal integrity deteriorates before it reaches the ignition control circuits, coil triggering problems will occur. Without a valid signal, the switching transistor(s) within the ICM will not be able to properly control ignition coil primary operation. The symptoms can be similar to those caused by a defective ICM, so it is important to verify sensor signal quality at the ICM.

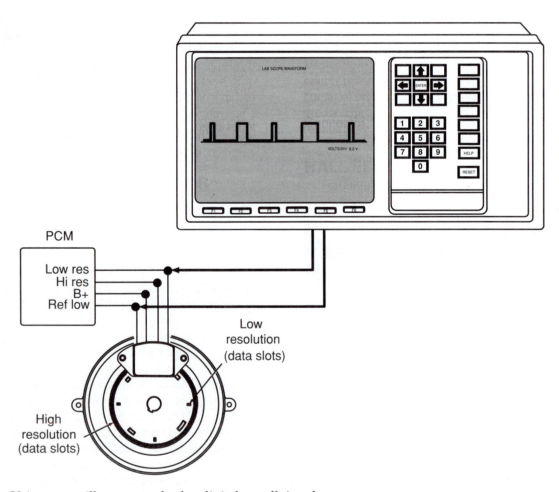

**Figure 8.** Using an oscilloscope to check a digital on-off signal.

If the system uses PCM-integrated ignition control circuits, the same basic troubleshooting procedures are followed. In this case, CKP and CMP signal quality is verified at the appropriate PCM terminals.

If valid signals are present at the ICM, test vehicle power and ground circuits to the ICM. A service manual is helpful in locating ICM connections. A DVOM set to measure low resistance is used to verify a good ignition ground is available at the ICM.

To verify the ICM is being supplied by vehicle power, a DVOM is used to measure battery voltage at the appropriate terminal. The voltage should be greater than 10.5V when the engine is cranked, and 12.6V when the ignition switch is at the key-on–engine-off position. Repair any defective connections before proceeding with any troubleshooting.

If all voltages are within specification, the ICM is most likely faulty. It should be replaced with an OEM component, using silicone thermal grease between the module and its mounting surface. The thermal grease will ensure adequate heat transfer from the module to protect it from heat. Always clean the old grease from the surfaces before adding new.

## VERIFYING BASE IGNITION TIMING

Base, or initial, timing can be verified using a timing light. The timing light is a stroboscopic xenon lamp that fires each time a signal is received by its inductive clamp **(Figure 9)**. The inductive clamp connects the high-voltage signal of cylinder number 1's ignition cable to the internal triggering circuits of the timing light. Cylinder number 1 is always used as the reference cylinder when checking or adjusting base, or initial, ignition timing.

Before checking base timing, make sure each of the following conditions is met:
- The engine is up to operating temperature.
- The curb idle rpm is within manufacturer's specifications.
- Electronic spark advance is disabled—usually by removing the timing advance (SPOUT) connector or entering "service mode" using a scan tool.
- You have located cylinder number 1 for connecting the timing light pickup.
- You have located the timing marks and can easily read them. The marks might need to be cleaned to allow you to read them accurately.

Consult the service manual for the correct timing verification procedure. Information regarding base timing is usually printed on the vehicle emissions control information (VECI) decal located in the engine compartment.

With the engine running and everyone clear from the moving engine components, point the timing light at the timing marks. The indicator should light up at the proper amount of ignition advance indicated in the service manual or listed on the VECI label.

## ADJUSTING BASE IGNITION TIMING

Before attempting to adjust ignition timing, consult the shop manual or VECI decal located in the engine compartment for specific procedures. Similar to verifying base ignition timing, electronic spark advance must be disabled before making any base timing adjustments. Follow the

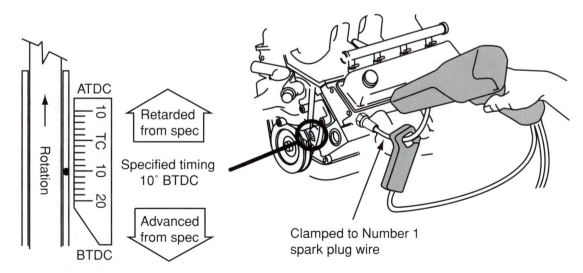

**Figure 9.**   The timing light fires each time a signal is received by the inductive pickup clamp. The clamp is attached around cylinder number 1's spark plug wire.

method outlined in the manual or printed on the VECI decal to disable electronic spark advance.

**Interesting Fact**

*If you have difficulty seeing the timing marks, apply a small amount of chalk or typewriter correction fluid to the marks and repeat the test. The light-colored compound will make the marks stand out when illuminated with the timing light, making them easier to read.*

## Distributor Ignition Systems

To adjust base ignition timing on DI systems, loosen the distributor holddown fastener. While pointing the timing light at the timing marks, rotate the distributor housing to adjust ignition timing. To advance ignition timing, rotate the distributor opposite the direction of rotor rotation. To retard ignition timing, rotate the distributor housing in the same direction that the rotor rotates. When the timing marks are correctly aligned, tighten the hold-down fastener securely.

## Electronic Distributorless Ignition Systems

Base ignition timing is usually nonadjustable on EI distributorless systems. In fact, many newer engines do not have a provision for checking base ignition timing. A scan tool is used to check spark advance.

If base timing is out of adjustment, it usually indicates a sensor problem or a broken trigger wheel. You should suspect a broken trigger wheel when base timing is significantly different than that called out in a service manual.

# Summary

- Always verify the complaint before proceeding with diagnosis.
- Always perform a thorough visual inspection of ignition components. Replace any that appear damaged or worn before continuing diagnosis of the ignition system.
- Intermittent concerns are difficult to diagnose. Use a logical troubleshooting process to isolate potentially defective components or connections.
- If an engine fails to start, verify that adequate spark is present using a spark tester.
- Causes of no spark or weak spark include loss of crankshaft position or camshaft position signals, a faulty ignition coil, a defective ICM, and poor connections.
- Ignition cable failures include excessive cable resistance and high-voltage breakdown, resulting in loss of spark or a weak spark.
- A test light can be used to check coil primary switching action.

- An ohmmeter can be used to check ignition coil primary and secondary resistances.
- Crankshaft position and camshaft position signals are best tested using an oscilloscope. A DVOM can sometimes be used to detect the presence of a crankshaft position or camshaft position signal.
- Ignition module tests include battery voltage and ground checks, as well as crankshaft position and camshaft position signal tests at the module. If all input signals are within specifications, the ignition module is likely defective.
- Base ignition timing can be checked with a timing light.
- Base ignition timing on DI systems is usually adjusted by loosening a holddown fastener and rotating the distributor.
- Ignition timing is normally not adjustable on EI distributorless systems.

# Review Questions

1. Technician A says that problems in the secondary high-voltage ignition system can be caused by a fault in the primary ignition system. Technician B says that an oscilloscope can be used to isolate primary ignition system concerns. Who is correct?
   A. Technician A
   B. Technician B
   C. Both A and B
   D. Neither A nor B

2. Technician A says that it is all right to check for spark by disconnecting the coil wire from the ignition coil and checking for a spark at the coil. Technician B says that a test light can be used to check spark. Who is correct?
   A. Technician A
   B. Technician B
   C. Both A and B
   D. Neither A nor B

3. All of the following can cause a no-spark condition *except:*
    A. An open ignition coil primary winding.
    B. A shorted ignition coil secondary winding.
    C. A faulty connection between the ignition coil (+) terminal and the ignition switch.
    D. An open radio suppression capacitor.
4. Technician A says that a timing light can be used to verify ignition coil primary voltage. Technician B says that a test light can be used to verify that the ignition coil primary (−) terminal is switching normally. Who is correct?
    A. Technician A
    B. Technician B
    C. Both A and B
    D. Neither A nor B
5. Technician A says that a test light connected between the ignition coil (−) terminal and the negative battery terminal will illuminate while the engine is cranked. Technician B says that a test light connected between the ignition coil (+) terminal and the negative battery terminal will flash. Who is correct?
    A. Technician A
    B. Technician B
    C. Both A and B
    D. Neither A nor B
6. Technician A measures 25 ohms across the ignition coil primary winding terminals and determines that the ignition coil is faulty. Technician B says that a coil primary resistance of 25 ohms is normal. Who is correct?
    A. Technician A
    B. Technician B
    C. Both A and B
    D. Neither A nor B
7. Technician A says that the output from an optical distributor is a sine wave signal. Technician B says that the output from an optical distributor is a digital on-off signal. Who is correct?
    A. Technician A
    B. Technician B
    C. Both A and B
    D. Neither A nor B
8. Technician A says that a DVOM set to measure resistance can be used to verify the integrity of an ignition module ground. Technician B says that a current probe should be used to check ground integrity. Who is correct?
    A. Technician A
    B. Technician B
    C. Both A and B
    D. Neither A nor B
9. Explain the process of verifying base ignition timing.
10. Technician A says that a timing light can be connected to any secondary ignition cable provided it uses an inductive pickup. Technician B says that the inductive clamp of a timing light must be connected to the coil wire on DI systems. Who is correct?
    A. Technician A
    B. Technician B
    C. Both A and B
    D. Neither A nor B

# Chapter 25

# Secondary Ignition System Diagnosis and Service

## Introduction

Testing the secondary ignition system often requires special instruments such as an ignition oscilloscope. Because of the high voltages involved, extra care must be used at all times to protect the technician and the equipment used to diagnose the system.

Secondary ignition patterns displayed on an ignition oscilloscope can reveal many subtle engine performance problems. However, basic tests will often isolate significant ignition problems that can severely reduce engine performance. The use of a DVOM can often isolate secondary ignition problems such as high-resistance ignition cables or ignition coil failures.

If the primary ignition system is operating properly, the technician should begin testing secondary components that can cause a no-spark, or weak-spark condition to occur. In particular, it is important to check for loose, corroded, or broken secondary ignition system components. For this reason, a thorough visual inspection is important.

### INSPECTING SECONDARY IGNITION WIRES

Faulty secondary ignition wires can cause hard engine starting and possibly even a no-start situation, especially in humid conditions. Before checking secondary ignition cable resistances, perform a visual inspection, paying particular attention to the condition of the jacket, spark plug boots, and coil or distributor terminals. If any burned spots are found, the cables should be replaced because arcing has likely occurred **(Figure 1)**.

Before continuing diagnosis, replace any cables that appear cut, torn, or otherwise damaged. Be especially mindful of spark plug boots that can easily rip if a boot puller tool is not used **(Figure 2)**.

Ignition cables should be replaced as a set. If one or two cables are found to be faulty, the others are soon to follow. The exception to this rule is in cases in which a single cable is damaged from a physical cause such as a tear in the insulation. In this case, it is acceptable to replace only the defective cable, provided that a source of single cables is available. Most ignition wires are sold only as a set.

Ignition cables used in hemispherical cylinder head engines often use long-reach boots. Internal arcing inside the spark plug tube can seriously affect ignition spark quality

*Interesting Fact*

*The first sign of deteriorating ignition cables can be extended crank times and a rough idle on rainy days. The problem often subsides after the vehicle is thoroughly warmed up.*

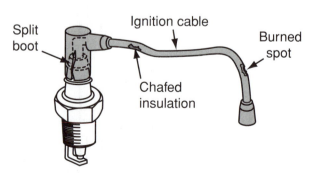

**Figure 1.** Ignition cables showing signs of burning or arcing must be replaced.

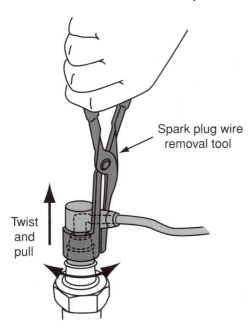

Spark plug wire removal tool

Twist and pull

Figure 2.   To minimize damaging spark plug boots, always use a boot puller.

> You Should Know
>
> *Never force a tight bend in an ignition cable, because the carbon filament might be weakened, causing premature cable failure.*

without visible signs of trouble **(Figure 3)**. Use extra care when inspecting this type of spark plug boot because internal arcing between the spark plug terminal and cylinder head often goes unnoticed.

## MEASURING SECONDARY IGNITION CABLE RESISTANCE

Secondary ignition cable resistance can be easily measured using an ohmmeter **(Figure 4)**. For carbon core

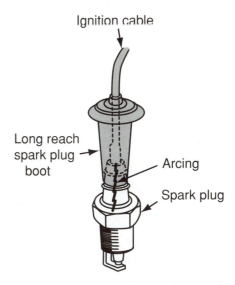

Ignition cable

Long reach spark plug boot

Arcing

Spark plug

Figure 3.   Internal arcing inside the engine often goes unnoticed.

cables, good ignition cables should measure less than 10,000 ohms per foot. Consult the manufacturer's service manual for exact specifications. The cables should be flexed slightly while making the measurements to ensure that there are no internal faults such as broken carbon filaments. Any cables that show increased resistance should be discarded and replaced with new cables.

> Interesting Fact
>
> *One of the best times to check for deteriorated ignition cables is at night. Light occurring from internal cable breakdown or arcing to the engine block or other components is easy to see. Be extremely careful of a running engine in the dark. Give your eyes several minutes to adjust to the new lighting conditions.*

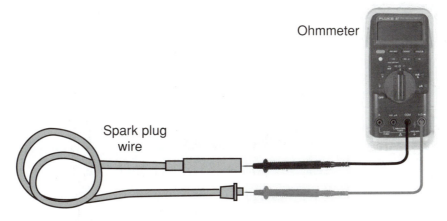

Ohmmeter

Spark plug wire

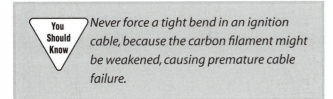

Figure 4.   Checking ignition cables using an ohmmeter.

*Deteriorating ignition cables are more sensitive to humidity and dirt. If the engine runs better after being steam cleaned, try wetting down the ignition cables with a spray bottle of water. The water causes a conductive path that shorts the spark energy to nearby objects, including adjacent ignition wires. If the engine begins to run rough or stalls, the ignition cables have failed. Use care to keep water away from connectors or electronic modules.*

## MEASURING SECONDARY IGNITION COIL RESISTANCE

Measuring coil resistance of the secondary winding is easily accomplished using an ohmmeter. Place one ohmmeter lead on the coil secondary tower and the other on either primary terminal and note the reading **(Figure 5)**.

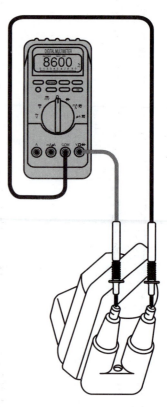

**Figure 6.** Checking a waste-spark coil using an ohmmeter.

Because the resistance of the primary winding is so low compared with the secondary winding, it does not matter which primary terminal you connect to. A value between 8,000 and 20,000 ohms is typical. Verify the specification for the particular ignition coil in the service manual.

In the case of EI systems using waste-spark ignition, the primary winding is completely isolated from the secondary winding. In this case, checking the secondary winding requires connecting the ohmmeter across both secondary coil towers **(Figure 6)**.

> **You Should Know** *Verify the engine's firing order if you suspect that anyone has tampered with the ignition system. Many technicians have been fooled trying to diagnose a misfire condition, yet the problem was traced to an incorrect firing order.*

## MANUALLY FIRING THE IGNITION COIL

To manually verify the coil is operating using a spark tester, first disconnect the ignition coil primary from the

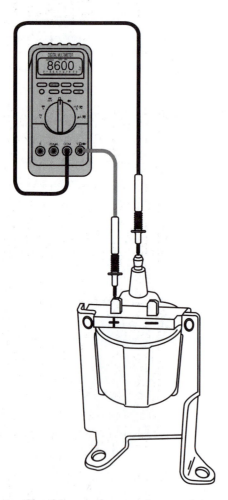

**Figure 5.** Checking coil secondary resistance using an ohmmeter.

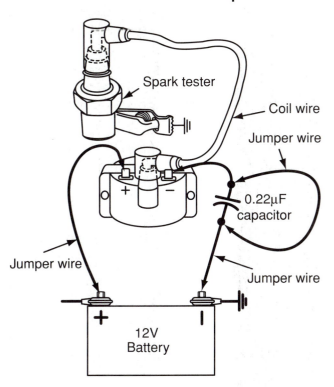

**Figure 7.** Using jumper wires and capacitor to fire the ignition coil into the spark tester.

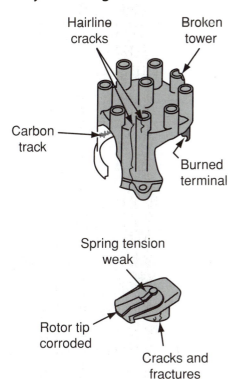

**Figure 8.** Periodically inspect the distributor cap and rotor for cracks, wear, corrosion, and carbon tracking.

harness. Obtain a jumper wire and connect the coil positive (+) terminal to the positive battery terminal. Connect a second jumper to the coil negative (−) terminal through a 0.22-μF (microfarad) capacitor **(Figure 7)**. Connect the other end of this jumper wire to a good ground. With a short jumper wire, short across the capacitor leads briefly. Upon disconnecting this jumper wire, the coil should fire.

## DISTRIBUTOR CAP AND ROTOR INSPECTION AND SERVICE

On vehicles equipped with an ignition distributor, the cap and rotor must be periodically inspected for wear, cracks, carbon tracking, and terminal corrosion **(Figure 8)**. If either component shows signs of deterioration, both the cap and rotor should be replaced. In some cases, the contact terminals can be cleaned, but it is usually better to replace than to reuse secondary ignition components.

The distributor cap and rotor should be routinely replaced according to the manufacturer's service intervals. As a general rule, the cap and rotor should be replaced at 3-year/36,000-mi. intervals to maintain optimum ignition capacity. This is particularly important in cold, damp conditions, where maximum spark energy is needed to start the engine.

## SPARK PLUG INSPECTION AND REPLACEMENT

Remove and inspect the spark plugs at regular intervals or when an engine performance problem suggests a misfire is occurring. Long-life spark plugs plated with platinum, silver, and iridium can provide a service life greater than 100,000 miles, but should never be left in the engine that long. Doing so can seriously impair your ability to remove the spark plug without resorting to machining, which is something to be avoided.

> **You Should Know** *The distributor cap and rotor should always be inspected whenever the spark plugs are removed for inspection or replacement. Do not forget to look for a retaining screw! Some rotors are held in place with a screw that must first be loosened before the rotor can be removed.*

Inspect the plugs for excessive electrode gap and other signs of plug failure or engine damage. **Figure 9** indicates a variety of engine and spark plug concerns that can impair ignition function. In some cases, severe internal engine

①*Normal condition*

Insulator nose grayish white or grayish yellow to brown. Engine is in order. Heat range of plug correct. Mixture setting and ignition timing are correct, no misfiring, cold-starting device functioning. No deposits from fuel additives containing lead or from alloying constituents in the engine oil. No overheating.

②*Sooted—carbon-fouled*

Insulator nose, electrodes, and spark-plug shell covered with velvet-like, dull black soot deposits.
*Cause:* Incorrect mixture setting (carburetor, fuel injection): mixture too rich, air filter very dirty, automatic choke not in order or manual choke pulled too long, mainly short-distance driving, spark plug too cold, heat-range code number too low.
*Effects:* Misfiring, difficult cold-starting.
*Remedy:* Adjust A/F mixture or choke device, check air filter.

③*Oil-fouled*

Insulator nose, electrodes, and spark-plug shell covered with shiny soot or carbon residues.
*Cause:* Too much oil in combustion chamber. Oil level too high, badly worn piston rings, cylinders, and valve guides. In two-stroke engines, too much oil in mixture.
*Effects:* Misfiring, difficult starting.
*Remedy:* Overhaul engine, adjust oil/fuel ratio (two-stroke engines), fit new spark plugs.

④*Lead fouling*

Insulator nose covered in places with brown/yellow glazing, which can have a greenish color.
*Cause:* Lead additives in fuel. Glazing results from high engine loading after extended part-load operation.
*Effects:* At high loads, the glazing becomes conductive and causes misfiring.
*Remedy:* Fit new spark plugs because cleaning the old ones is pointless.

⑤*Pronounced lead fouling*

Insulator nose covered in places with thick brown/yellow glazing, which can have a greenish color.
*Cause:* Lead additives in fuel. Glazing results from high engine loading after extended part-load operation.
*Effects:* At high loads, the glazing becomes conductive and causes misfiring.
*Remedy:* Fit new spark plugs because cleaning the old ones is pointless.

⑥*Formation of ash*

Heavy ash deposits on the insulator nose resulting from oil and fuel additives, in the scavenging area, and on the ground electrode. The structure of the ash is loose to cinder-like.
*Cause:* Alloying constituents, particularly from engine oil, can deposit this ash in the combustion chamber and on the spark-plug face.
*Effects:* Can lead to autoignition with loss of power and possible engine damage.
*Remedy:* Repair the engine. Fit new spark plugs. Possibly change engine-oil type.

**Figure 9.** Reading spark plugs to determine engine problems. (*continued on following page*)

⑦ *Center electrode covered with melted deposits*

Melted deposits on center electrode. Insulator tip blistered, spongy, and soft.

*Cause:* Overheating caused by autoignition. For instance, due to ignition being too far advanced, combustion deposits in the combustion chamber, defective valves, defective ignition distributor, poor-quality fuel. Possibly spark-plug heat-range value too low.

*Effects:* Misfiring, loss of power (engine damage).

*Remedy:* Check the engine, ignition, and mixture-formation system. Fit new spark plugs with correct heat-range code number.

⑧ *Partially melted center electrode*

Center electrode has melted and ground electrode is severely damaged.

*Cause:* Overheating caused by autoignition. For instance, due to ignition being too far advanced, combustion deposits in the combustion chamber, defective valves, defective ignition distributor, poor-quality fuel.

*Effects:* Misfiring, loss of power (engine damage). Insulator-nose fracture possible due to overheated center electrode.

*Remedy:* Check the engine, ignition, and mixture-formation system. Fit new spark plugs.

⑨ *Partially melted electrode*

Cauliflower-like appearance of the electrodes. Possible deposit of materials not originating from the spark plug.

*Cause:* Overheating caused by autoignition. For instance, due to ignition being too far advanced, combustion deposits in the combustion chamber, defective valves, defective ignition distributor, poor-quality fuel.

⑨ *Partially melted electrode (continued)*

*Effects:* Power loss becomes noticeable before total failure occurs (engine damage).

*Remedy:* Check engine and mixture-formation system. Fit new spark plugs.

⑩ *Heavy wear on center electrode*

*Cause:* Spark-plug exchange interval has been exceeded.

*Effects:* Misfiring, particularly during acceleration (ignition voltage no longer sufficient for the large electrode gap). Poor starting.

*Remedy:* Fit new spark plugs.

⑪ *Heavy wear on ground electrode*

*Cause:* Aggressive fuel and oil additives. Unfavorable flow conditions in combustion chamber, possibly as a result of combustion deposits. Engine knock. Overheating has not taken place.

*Effects:* Misfiring, particularly during acceleration (ignition voltage no longer sufficient for the large electrode gap). Poor starting.

*Remedy:* Fit new spark plugs.

⑫ *Insulator-nose fracture*

*Cause:* Mechanical damage (spark plug has been dropped, or bad handling has put pressure on the center electrode). In exceptional cases, deposits between the insulator nose and the center electrode, as well as center-electrode corrosion, can cause the insulator nose to fracture (this applies particularly for excessively long periods of use).

*Effects:* Misfiring, spark arcs over at a point that is inaccessible for the fresh charge of A/F mixture.

*Remedy:* Fit new spark plugs.

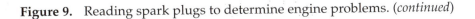

**Figure 9.** Reading spark plugs to determine engine problems. (*continued*)

damage or wear might have occurred, causing the spark plug to react to the damage. In these instances, a major engine overhaul is indicated.

> **You Should Know**  *Always make sure the cylinder head is cool before attempting to remove any spark plugs. Failure to do so can lock the spark plug within the threads, requiring that you remove the head for machining. This caution is especially important when servicing aluminum cylinder heads.*

Use compressed air to blow dirt from around the spark plug before removing the plug. Debris can enter the spark plug hole, causing thread damage or damage within the cylinder.

## To Regap or Replace?

Because the labor to remove and inspect the spark plugs is the same as that for installing new spark plugs, there is little reason to reuse the old plugs. The cost of installing new plugs is usually small compared with the labor required to accomplish the task.

Cleaning and regapping spark plugs is usually not cost-effective because the old plugs will never last as long as new ones. Never attempt to regap used long-life spark plugs. The plating material can be broken away from the side electrode, resulting in rapid electrode wear.

## TESTING THE IGNITION SYSTEM USING AN OSCILLOSCOPE

Using an ignition oscilloscope to check secondary ignition is an effective method to rule out a variety of ignition concerns. Special inductive probes must be used to prevent scope damage. For this reason, most automotive ignition scopes have secondary high-voltage leads that are designed specifically to be used with the instrument. **Figure 10** shows a typical ignition scope hookup for a DI system.

### Analyzing the Secondary Waveform

The secondary ignition waveform is a complex signal that can be broken down into several parts. The signal consists of portions that go above and below the ground reference trace.

The vertical deflection of the scope represents voltage. The horizontal axis represents time, usually measured in milliseconds. Because the signal changes over time, the oscilloscope is the ideal instrument to display ignition events. A typical secondary ignition pattern is shown in **Figure 11**.

### Firing Line

The firing line represents the voltage across the plug electrodes at the instant the plug fires. The firing line varies depending upon many factors, including compression ratio, electrode gap spacing, and air/fuel mixture. Excessive resistance in the spark plug ignition cable can cause the firing line to increase.

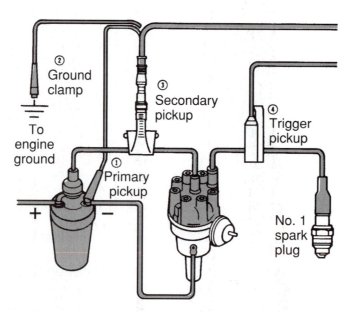

**Figure 10.** Typical ignition test lead hookup using inductive secondary ignition probes.

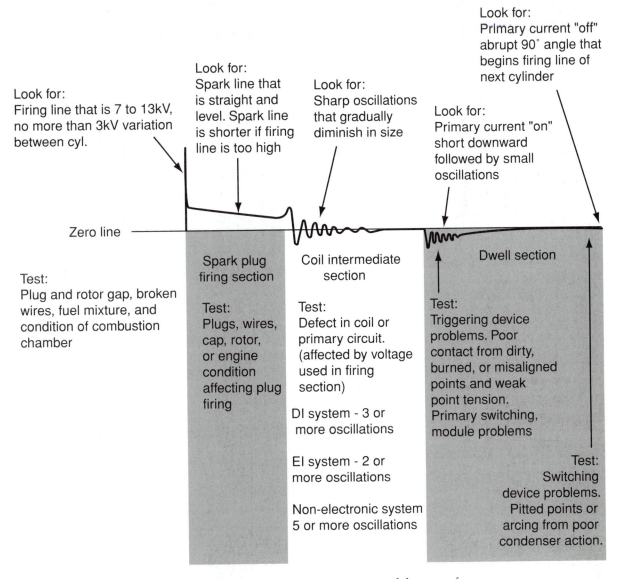

**Figure 11.** Secondary ignition pattern showing various components of the waveform.

The variation between cylinders should be less than 3kV. If a larger variation is noted, check the spark plug gap and ignition cable resistance first. A low-flowing fuel injector can cause an increase in the firing line. This is because a lean air/fuel charge causes greater resistance in the region near the electrodes.

In some cases, abnormally high firing lines are noted on all cylinders. This can be caused by excessive spark plug wear on all cylinders, a worn distributor cap and rotor, or a coil wire that has excessive resistance. Eliminate each possible cause in a systematic fashion until the firing line returns to normal values.

## The Spark Line

The spark line is a short line segment that begins at the firing line and continues for about 1 msec. The amplitude

of this section of the waveform represents the amount of voltage required to sustain the arc across the spark plug electrodes. Typically, this voltage is about 1.5–2.5kV, depending upon the system. The duration of the spark should be between 0.8 and 2.0 msec. The value is somewhat shortened for gaseous fuel vehicles running on natural gas, for example.

## Coil Oscillations

Even after the spark has quenched, or stopped, the ignition coil still contains some stored energy. The energy dissipates in the ignition coil windings, including the entire secondary high-voltage circuit. Because of the inductive behavior of the ignition coil, the decaying signal occurs in the form of oscillations. These oscillations are called "ringing" because the signal tends to echo back and forth like sound waves in a large auditorium. Eventually, the energy is depleted and the ringing stops.

Depending upon the ignition system, two to five coil oscillations are normal. Fewer oscillations indicate shorted turns in the ignition coil. In this case, the coil should be replaced.

## Coil Charging Point

After the oscillations decay, the coil is depleted of any stored energy in the magnetic field, as indicated by the signal resting at the 0V level for a brief interval. When the coil primary is energized by the switching device (usually an ICM), the coil does not charge instantaneously. This is because the inductance of the ignition coil opposes any change in current. Coil charging takes place slowly. Some ringing of the waveform occurs because of the inductive effects of the primary winding. These soon die out as the coil charges to capacity. Coil charging takes approximately 3 msec, depending upon the specific system.

## Dwell Time

Dwell time is the time between when coil charging begins and ends. When the switching device within the ICM opens, the primary current stops and the magnetic field collapses. This results in another spark event.

## WASTE-SPARK IGNITION SYSTEMS

When testing distributorless EI systems, a special oscilloscope adapter is required (**Figure 12**). The adapter allows all cylinders to be displayed according to the firing order. Because half of the spark plugs fire in the forward direction and the remaining half fire in the reverse direction, connecting the cables with the proper polarity is important. The signal pattern should resemble that of **Figure 13**.

## COIL-ON-PLUG IGNITION SYSTEMS

Unfortunately, there is no access to the secondary high-voltage signal in COP ignition systems. Because of this, only primary oscilloscope patterns can be used to diagnose ignition system concerns. It is interesting to note that adequate diagnostic information is available using only primary circuit waveforms. This is because signals occurring on the secondary high-voltage ignition system are reflected back to the primary low-voltage side, where they are easily analyzed.

## COIL-NEAR-PLUG IGNITION SYSTEMS

Because a secondary high-voltage cable is used with CNP ignition systems, conventional secondary waveform

**Figure 12.**    Special distributorless ignition parade adaptor for waste-spark ignition systems.

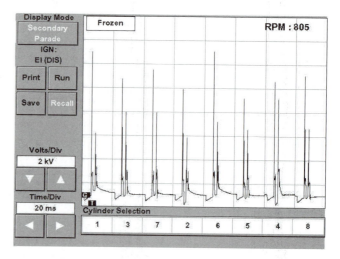

**Figure 13.**    Secondary ignition waveform from a distributorless waste-spark ignition system.

analysis is possible. Unfortunately, each cylinder must be individually tested because no ignition distributor is used.

## INTERPRETING THE RESULTS

If the secondary ignition waveform deviates from the ideal signal, problems ranging from ignition system faults to mechanical problems are to blame. With adequate experience, a technician can even pinpoint sticking intake valves. Interpreting waveform variations in the secondary ignition system is a powerful tool in diagnosing ignition system concerns.

## ON-BOARD DIAGNOSTICS II MISFIRE MONITORS

Vehicles equipped with on-board diagnostics II (OBDII) emissions control system can alert the driver to potentially damaging ignition misfire. The "check engine" warning lamp (MIL) will illuminate if ignition misfire is detected. In the case of a catalyst-damaging misfire, the MIL will flash to warn the driver of a serious failure in the engine control system.

## Crankshaft Acceleration Misfire Detection

The OBDII system monitors ignition misfire by checking for crankshaft acceleration. Each engine power stroke contributes a certain amount of crankshaft acceleration. The PCM monitors the CKP or CMP sensors, which supply crankshaft acceleration information. Using sophisticated sampling and processing circuitry within the PCM, the OBDII

system is able to determine if a cylinder misfire event occurred.

## Ion-Sensing Misfire Detection

Improved misfire detection can be achieved by using an ion-sensing circuit instead of crankshaft acceleration information. Ion sensing can be used at any crankshaft speed, making it more suitable for detecting engine misfires at high engine speeds and heavier engine loads.

Ion-sensing misfire detection uses a DC bias voltage applied to the spark plug electrodes. The bias voltage is developed from the energy delivered to the spark plug to cause an ignition event. During the combustion process, free ions are released. Using signal processing, the engine's combustion characteristics can be mapped out to detect cylinder misfire.

No ions will be released if combustion does not occur, resulting in zero ion current. The PCM is calibrated to recognize this unique signal, triggering a misfire code in the OBDII system. Because monitoring cannot take place while the spark is occurring, the DC bias voltage is not applied until the spark event ends.

## Summary

- Testing the secondary ignition system often requires special instruments such as an ignition oscilloscope.
- Faulty secondary ignition cables can cause hard engine starting and possibly even a no-start condition.
- Secondary ignition cable resistance can be easily measured using an ohmmeter.
- Measuring coil resistance of the secondary winding is easily accomplished using an ohmmeter.
- The ignition coil can be manually activated using jumper leads and a spark tester.
- On vehicles equipped with an ignition distributor, the cap and rotor must be periodically inspected for wear, cracks, carbon tracking, and terminal corrosion.

- Remove and inspect the spark plugs at regular intervals or when an engine performance problem suggests a misfire is occurring.
- Using an ignition oscilloscope to check secondary ignition is an effective method to rule out a variety of ignition concerns.
- The secondary ignition waveform is a complex signal that can be broken down into several components, including firing line, spark line, and dwell time.
- When testing distributorless EI systems, a special oscilloscope adapter is required.
- The OBDII system monitors cylinder misfire by measuring crankshaft acceleration or ion current produced by the free ions created during the combustion event.

## Review Questions

1. Technician A says that an ignition oscilloscope can be used to check ignition secondary high-voltage signals. Technician B says that an ignition scope can be used only to make primary low-voltage measurements of the ignition system. Who is correct?
   A. Technician A
   B. Technician B
   C. Both A and B
   D. Neither A nor B

2. Technician A says that any test probe can be used to verify correct ignition secondary operation. Technician B says that only specialized inductive probes should be used when taking high-voltage secondary measurement. Who is correct?
   A. Technician A
   B. Technician B
   C. Both A and B
   D. Neither A nor B

3. Technician A says that a voltmeter can be used to check secondary high-voltage ignition cables. Technician B says that only an ohmmeter can be used to check the resistance of secondary high-voltage ignition cables. Who is correct?
   A. Technician A
   B. Technician B
   C. Both A and B
   D. Neither A nor B

4. Technician A says that the ignition coil can be manually fired to test its operation. Technician B says that if there is no spark, the ignition coil should be replaced. Who is correct?
   A. Technician A
   B. Technician B
   C. Both A and B
   D. Neither A nor B

5. Technician A says that ignition coil secondary resistance checks are meaningless. Technician B says that ignition coil secondary resistance can be checked using an ohmmeter. Who is correct?
   A. Technician A
   B. Technician B
   C. Both A and B
   D. Neither A nor B

6. Technician A says that long-life spark plugs can remain in an engine for at least 100,000 mi. Technician B says that reusing old spark plugs is not cost-effective. Who is correct?
   A. Technician A
   B. Technician B
   C. Both A and B
   D. Neither A nor B

7. Technician A says that used platinum spark plugs should always be regapped. Technician B says that used spark plugs should never be regapped. Who is correct?
   A. Technician A
   B. Technician B
   C. Both A and B
   D. Neither A nor B

8. Technician A says that an ignition oscilloscope is useful in locating spark plugs with excessively wide electrode gaps. Technician B says that wide electrode gaps reduce the firing line. Who is correct?
   A. Technician A
   B. Technician B
   C. Both A and B
   D. Neither A nor B

9. Technician A says that dwell time refers to the time that the ignition coil is charging. Technician B says that "ringing" in the secondary circuit occurs because of the collapsing magnetic field. Who is correct?
   A. Technician A
   B. Technician B
   C. Both A and B
   D. Neither A nor B

10. Technician A says that COP ignition systems can be diagnosed using secondary ignition waveforms. Technician B says that secondary waveforms cannot be used to diagnose ignition-related concerns. Who is correct?
   A. Technician A
   B. Technician B
   C. Both A and B
   D. Neither A nor B

# Section 5

## Fuel and Induction

## SECTION OBJECTIVES

After you have read, studied, and practiced the contents of this section, you should be able to:

- Describe what makes gasoline a desirable motor fuel.
- Explain what causes preignition and detonation. Understand the detrimental effects of detonation.
- Describe the construction and operation of primary fuel delivery systems.
- Explain how air is metered into the engine.
- Explain how exhaust gases are expelled from the engine.
- Explain how power can be increased using external means of pressurization.
- Explain the basic concepts of fuel injection systems.
- Describe the construction and operation characteristics and operating modes of basic throttle body fuel injection.
- Describe the construction and operation of various types of port fuel injection systems and how they differ from TBI and from one another.
- Perform basic fuel injection tests.
- Analyze test results and form a diagnostic conclusion.
- Perform basic induction system testing and service.
- Diagnose exhaust system restrictions and noises.

**Interesting Fact** One of the first mass-produced fuel injection systems was introduced in 1957 by General Motors. It was a complete mechanical system that was used in both Chevrolets and Pontiacs. It never gained wide acceptance and was discontinued in 1959, except in Corvettes.

# Introduction
# to Fuel
# Systems

## Introduction

The modern vehicle, as it has for decades, relies on the use of an internal combustion engine as its sole power source. As we have learned, the internal combustion engine relies on a compressed fuel and air mixture ignited at the proper time to produce combustion. This may seem simple but, as we will see, is slightly more complex and depends on the interaction of a number of components to make this happen.

### GASOLINE

The most popular fuel that is used in the automobile today is gasoline. Gasoline is a hydrocarbon-based material that is a by-product of **crude oil**. In its natural state in which it is removed from beneath the surface of the earth, crude oil is a complex mixture of many different hydrocarbon compounds. Products derived from the refining process of crude oil are gasoline, diesel, kerosene, lubricating oils, plastics, and many other products that you encounter on a daily basis.

> **You Should Know** *Gasoline is a highly combustible chemical. You should avoid creating any sparks or heat when working around gasoline. Gasoline should be stored only in well-ventilated areas and in approved fuel containers. Under no circumstance should gasoline ever be stored in a glass container.*

Gasoline is a compound that is comprised of hydrogen and carbon molecules, thus the name hydrocarbon. Gasoline is an attractive motor fuel because in its purest refined state it possesses excellent vaporization qualities. However, to perform well as a motor fuel, a chemical must have specific qualities, and gasoline is easily modified to meet those quality needs. The basic qualities that are required of gasoline are:

- Volatility
- Antiknock quality
- Deposit control

### VOLATILITY

Volatility is a fuel's ability to vaporize and is a critical element of gasoline composition. Volatility is a critical element for a motor fuel because for a fuel to burn completely and efficiently within the engine, the fuel must be fully vaporized. Volatility is easily modified and is often changed to meet various requirements.

Several environmental elements affect the rate at which fuel vaporizes. Two of those elements are temperature and altitude. Fuel blends are optimized for maximum performance in specific operating conditions in which the gasoline is sold. For example, if the temperature of a specific fuel is raised, the fuel will vaporize more easily. The same fuel will vaporize more slowly if the temperature is lowered. Altitude will have the same effects, with lower altitude causing slower vaporization and high altitude causing faster vaporization. Because of these differences, fuel blends vary from region to region and from season to season.

Winter and summer blended fuels are blended in order to readily vaporize in the expected ambient temperatures for the particular season in that region. Because vaporization rates are so tightly calibrated, it is common for a vehicle to experience some temporary driveability concerns caused by fuel volatility issues if temperatures become unseasonable.

Fuels that are intended for cold weather use are blended to vaporize easily in low temperatures. If a vehicle with winter fuel is exposed to temperatures that are greater than what the fuel was blended for, the fuel will have a tendency to partially boil in the fuel lines and turn to a vapor before it reaches the fuel injectors, causing a lean condition. This condition is commonly known as vapor lock and can cause rough idle, stalling, or hesitation.

> **You Should Know** *Extreme caution should be exercised when using portable lighting when working around gasoline. Drop light bulbs are prone to bursting when accidentally dropped. If this occurs around a source of gasoline vapors, a fire or explosion will occur.*

Summer fuels are blended to be more difficult to vaporize. This prevents vapor lock conditions during the hot summer months. Because summer fuel is more difficult to vaporize if it were used in temperature conditions lower than intended, it will be slow to vaporize, and complete combustion will not occur. This condition will generally manifest itself as longer-than-normal cranking times before starting, called **extended crank**. This can also be characterized by a **cold start stall**. Cold start stall is a condition in which the engine might start properly but then stalls almost immediately. A rough idle will often follow both of these conditions after the engine has started. When these specific concerns and conditions are present, a change in fuel should be recommended before extensive diagnostics take place.

## DETONATION AND PREIGNITION

For combustion to occur, the air/fuel mixture must be compressed. Generally, the more that we can compress this mixture, the more efficient and powerful the combustion process will be. The most common way to increase the pressure within the cylinder is by increasing the engine's mechanical compression ratio. The limiting factor in this case is the fuel itself. As we studied earlier, the higher the pressure upon a gas or liquid becomes, the higher the temperature will be. In the case of gasoline, the pressure caused by compression combined with the natural heat that builds in the engine can be enough to ignite the mixture without the aid of a spark from the spark plugs. This is called **preignition**.

Combustion within the cylinder does not occur instantly. Ideally, what happens inside of the engine is that combustion will start with a spark from the spark plug and create a **flame front**. The flame front is the leading edge of the burning air/fuel mixture. Once ignited, the flame front expands smoothly across the cylinder, consuming all of the air in fuel in the cylinder in the process.

In a typical engine using a wedge-shaped combustion chamber, a small part of the air/fuel mixture is compressed more on the side opposite the spark plug. What can occur is that before the flame front has had time to consume the entire mixture, the heat and pressure in that portion of the cylinder causes combustion and the formation of a second flame front that travels in the opposite direction. The two flame fronts collide and create shock waves that are heard as a pinging or knocking inside the engine. This is called **detonation (Figure 1)**. When detonation occurs, the pressure and heat within the cylinder rise sharply and can cause severe damage to the pistons and cylinder head gaskets. Short intermittent bursts of detonation are normal when the engine is placed under a heavy load condition, such as climbing a long steady grade or during rapid acceleration. These bursts are not dangerous unless they are allowed to continue for a prolonged amount of time. Several possible causes for preignition and detonation are:

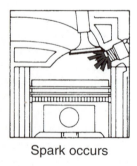
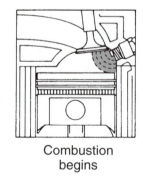

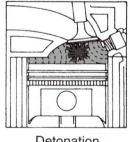
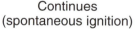

| Spark occurs | Combustion begins | Continues (spontaneous ignition) | Detonation (flame fronts collide) |

**Figure 1.**   Stages of combustion leading to detonation.

- *Lean air/fuel mixtures.* Because a lean mixture burns at a slower rate but at a higher temperature, the extra heat within the cylinder often is enough to induce detonation.
- *Excessive carbon deposits.* Carbon can cause detonation in two ways. The most common carbon occurrence happens when the sharp edges of the carbon flakes heat up to the point that they will begin to glow and act as an ember within the cylinder to start ignition prematurely. If carbon buildup is severe, it can raise the mechanical compression ratio in the engine.
- *Engine temperature.* If the engine enters a situation of overheating, the temperature of the engine parts can become great enough to start preignition.
- *Mechanical compression ratio.* The mechanical compression of the engine will cause an excessive amount of heat to be built within the cylinder and cause preignition.

---

**Interesting Fact** *A hemispherical-shaped combustion chamber is less prone to detonation for two reasons. The first reason is that the chamber's shape allows the spark plug to be placed near the center of the cylinder, allowing combustion to start in the middle (**Figure 2**). The second reason is that the shape of the chamber allows for more even compression of the air/fuel mixture.*

---

- *Overadvanced timing.* Overadvanced timing can cause combustion to occur before the piston has reached TDC. Detonation occurs when the piston, still moving

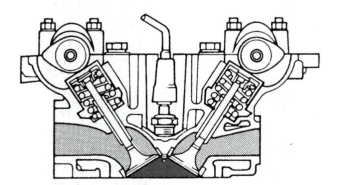

**Figure 2.** Cross-sectional area of a hemispherical combustion chamber. Notice the shape of the chamber and location of the spark plug.

upward, encounters the rapidly expanding combustion process created by combustion.

## OCTANE RATING

To combat the occurrence of preignition, fuel manufacturers blend additives with the fuel to enhance the fuel's antiknock quality. The antiknock quality is represented by the octane number that is found at the gas pump. This number, in simple terms, is the ability of gasoline to resist ignition. The higher the octane, the more difficult it is for the fuel to burn and preignite. The octane number of the gasoline has no effect on economy or efficiency. The general mistaken belief is that the higher the octane number that a fuel has, the more performance that can be expected from the engine. In reality, the opposite is true. Using a fuel with a higher octane number than is actually needed will cause combustion to slow down and can slightly decrease performance. Additionally, using a fuel with a higher octane than is recommended by the manufacturer can cause driveability symptoms in cold weather identical to those found when using a summer blended fuel in cold weather. An engine in good condition will need a gasoline with a specific octane number to keep detonation from occurring. This is typically an octane number between 87 and 92.

---

**You Should Know** *The recommended octane number for a vehicle can be found in the owner's manual.*

---

The octane number of a fuel is measured using a one-cylinder test engine. A quantity of fuel with an unknown octane rating is run in the engine and the severity of the knock is measured. A second test is run with various proportions of two fuels with known knock qualities, heptane and isooctane. Heptane, which has an octane rating of 100, will not knock when burned in an engine. Isooctane has an octane rating of zero and knocks severely when burned in an engine. Various amounts of these two chemicals are combined and run in the test engine to duplicate intensity of the knock produced by the test fuel. The percentage of heptane that was used is the octane number assigned to the test fuel. For example, a mixture of 82 percent heptane and 18 percent isooctane produced a knock of identical intensity to that of the test fuel. The fuel is assigned an 82 octane rating.

There are two methods to measuring octane. Both use the same basic procedure but test parameters vary. These are the research octane number (RON) and the motor octane number (MON). The number that is displayed on the fuel pump is an average of the results achieved by these methods. This is the formula:

$$RON + MON/2$$

An abbreviation of this formula,

$$(R + M)/2,$$

is often found on the pump to indicate that this method was used to determine the octane rating of the fuel. The (R + M)/2 formula is sometimes called the **road octane** number. Adding solvents, such as toluene and xylene, and alcohols, such as ethanol and methanol, can be one way to raise the octane rating in gasoline. Various other chemicals are added to gasoline to not only raise octane, but prevent freezing, and control carbon and deposit formation as well.

*Interesting Fact*

*Before gasoline was discovered to be useful as a motor fuel, it was an unwanted by-product left from the production of lubricants and kerosene.*

## FUEL DELIVERY

To get the full benefit from a charge of gasoline, it must be vaporized and mixed with air. When properly mixed, the fuel vapor is suspended within the air stream as it travels toward the cylinder **(Figure 3)**. If the fuel is not fully vaporized, the fuel will not remain suspended and can puddle in

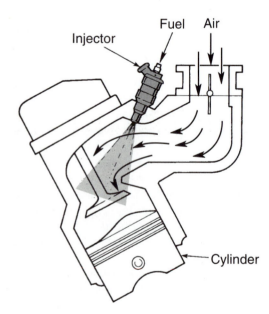

**Figure 3.** The fuel must be mixed with air for efficient combustion to occur.

the intake manifold. This can result in a failure of fuel to reach the cylinder, causing a lean condition in some cylinders.

For the fuel to burn efficiently, the fuel and air must be mixed in the proper amounts, called an air/fuel ratio (A/F ratio). The air/fuel ratio is the numerical comparison of the amount of air to the amount of fuel, both measured by weight. It has been found that a **stoichiometric** ratio of 14.7 parts air mixed with one part fuel (14.7:1) is the most desirable ratio attainable for good engine operation and acceptable tailpipe emissions.

## Delivery Methods

Since the inception of the automobile, there have been two primary methods of fuel delivery, the carburetor and fuel injection. In the last 25 years, fuel delivery system advancements have increased ten-fold over what was experienced the first 100 years. In 1980, the fuel delivery method of choice was the carburetor. By 1990, the carburetor was all but gone from the new car market. By this time, **electronic fuel injection (EFI)** had established its dominance as the fuel system of the future.

## Carburetor

The carburetor is a mechanical device that sat on the top of the engine on an intake manifold. The intake manifold provides a path for air and fuel to be evenly distributed to the cylinders. The carburetor was charged with controlling the amount of air entering the engine and mixing that air with fuel in the proper ratio. Fuel was drawn into the engine using a venturi effect. Basic operation of the carburetor is shown in **Figure 4**.

## Fuel Injection

Electronic fuel injection (EFI) is the most efficient method of fuel control that has ever been mass produced for automotive use. Electronic fuel injection uses an electrically actuated injector to precisely control the amount of fuel that is allowed into the engine. By using information from various sensors about engine operating conditions, the injection system can make very small and accurate adjustments to keep the A/F ratio very close to stoichiometric. A simple fuel injection system is shown in **Figure 5**. There are two primary types of fuel injection systems, **throttle body injection (TBI)** and **port fuel injection (PFI)**. TBI systems resemble carburetors. The fuel injectors are installed in the throttle body, and the air and fuel mix within the throttle body and intake manifold. PFI systems have individual injectors for each cylinder; these injectors are located near the entrance of the cylinder head port.

## PRIMARY FUEL DELIVERY

The method of supplying fuel to the carburetor or injection system is practically the same on all vehicles

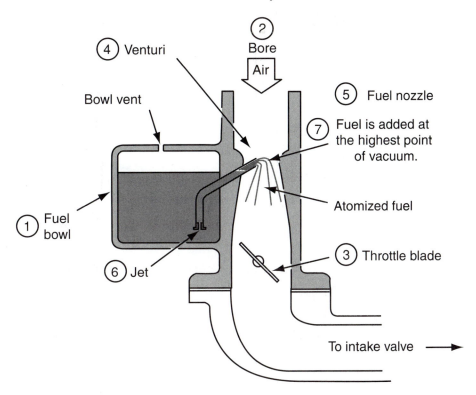

**Figure 4.**  The basic systems of a carburetor. (1) Fuel from the fuel tank is pumped into a small fuel storage area called a fuel bowl. A float system similar to that found in a toilet tank is used to control the fuel level in the bowl. The fuel bowl is vented to atmospheric air. (2) One to four throttle bores are at the center of the carburetor. This is the where the air passes into the engine. (3) At the bottom of each bore is a blade that is connected to the accelerator pedal and controls the amount of air that flows into the engine. (4) Near the center of each throttle bore, the passage narrows. This is called a venturi. At this point, the air flowing into the engine speeds up dramatically. (5) Extending into each venturi is a tube that extends from the bottom of the fuel bowl to the center of the venturi; this is called a fuel nozzle. As the air speeds past the end of the nozzle, a low-pressure area is created. This principle is similar to what happens when you drink through a straw. The air pressure acting on the fuel in the bowl forces the fuel out of the nozzle in a fine mist into the passing air stream and into the engine. (6) The amount of fuel introduced into the air stream is controlled by an orifice called a jet, located at the opening of the fuel nozzle in the fuel bowl. The size of this jet will control the air/fuel ratio. (7) As the volume of air flowing through the throttle bore changes, so does the amount of fuel that passes through the fuel nozzle.

regardless of the type of fuel system used. The primary fuel delivery system consists of the fuel tank, fuel lines, fuel pump, and fuel filters. Each of the components is essential to fuel delivery.

## FUEL TANK

As the name implies, the fuel tank is the storage vessel that is used to store a quantity of fuel that is to be used in the engine. Most fuel tanks have a capacity of between 12 and 40 gallons. The fuel tank may be located in a variety of locations on the vehicle but typically the tank is found near the rear of the vehicle beneath the trunk or near the center of the vehicle beneath the trunk and rear seat. The tanks on

pickups are usually located near the bedsides between the rear axle and the cab.

Fuel tanks have been made from both coated stamped steel and plastic materials. Plastic is the material of choice for most late-model vehicles because plastic tanks are lighter, will not dent, are puncture resistant, and will not rust or corrode.

Fuel tanks might be equipped with internal baffling to provide slosh resistance for the fuel, helping to eliminate the noise caused by sloshing fuel. Baffling will also stabilize fuel level readings while driving. Baffles are also used to keep the fuel pump submerged in a consistent amount of fuel to keep the fuel pump from picking up air bubbles and debris and introducing them into the fuel stream.

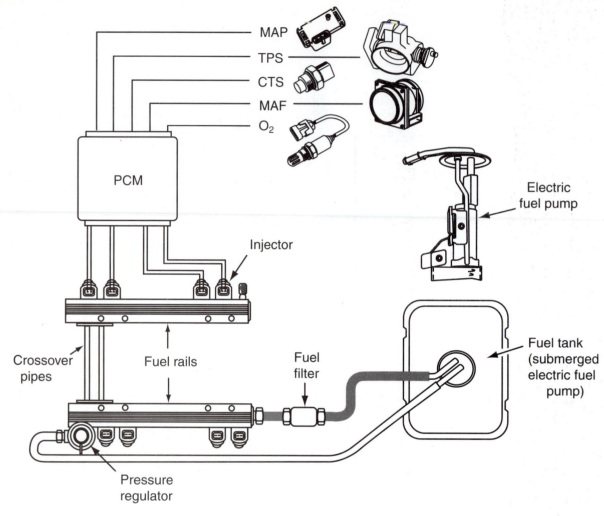

**Figure 5.** A simple fuel injection system.

Most fuel tanks have provisions for a fuel level sensor and fuel pump to be installed within the tank **(Figure 6)**. As the emission laws have become more stringent, fuel tanks have been redesigned to meet the growing demands of fuel vapor capture. Some of the elements relating to the tank are measures that make the fuel tank able to maintain a complete seal against escaping vapors. The fuel filler necks that are used to fill the tank have also been redesigned to prevent the escape of fuel and its vapors while filling the tank. Additionally, the fuel tank is attached to a vapor recovery system. These systems are covered in depth in Section 6.

The filler neck of the fuel tank is topped with a fuel tank cap. The cap uses a seal at the filler neck to ensure a proper airtight seal. Many fuel tank caps are equipped with a pressure relief and a vacuum valve. The pressure relief valve is installed to allow the cap to relieve tank pressure in the event that pressure should become too great. The vacuum valve is used to pull air into the tank to fill the space as fuel

is used or as a temperature decrease causes the fuel to contract **(Figure 7)**.

Fuel caps have become an integral part of the vapor recovery system, and a loose gas cap can even result in the illumination of the MIL. Some states have even mandated gas cap pressure testing as part of an annual safety inspection.

## FUEL SENDER

The fuel sender unit is a component that is usually mounted into the top of the fuel tank. The fuel sender provides a mounting location for the fuel level sensor unit, fuel pump, inlet screen, and the fuel line connections. The fuel sending unit is sealed to the top of the tanks using an O-ring-type seal or a boot that surrounds the tank opening. The sender can be held in place by a large threaded nut or a cam-locking device.

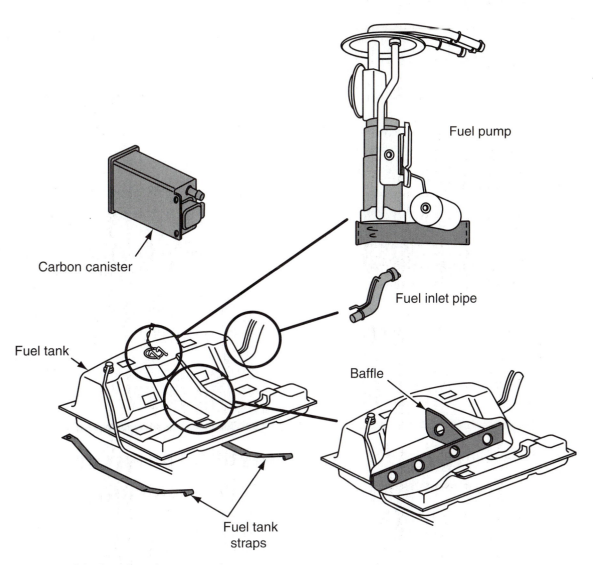

**Figure 6.** A typical fuel tank and associated components.

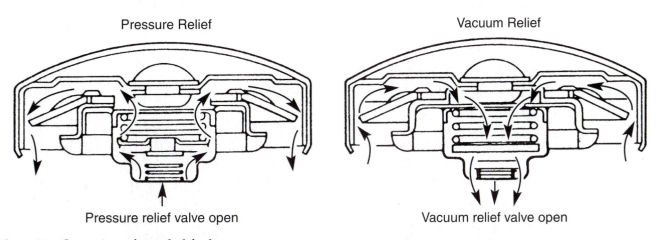

**Figure 7.** Operation of a sealed fuel cap.

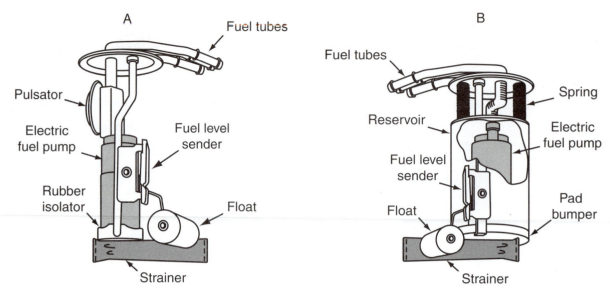

**Figure 8.** (A) A typical fuel pump and sender combination. (B) A typical modular sending unit.

Modular fuel senders have become commonplace in many modern vehicles. These units differ from other sending units based on their design. The modular assembly encloses the fuel pump within a cylindrical container. The construction of the modular sender creates an internal sump to ensure that the pump has a constant supply of fuel. Because the fuel pump is provided with a fuel sump within the sender, many tanks using this type of sender will have very little or no internal tank baffling. Additionally, the fuel pump and sender must be replaced as a unit and may not be serviced separately. A comparison of these senders is shown in **Figure 8**.

## FUEL PUMP

The fuel pump is responsible for transferring fuel from the tank up to the carburetor or to the fuel injection system. In most applications, a carburetor fuel system will use a mechanical fuel pump, and an electronic fuel injection system will use an electric pump.

### Pressure and Volume

Pressure is an amount of force applied to an object or substance. Volume is a measurement of an amount. Fuel pump pressure is the pump's ability to apply force to the fuel. Fuel pump volume is the pump's ability to deliver a specific amount of fuel. Fuel pressure and volume are directly related. It takes a specific amount of pressure to push a specific quantity of fuel. If pressure is low, relative fuel volume will be affected. If the pump is unable to deliver an adequate volume, as soon as its ability to deliver volume is exceeded the pressure will drop.

The carburetor operates at a pressure of 4–8 psi. This system needs only enough volume to keep the fuel bowl full and needs only enough pressure to open the needle valve that controls fuel flow into the bowl and to overcome the weight of the fuel column in the fuel line under acceleration. Any extra pressure would cause the fuel to overcome the fuel control mechanism and cause the bowl to overflow.

 *Interesting Fact* *As a vehicle accelerates, the weight of the fuel within the line is pushed back toward the fuel tank. Although not a major concern on a typical passenger vehicle, it is a factor that must be considered when assembling a high-performance fuel system.*

Fuel injection systems operate in ranges as low as 9–12 psi for a TBI unit or in ranges from 30 to 60 psi for an EFI system. Because the EFI system does not have a fuel storage system, the fuel system must supply enough fuel volume to meet engine demands and keep the system under pressure under the heaviest demands. If the fuel pressure drops, even for a short time, a lean condition or stalling will result.

Because proper fuel atomization depends directly on the amount of pressure that is used to force the fuel through the injector, fuel pressure is critical to the operation of the fuel injection system. If fuel pressure is too low, not enough fuel will be delivered to the engine. If the pressure is too great, an excessive amount of fuel will be delivered to the engine, resulting in a rich condition.

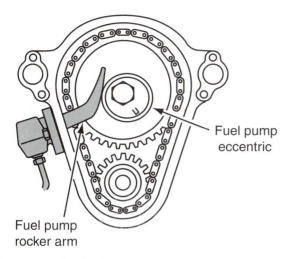

**Figure 9.** The fuel pump eccentric is bolted to the front of the camshaft.

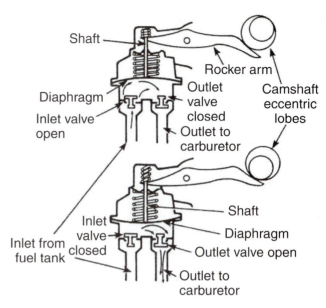

**Figure 10.** Operation of a mechanical fuel pump.

## MECHANICAL FUEL PUMP

The mechanical fuel pump is a diaphragm-type pump. The fuel pump is operated by the expansion and contraction of the diaphragm. Attached to the upper portion of the chamber is the housing that attaches the pump assembly to the engine and supports the pump arm. The pump arm rides on the top of the diaphragm inside the pump while the other end is inserted into the engine. Inside the engine, the pump arm rides against a lobe on the camshaft designed to activate the fuel pump. The arm can also ride against an eccentric that is bolted to the front of the camshaft. The eccentric has an offset mounting hole that, when mounted to the camshaft, has the same effect of a camshaft lobe **(Figure 9)**. Some fuel pumps operate using a push rod that contacts the camshaft on one end and the fuel pump arm on the other. The action of the pump is still the same.

We begin looking at the fuel pump operation with the pump arm in its natural uncompressed state. In this position, the diaphragm is collapsed toward the bottom of the chamber. As the camshaft rotates, the lobe acts upon the arm, and a spring under the diaphragm pushes upward against the arm. As the diaphragm moves upward, a low-pressure area is created within the lower chamber. The lower pressure causes the inlet valve to lift and causes the outlet valve to seal tighter. The fuel from the tank flows through the inlet hose past the inlet valve and into the chamber. As the cam continues to rotate, the arm rides against the base circle of the lobe, and the diaphragm is pushed down toward the bottom of the chamber. As the diaphragm is pressed toward the bottom of the chamber, the fuel trapped inside is pressurized. The resulting pressure forces the intake valve to seal and the outlet valve to open. The fuel flows through the open outlet valve and to the fuel system **(Figure 10)**.

## ELECTRIC FUEL PUMP

The electric pump that is described in this section is typically located within the fuel tank and is submerged in fuel. The major components that are contained within the pump assembly are the pump, motor, check valve, and pressure relief valve. Typical output pressures can range from 9 psi to over 60 psi, depending upon the specific application.

The pump assembly is a cylindrical unit that has an inlet port on the bottom and an outlet port on the top. The pump is driven by a small DC electric motor. A pump assembly is fitted to the end of the motor assembly and is driven by the armature. There are several different styles of pump assemblies; however, each one works essentially the same from a basic diagnostic standpoint.

> **You Should Know** *Fuel pump inlet screens become brittle over time and are easily torn. Because of this, the screen should be replaced with a new one anytime that it is removed.*

When the motor is energized, it spins the pump assembly. The pump creates a low-pressure area at the pump inlet. The fuel that surrounds the pump moves in to fill this low-pressure area. Before the fuel enters the pump, it is passed through a mesh screen that covers the inlet. This screen is in place only to keep out large debris that could damage the pump. (Refer to Figure 8.) Once the fuel has entered the pump, it is pressurized and pushed upward across the pump motor. This acts to cool the motor. The fuel

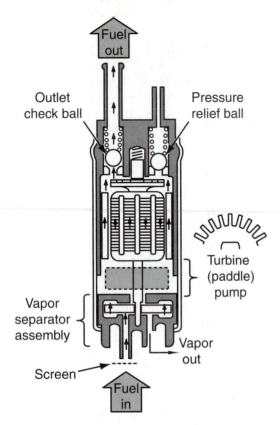

**Figure 11.**   Typical electric fuel pump.

then passes through a check valve and then the outlet of the pump and on to the fuel filter and into the fuel system **(Figure 11)**.

The check valve is an important element to satisfactory operation of the fuel system. The fuel pump is required to pull the fuel from the bottom of the tank and transport it to the top of the engine. The fuel will typically have to rise approximately 26 inches from the pump inlet to the engine and can often cover a distance of 12–18 feet. Because of the rise and length, when the pump stops running the tendency of the fuel is to drain back into the tank. This can lead to the entire fuel line being emptied of fuel and will end up in a large amount of air being present in the line. This will lead to driveability concerns such as an extended crank and stalling on startup. The check valve is placed in the top of the pump to prevent the drain back of fuel into the fuel tank when the pump is not running. This keeps the outlet line full and makes fuel immediately available when the pump is switched on.

Another component that is located in the top of the pump assembly is the pressure relief valve. The pressure relief valve is a spring-loaded valve designed to release fuel pump pressure in the event that an outlet line should become severely restricted. A restricted fuel supply line will cause the fuel pressure to rise significantly within the pump and will ultimately cause pump damage. The most common condition causing a restriction in the outlet line is a restricted fuel filter.

*A severe restriction in the fuel filter can exist for a long period of time before the customer notices any driveability concerns. By this time, it is likely that the fuel pump has been damaged and its service life shortened.*

## FUEL PUMP CIRCUITS

For the electric fuel pump to work, we must be able to supply both power and ground to the electric motor. Every manufacturer will have a specific fuel pump circuit but even these will vary among models **(Figure 12)**.

The fuel pump circuit often has a dedicated fuse. Voltage is supplied to terminal 30 at the fuel pump relay by the fuel pump fuse. When voltage is supplied by the PCM to terminal 86, a magnetic field is created within the relay coil, causing the switch contacts to close. When the switch contacts close, voltage is allowed to flow from terminal 30

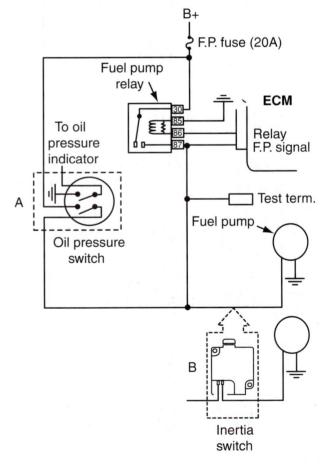

**Figure 12.**   A basic electric fuel pump circuit. (A) A diagram of an oil pressure switch installation. (B) A diagram illustrating the installation of an inertia switch.

through the relay contacts and out terminal 87 to the fuel pump. In many systems, the PCM will supply a voltage to terminal 86 when the ignition is first turned on and will last about 2 seconds. This primes the fuel line to ensure that adequate fuel is available to the fuel system. Once the PCM senses an rpm signal, the PCM will then supply 12V to terminal 86 as long as an rpm signal is present. Notice that the PCM can also monitor voltage by a parallel circuit that is spliced into the battery side of the fuel pump circuit.

Part A of the diagram in Figure 12 illustrates a common variation to the basic diagram. In this application, an oil pressure switch is used as a backup to the relay. The oil pressure switch is wired in parallel to the relay's fuel pump feed circuits. When oil pressure reaches a predetermined amount, the contacts within the fuel pump switch close and allow voltage to flow to the pump. This will act as a backup in the event that the fuel pressure relay or associated circuit is inoperative. Because oil pressure will take a few seconds to build up, a circuit that has a faulty relay might experience an extended crank condition until the oil pressure is sufficient to complete the circuit.

Part B in Figure 12 represents another popular variation found in many vehicles. An inertia fuel shutoff switch is installed in series between the relay and the fuel pump. In the event that the vehicle is in a collision of sufficient force, the inertia switch opens the fuel pump circuit, turning off the fuel pump. Once the inertia switch has been tripped, the switch must be manually reset by pushing a button on top of the switch. These switches may be located either in the trunk of a vehicle or under or behind the instrument panel and should be checked whenever a no-start condition exists on a vehicle that has this switch installed.

 *Interesting Fact*

*Keeping the fuel tank as full as possible will extend the life of an electric fuel pump by allowing the pump to run cooler. In addition, the more fuel that is present in the tank, the less space for air to reside and condense to form moisture within the tank.*

## FUEL FILTERS

As the fuel makes its way forward from the tank through the pump, it is passed through a dense filter. The fuel filter is designed to trap very small dirt particles that were able to pass through the fuel pump inlet screen. These filters are typically constructed of a dense paper element contained in a plastic or metal canister. **Figure 13** shows an example of a typical fuel injection fuel filter. When used in fuel injection systems, these filters will be located in the fuel line between the fuel pump and the fuel system and can be found underneath the vehicle or in the engine

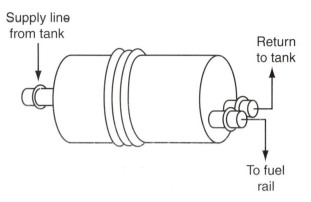

**Figure 13.** A typical fuel filter found in a fuel injection system.

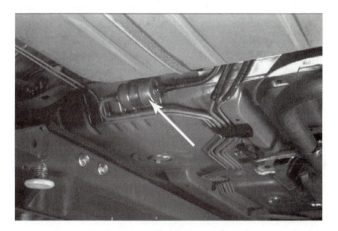

**Figure 14.** A typical fuel filter location.

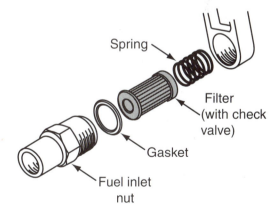

**Figure 15.** The fuel filter for a carburetor is located in the fuel inlet.

compartment **(Figure 14)**. A carburetor system will typically have the filter located in the fuel inlet **(Figure 15)**.

## FUEL LINES

Often overlooked parts of the fuel system are the fuel lines. Although the fuel lines do not give much trouble, it is important to have a basic understanding of fuel line construction. The fuel lines connect all of the major components of the fuel system together. This series of lines allows the fuel that we put into the tank to reach the engine's cylinders.

Fuel lines are primarily made from three materials: steel, rubber, and nylon. Steel fuel lines are constructed of coated double-wall tubing in varying sizes but typically $5/16$ or $3/8$ inch inner diameter. These lines are connected to each other or other components using various fittings, primarily a double flare or O-ring. Snap connect fittings are popular as well, especially when connecting a flexible line to a steel line. Some snap connect fittings allow service without the aid of any tools. Rigid steel fuel lines are usually secured to the frame rails and run the length of the vehicle. In areas where the lines are connected to components that have some movement, a rubber or nylon juncture is often used. If the fuel line material is in good condition, repairs can be made to steel lines by using typical double-flare splicing methods such as a double-flare union. **Figure 16** shows several different methods of connecting lines and hoses.

Rubber lines come in a variety of sizes and are used in those applications in which a solid steel line must connect to a flexible object, such as the engine. Because rubber is prone to deterioration, manufacturers try to limit the amount of rubber hose that is used on a vehicle. There are several different types of rubber hose and the type should be selected according to application. Only those hoses designated for use as fuel lines should be used in a fuel system. Hoses are also built to withstand certain internal pressures. For example, a common $3/8$-inch-wide hose designed to carry 9 psi to a carburetor would burst when used in a high-pressure fuel injection system. Because of this, fuel line selection is critical when replacing rubber hoses. Rubber hoses are usually connected to components using spring and worm drive clamps or using special O-ring- or flare-type fittings attached from the factory. No attempt to repair rubber lines should be made. Worn or damaged hose sections should be discarded and replaced with identical components.

Nylon lines are commonplace on modern vehicles because they are lightweight, somewhat flexible, strong, and corrosion resistant. These types of lines use the quick

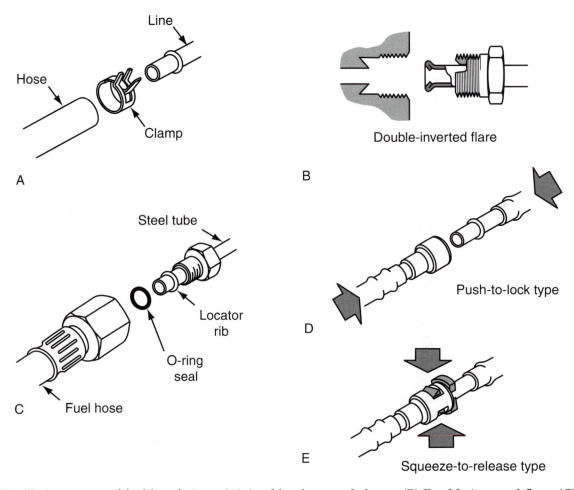

**Figure 16.**   Various types of fuel line fittings. (A) A rubber hose and clamp. (B) Double-inverted flare. (C) O-ring-style fitting. (D) A push-to-lock fuel line fitting. (E) A squeeze-to-release fuel line fitting.

connect fittings to connect to other lines and components. It is commonplace to find this type of line installed from the fuel tank to the engine. Because nylon line is somewhat flexible, rubber-connecting hoses are generally not required when connecting to components that have slight movements. Nylon fuel lines may be repaired using special service fittings if damaged. However, some manufacturers recommend line replacement in the event of damage.

## Summary

- Gasoline is comprised of hydrogen and carbon molecules. Gasoline is derived from the refining of crude oil.
- Volatility is gasoline's ability to vaporize. Temperature and altitude can affect a fuel's ability to vaporize.
- The fuel vaporization threshold is changed as the seasons change. A fuel's volatility can cause driveability concerns.
- Preignition takes place when the air/fuel mixture ignites without the aid of spark from the ignition system. Preignition will ultimately lead to detonation.
- Detonation occurs when two opposing flame fronts collide. Detonation, if left unchecked, can cause serious engine damage. Short bursts of detonation are a normal condition in many vehicles.
- The octane rating is a fuel's ability to resist combustion. Using a higher octane than the engine needs to resist preignition can cause some driveability concerns and can decrease engine performance.
- The fuel system meters both air and fuel. Fuel must be vaporized and mixed with air to properly burn. Most modern vehicles strive to maintain as close to a 14.7:1 A/F ratio as possible.
- The carburetor and fuel injection are the two most common fuel delivery methods that have been used in vehicles.
- The primary fuel delivery system stores fuel, measures the quantity of fuel, and delivers it to the carburetor or fuel injection system at the engine. The fuel tank stores fuel and has provisions for a fuel level sensor and fuel pump mounting.
- The two types of fuel pumps are electric and mechanical. Mechanical fuel pumps operate from a direct connection to the engine's camshaft. The mechanical pump uses a flexible diaphragm and two valves to move fuel.
- The electric fuel pump is controlled through an electrical circuit. Most electric fuel pumps are located inside the fuel tank. A screen is located at the fuel pump inlet to prevent debris from damaging the pump.
- Fuel filters are installed to trap any debris that might have passed through the fuel pump inlet screen.

## Review Questions

1. Technician A says that gasoline has excellent vaporization qualities. Technician B says that gasoline is popular as a motor fuel because it needs no modification to be used as a fuel in an engine. Who is correct?
   - A. Technician A
   - B. Technician B
   - C. Both A and B
   - D. Neither A nor B

2. Technician A says that a customer should be advised to add a commercially available octane booster to the gasoline to enhance the cold start performance of their engine. Technician B says that temporary preignition symptoms might be persistent if unseasonably warm temperatures are present during the winter months. Who is correct?
   - A. Technician A
   - B. Technician B
   - C. Both A and B
   - D. Neither A nor B

3. Technician A says that a thermostat that has stuck in the closed position can cause serious detonation. Technician B says that a restricted fuel filter could cause detonation. Who is correct?
   - A. Technician A
   - B. Technician B
   - C. Both A and B
   - D. Neither A nor B

4. Technician A says that a restricted air filter in a fuel-injected engine will cause the air/fuel ratio to be leaner than stoichiometric. Technician B says that a piece of trash stuck in the carburetor jet will cause a leaner than stoichiometric air/fuel ratio. Who is correct?
   - A. Technician A
   - B. Technician B
   - C. Both A and B
   - D. Neither A nor B

5. Technician A says that a torn fuel cap seal can cause the MIL to illuminate. Technician B says that a fuel tank cap that has a pressure relief valve that is stuck in the open position can cause the MIL to illuminate. Who is correct?
   A. Technician A
   B. Technician B
   C. Both A and B
   D. Neither A nor B

6. An engine with a mechanical fuel pump has no fuel pressure. Technician A says that a worn cam lobe could cause this concern. Technician B says that a ruptured diaphragm could cause this concern. Who is correct?
   A. Technician A
   B. Technician B
   C. Both A and B
   D. Neither A nor B

7. A vehicle with an electric fuel pump has a low fuel pressure concern. Technician A says that a stuck open pressure relief valve could be the cause. Technician B says that a tripped inertia fuel cutoff switch could be the cause. Who is correct?
   A. Technician A
   B. Technician B
   C. Both A and B
   D. Neither A nor B

8. A vehicle with an electric fuel pump has a pinched outlet line leading to the engine. Technician A says that this could damage the fuel pump if left uncorrected. Technician B says that insufficient fuel volume will be present at the fuel injection system. Who is correct?
   A. Technician A
   B. Technician B
   C. Both A and B
   D. Neither A nor B

9. Technician A says that a clogged fuel filter will cause a leaner than stoichiometric fuel ratio. Technician B says that a restricted fuel filter could cause the fuel by-pass valve in an electric fuel pump to open. Who is correct?
   A. Technician A
   B. Technician B
   C. Both A and B
   D. Neither A nor B

10. Technician A says that a torn fuel pump inlet screen should be replaced. Technician B says that a clogged fuel pump inlet screen can cause fuel pressure to be lower than normal. Who is correct?
    A. Technician A
    B. Technician B
    C. Both A and B
    D. Neither A nor B

# Chapter 27

# Air Induction and Exhaust Systems

## Introduction

This chapter explains how airflow into and out of the engine is carried out. Because an engine is primarily an air pump, the way in which the airflow is managed is critical to an engine's performance. The air induction system is responsible for introducing an adequate supply of fresh, clean air into the engine. In this chapter, we look at this system from the air inlet to the cylinder head. The exhaust system is equally critical. It is responsible for removing burned gases from the engine and expelling them into the atmosphere. Here, we examine the exhaust system from the cylinder head to the tailpipe. Additionally, this chapter briefly explores the operation of forced induction methods, used to boost power within the engine. Additional elements dealing with the emissions aspect of the exhaust system is covered in Section 6.

### VOLUMETRIC EFFICIENCY

As engine sizes have decreased, demands for performance have increased. Manufacturers have sought ways to increase the **volumetric efficiency (VE)** of their engines. The VE is a comparison of how much air the cylinders can hold when completely filled compared with how much is being drawn in. The VE is actually a measurement of how well the engine is able to bring in air and expel air, or "breathe." When examining VE, every component, starting at the inlet of the air cleaner housing to the end of the tailpipe, will affect the efficiency of the engine. Understand that volumetric efficiency considers only the amount of air that the engine is moving. However, the more air that we can pack into the engine, the more fuel we can effectively

burn and the more power we can make. Factors that affect volumetric efficiency are engine rpm and the number of restrictions within the engine. A restriction can be as small as a bump in the intake tract or a sharp bend in the exhaust.

**Interesting Fact**

*The faster the engine spins, the more difficult it is to achieve high volumetric efficiency numbers because the cylinders have less time to completely fill up. Other factors that affect VE are how well the air can flow into the engine and how well the exhaust gases can exit. These factors are what performance-engine builders look at when building engines. They make modifications and add parts that will help the engine breathe better.*

### BACKPRESSURE

Backpressure occurs when the exit of exhaust gases from the engine is slowed. Some backpressure within the engine is desirable. Much of an engine's normal backpressure is created within the exhaust manifold. The basic design of the manifold dumps each cylinder runner into a common tube. As the engine runs, this tube is filled with more exhaust gas than it can move. Each time an exhaust valve opens, the pressure in the manifold slows the exit of the gases from the exhaust port. Backpressure is further created by each component in the exhaust system.

When backpressure is modified, either increasing or decreasing the flow, flow characteristics that take place within the cylinder head are altered. As we learned in Chapter 8, on the intake stroke, both the intake and exhaust valves are open, and the exhaust gas that exits the cylinder helps draw in the intake charge. If backpressure in the exhaust is reduced, the exhaust gases are able to move much faster. As the exhaust gases speed up, they end up pulling an excessive amount of the air/fuel mixture out of the cylinder and into the exhaust. This typically decreases the low-speed engine performance but can increase higher speed engine performance. Furthermore, a decrease of engine backpressure can alter the operation of the EGR system. This occurs because the decrease in exhaust pressure directly affects the volume of exhaust gases flowing into the EGR ports. In some cases, this condition can be severe enough to cause illumination of the MIL.

When exhaust backpressure is increased, the opposite effects occur. The exhaust gases are not allowed to leave the exhaust manifold. When the exhaust valve opens, those gases are not allowed to leave the cylinder, thus keeping a full charge of the fresh air/fuel mixture from entering through the intake valve. In cases of severe restriction, the exhaust gases might try to revert into the intake manifold when the intake valve opens. This often causes a backfire in the intake manifold. This condition commonly occurs when an exhaust pipe is pinched or when severe catalytic converter damage has occurred. A clogged exhaust will result in a severe loss of power.

## AIR INDUCTION SYSTEMS

As we begin our study of intake systems, let us look at how the air is brought into the engine. For the most part, air is forced into the engine at atmospheric pressure or slightly above, as provided by a slight ram effect from the location of the fresh air inlet. As the pistons move down in their respective bores with the intake valve open, a low-pressure area is created within that cylinder. The air from the induction system rushes to fill that void created by the piston. Atomized fuel is then mixed with the incoming air between the air inlet and the intake valve.

## Intake Manifold

The intake manifold is the engine component that is responsible for efficiently moving the air/fuel mixture to the cylinder head ports. The intake manifold will also have dedicated passages to recirculate burned exhaust gases in order to reduce emissions. In addition, the intake manifold might also provide crossover passages for coolant flow that allow the transfer of engine coolant from one cylinder head to another on a V-type engine. There are several styles of intake manifolds, and they will greatly differ, depending on the particular type of fuel injection system that is used **(Figure 1)**.

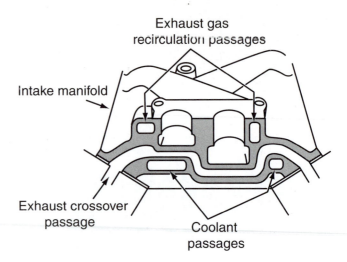

**Figure 1.**   A cutaway of a typical manifold found on a V-type engine using a TBI or carburetor.

## Throttle Body

The throttle body is the component that controls the exact amount of air that is allowed to flow into the engine. It controls the amount of air by using a moveable plate called a throttle plate or throttle blade. The throttle bore of the throttle body is a round hole bored into a plate that is mounted to the manifold inlet. A shaft extends through a perpendicular hole drilled across the throttle bore, and the throttle plate is bolted to the shaft, called the throttle shaft. As the shaft is allowed to rotate, the blade is moved to block the throttle bore to varying degrees. This changes the amount of air that is allowed to pass through the throttle body into the engine **(Figure 2)**. The driver of the vehicle has complete control of the throttle shaft movement using a cable connected between the throttle shaft and accelerator pedal. In drive-by-wire systems, the throttle shaft is connected to the TAC motor.

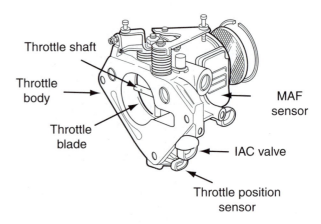

**Figure 2.**   A typical throttle body found on a PFI engine.

Depending on the particular application, the throttle body can have from one to four throttle bores. In both a carburetor and a throttle body fuel injection system, the fuel passes directly through the throttle bore and throttle plates. In these systems, the size of the throttle bore directly affects the velocity of the air passing through and, therefore, also affects the atomization of the fuel. In port fuel systems where fuel is introduced above the intake valve, the throttle body controls only the flow of air. Most vehicles produced today are port fuel injected.

The throttle body is an extremely important component on any type of fuel system, and the appearance and number of functions will vary among different fuel delivery systems. The throttle will, however, have provisions to control base idle speed using passages that allow air to bypass the throttle blades and move into the engine. The amount of air flowing in these passages is controlled using an IAC valve. The throttle body will additionally have provisions for the installation of the TP sensor.

## Air Intake Systems

As air intake systems exist in modern vehicles today, cool fresh air is introduced from somewhere around the front of the vehicle, either near the radiator core support or sometimes behind the fenders. The fresh air is piped to a remote air cleaner assembly. The remote air cleaner assembly can be located just about anywhere under the hood. It typically consists of a lower housing and a cover. The air filter element is placed between the upper and lower components, forming an airtight seal between the two halves; this keeps dirt from passing through the seams. Most air filters are made of a pleated paper material. The ends of the pleats are molded into a flexible material similar to foam rubber to seal out any unfiltered air.

As the air passes from one side of the housing, through the filter, and into the other side of the housing, a tube then connects the air cleaner assembly to the engine's throttle body, where the air is metered into the engine. The air is routed from the fresh air inlet to the air cleaner housing to the throttle body using large rubber ducts. These are molded assemblies with some flexibility to allow for engine movement and easy assembly and disassembly. The IAT sensor is typically located either in the air cleaner housing or in the tube between the housing and throttle body. The flexibility allowed with remote air cleaner housings has allowed the manufacturers to greatly increase the aerodynamic design possibilities associated with the engine compartment of the vehicle (**Figure 3**).

## Mass Air Flow Sensor

In those systems that use a MAF sensor, it will be placed between the air cleaner and the throttle body, as shown in Figure 3. In some cases, the MAF sensor may be attached to the air cleaner housing. Other configurations have the MAF sensor attached to the throttle body; an example of this is shown in Figure 2. Using any of these locations ensures that only filtered air will pass through the sensor, preventing possible damage and contamination

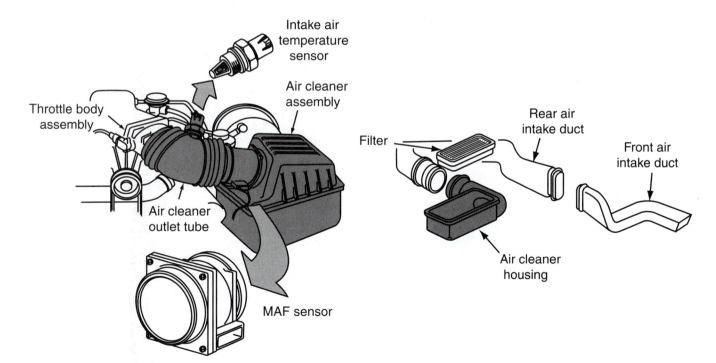

**Figure 3.**   Remote air cleaner assembly.

from debris. Additionally, these locations allow the MAF sensor to measure all of the air that is entering the engine. Because the MAF sensor needs to measure all of the air entering to operate properly, it is imperative that the connecting hoses are in good condition and clamps are tight. If air is allowed to enter the air stream behind the sensor, an incorrect reading will be observed and air/fuel mixtures and ignition timing will be incorrect for the engine conditions.

## EXHAUST SYSTEM

The exhaust system is used to remove all of the burned gases from the engine and expel them into the atmosphere. The exhaust system begins at the exhaust port of the cylinder head. Each time an engine exhaust valve opens, the burned exhaust gases are pushed into the exhaust port as an exhaust pulse. These pulses are essentially what the exhaust system is controlling. An exhaust manifold is bolted to the exhaust ports of the cylinder head. Each cylinder head exhaust port will have one corresponding manifold runner. Each of the individual runners is connected into one common runner and then into the exhaust system.

Typical manifolds use a cast iron design, with each of the manifold runners cast together in one unit. This type of manifold construction is economical to manufacture and extremely durable. Another common design of manifold uses steel tubing welded to a flange that bolts to the cylinder head, with each individual tube welded into a common tube. The exhaust manifolds often provide provisions for $O_2$Ss to be installed. Other features of the exhaust manifolds are provisions for the injection of fresh air used in some emissions control systems (**Figure 4**).

When the gases leave the manifold, they will enter an exhaust down pipe. In some V-type engines, the gases might flow into a crossover pipe sometimes called a "Y" pipe because its shape resembles the letter Y. A crossover pipe is a pipe that directs gases from both cylinder banks into one common exhaust pipe before they enter the rest of the exhaust system. As the gases leave the exhaust pipe or crossover, they might be passed into an intermediate pipe or directly into the catalytic converter. The intermediate pipe is simply a connection between the exhaust down pipe and the catalytic converter.

> **Interesting Fact**
> An exhaust header is similar to a manifold except that the individual tubes never join one another until they enter the exhaust pipe. Headers increase engine performance by reducing backpressure.

Once the gases leave the exhaust pipe, they then flow into the **catalytic converter**. The converter, as it is often referred to, converts the harmful pollutants that are a byproduct of combustion and reduces the pollutants to acceptable levels. Full converter function is covered extensively in Section 6. Catalytic converters were once thought to be highly restrictive and greatly reduced engine performance. However, many modern converter designs are less restrictive than the muffler itself. Removal of the converter is illegal and can reduce the performance of the engine.

Once the gases have been converted, they will exit the converter and flow through the exhaust pipe and into the muffler. The sole purpose of the muffler is to reduce exhaust pulses and thus the amount of noise emitted from the engine.

> **Interesting Fact**
> Some V-type engines use two parallel but completely separate exhaust systems. This is called dual exhaust. Dual exhausts reduce the amount of backpressure in the engine to allow better breathing qualities. Dual exhausts are often used in applications that use high output engines.

There are two predominant designs of mufflers: reverse flow and straight through. Reverse-flow mufflers cause the exhaust gases to change direction within the muffler, canceling out the exhaust pulses. Mufflers are also equipped with a small drain hole located at the rear of the muffler to allow condensation that naturally builds within the muffler to drain out.

Exhaust ports

Runners

Cylinder head

Gasket

Exhaust manifold

**Figure 4.**  Typical exhaust manifold and mounting diagram.

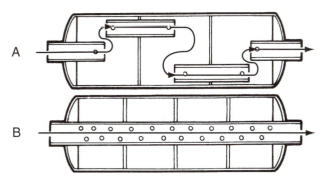

**Figure 5.** (A) Reverse-flow muffler. (B) Straight flow-through design.

A straight-through muffler is made from a perforated pipe surrounded by a sound-deadening material such as fiberglass. The gases are allowed to pass unobstructed through the pipe but as the exhaust pulses move through the pipe, they naturally expand outward through the perforations and are absorbed by the deadening material. Traditionally, straight-through designs are much louder than their reverse-flow counterparts **(Figure 5)**. Muffler locations are usually near the center or toward the rear of the vehicle. The farther the muffler is located from the engine, the cooler the muffler is able to operate, extending the life of the muffler.

After the gases leave the muffler, they might exit into another component called a resonator. A resonator is essentially a straight-through muffler designed to further reduce the noise emitted from the exhaust. Not all vehicles are equipped with resonators. The tailpipe is the last component of the exhaust system and the exit point for

the exhaust gases. The tailpipe may be connected to the resonator or muffler and is of sufficient length to allow all of the exhaust gases to exit from underneath the vehicle.

> **You Should Know** *Whenever it is necessary to operate the engine of a vehicle, it should be run in a well-ventilated area where exhaust gases can escape. When working indoors, an approved exhaust evacuation system should be used to prevent the possibility of carbon monoxide poisoning. In addition to carbon monoxide dangers, exhaust gases of an operating vehicle can often be irritating to the eyes, nose, and throat.*

The exhaust system is supported across the underside of the vehicle using rubber hangers designed to isolate the vibration of the exhaust system from the vehicle's body. Heat protection is provided to the body of the vehicle using heat shields placed between the system and the body. The heat shields are fastened to the underside of the body using rivets or screws **(Figure 6)**.

Any damage to the exhaust system that allows exhaust gases to remain under the vehicle can be very dangerous. Gases that exit underneath the vehicle can become trapped and can find their way into the vehicle and fill the passenger compartment with **carbon monoxide**. Carbon monoxide is a colorless odorless gas that is present in all engine exhaust gases. When inhaled by humans, the gas will displace oxygen within the lungs. If a person is exposed to sufficient amounts of carbon monoxide, death can occur.

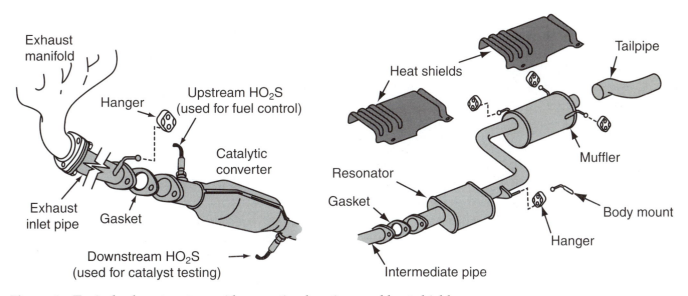

**Figure 6.** Typical exhaust system with mounting locations and heat shields.

# FORCED INDUCTION

A **naturally aspirated** engine depends on the natural flow of the atmospheric pressure of air to fill the void created within the cylinders. Within an operating engine, every component in the induction tract between the air cleaner inlet and the back of the intake valve creates a restriction to airflow. Each restriction is small but when the sum of all restrictions is realized, the restrictions are able to greatly decrease airflow and keep the engine from operating at full capacity. A means of forcing air into the engine can overcome the detrimental effects of these small restrictions and will force air into the cylinders. This is called forced induction.

Forced induction is a method of filling the engine's cylinders using a pump to force air through the induction system. Forced induction creates pressure in the intake system to push more air into the cylinders. This is called **boost pressure**. By creating boost pressure in the engine, we can force enough air into the cylinder so that the engine can operate above 100 percent volumetric efficiency. For example, a single cylinder capacity might be 0.5L. Under ideal conditions, atmospheric pressure will not be able to completely fill the cylinder; 70–80 percent is typical. However, the same engine using forced induction might be able to pack 0.6L of air into the 0.5L space, making the cylinder operate at 120 percent efficiency. Supercharging and turbocharging are the two most widely used methods of forced induction.

Adding additional air is only part of the equation to increasing an engine's output. To realize any benefit from the additional air, the amount of fuel must also be increased. Other factors that must be considered when forced induction is applied are that the extra pressure applied to the engine increases the heat of the air and the dynamic compression within the engine. These two factors can lead to preignition and detonation. To counter these effects, the ignition timing advance and static compression must also be considered when forced induction is applied to the engine.

# SUPERCHARGER

The supercharger is a belt-driven pump that is driven directly off the crankshaft. As the engine spins, so does the supercharger. This makes some boost available at the lowest of engine speeds. However, because the supercharger is belt-driven it accounts for some amount of **parasitic loss**. Parasitic loss is the amount of power required to operate an accessory. In the case of a supercharger, the parasitic loss is outweighed by the increased performance.

The most popular design of a supercharger is the roots type, which uses two rotors contained within a housing. The rotors typically have three lobes that interlock with one another as they spin. Although the rotors do not touch

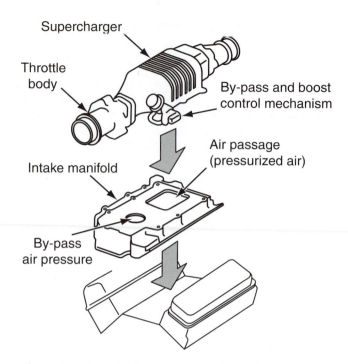

**Figure 7.**   A typical supercharger installation.

one another, they do have very close clearances. The gear drive mechanism located at the pulley end of the supercharger keeps the rotors separated. The design of the drive gears causes the rotors to spin in opposite directions. When in operation, each rotor spins toward the center of the housing, trapping air between the outside of the housing and the rotor. This action compresses the air and forces it down into the engine. The supercharger can be located on top of the engine, essentially becoming part of the engine's structure, or may be mounted as an accessory to the engine. A typical supercharger installation is shown in **Figure 7**.

> **Interesting Fact**
> The rotors within the supercharger are often helical cut. Lobes on helical rotors have a twisted appearance. By twisting the lobes, they are able to engage one another gradually as they spin. This method of engagement greatly reduces the traditional whining noise that is associated with a supercharger.

As the rotors spin, a low-pressure area is created at the top of the lobes. This is the inlet. Air from the induction system rushes in to fill the low-pressure area created by the rotors. As the air is pulled into the supercharger, it is compressed and forced through the outlet. From the outlet, the

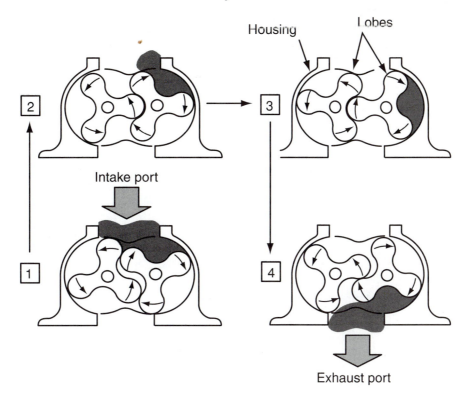

**Figure 8.** Basic supercharger operation.

air is passed into the engine through the intake manifold. Simple supercharger operation is shown in **Figure 8**.

The desired amount of boost is typically 7–16psi more than what is provided by the atmosphere. A by-pass valve is used to control the amount of boost produced by the supercharger during deceleration, high engine load situations, and conditions in which no boost is desired. When opened, the valve diverts some of the compressed air provided by the supercharger back to the inlet. When the valve is closed, all of the compressed air is allowed into the engine.

The by-pass valve is controlled by an actuator connected near the outlet of the supercharger. A diaphragm separates the two sides of the actuator. A spring is located on one side, forcing the diaphragm in a direction to close the valve. A vacuum hose is also connected to the spring side of the actuator. A boost pressure hose is connected to the opposite side of the actuator.

When vacuum in the engine is high, it is strong enough to overcome spring pressure and open the by-pass valve. When vacuum drops below a predetermined level, the spring overcomes the vacuum in the chamber and causes the valve to close, causing a rise in boost pressure in the intake manifold. As the boost pressure begins to rise, it begins to act upon the opposite side of the actuator and push the diaphragm against spring pressure to open the valve. Application of boost pressure supplied to the actuator can be controlled by the PCM using a solenoid. This

allows the PCM to precisely control the amount of boost available in the engine. A typical by-pass valve installation is illustrated in the cutaway view of a supercharger in **Figure 9**.

## TURBOCHARGER

Turbocharging is another method of forced induction. The turbo, as a turbocharger is commonly referred to, works by using the pressure and movement of the exhaust gases to spin a compressor wheel in order to pressurize the intake charge. Because the amount of boost is dependent on the speed of exhaust flow, boost is delayed until the engine produces a significant amount of exhaust flow to speed up the turbine, this is called **turbo lag**. Unlike the supercharger, the turbo adds no parasitic load on the engine.

The turbo uses a compressor wheel and turbine wheel mounted on a common shaft. Each of the wheels is located within its own sealed housing. Exhaust gases are routed through the turbine side of the turbo and then back into the exhaust. As the gases pass through the turbine portion of the housing, they exert a force on the turbine wheel and cause it and the shaft to spin **(Figure 10)**. The faster the engine turns, the faster the exhaust gases move and the faster the turbine wheel spins. Because the turbine wheel and the compressor wheel are located on the same shaft, they spin at the same speed.

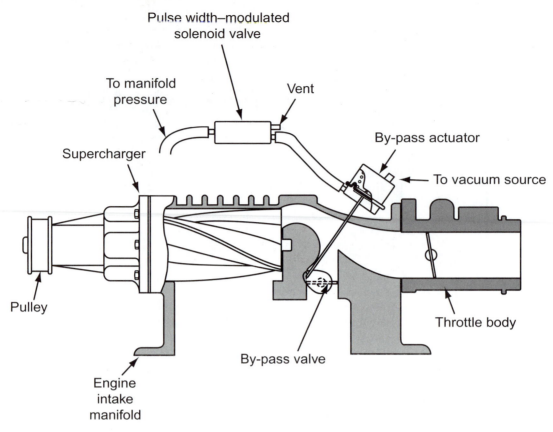

**Figure 9.**   A cutaway view of a supercharger.

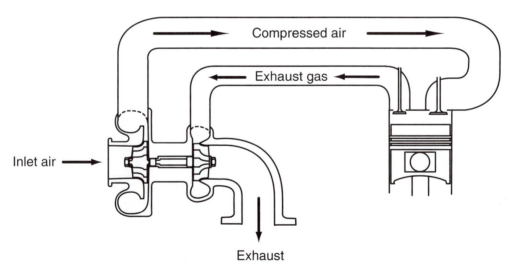

**Figure 10.**   Basic turbocharger operation.

The compressor side of the turbo is connected to the air induction system on the inlet and the throttle body on the other side. As the compressor wheel begins spinning, a low-pressure area is created at the turbo inlet. The air that fills this low-pressure area is compressed and forced through the outlet of the turbo and through the throttle body of the engine. The amount of boost created is directly related to the speed of the compressor wheel **(Figure 11)**.

To control the boost pressure of the turbo, it is equipped with a **wastegate**. A wastegate is similar to a by-pass valve

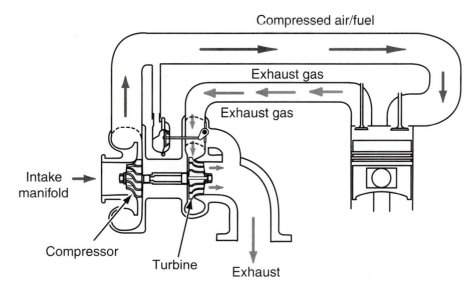

**Figure 11.** Exhaust gas and airflow movement through a turbocharger.

used on the supercharger. The wastegate is a butterfly valve that is activated by a pressure-sensitive actuator. Inside the actuator is a diaphragm that is connected directly to the butterfly valve. A spring within the actuator acts on the diaphragm. In its normal position, the spring forces the diaphragm and the butterfly valve to the closed position. This directs all of the exhaust gas to move through the turbine wheel. A hose is connected to the actuator opposite of the diaphragm. The opposite end of the hose is connected downstream on the pressure side of the turbo; this hose is

exposed to the same pressure as the intake manifold of the engine. As boost pressure approaches a predetermined level, the pressure supplied to the actuator begins to raise and force the diaphragm against the spring pressure; this opens the wastegate valve and allows some of the exhaust pressure to bypass the turbine wheel and causes the turbine wheel to slow down **(Figure 12)**. By slowing the turbine, we reduce the boost in the engine. The wastegate pressure might be controlled by a PCM-activated solenoid in some applications.

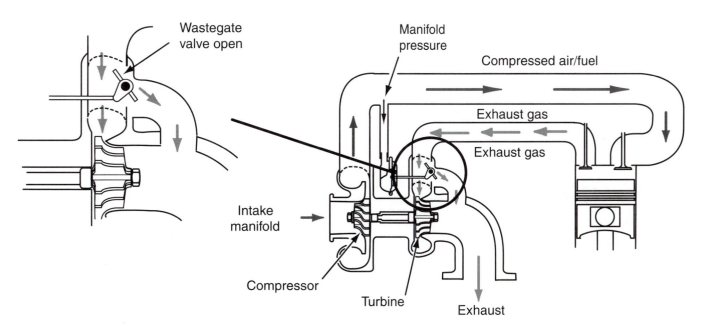

**Figure 12.** Basic wastegate operation.

## Intercoolers

Pressurized air as it leaves a supercharger or a turbo is very hot. An intercooler is installed in some applications to cool down the air after it leaves the turbocharger, before it enters the engine. An intercooler is similar to a radiator and is designed to remove heat from the pressurized air. The pressurized air is forced through the tubes of the cooler the same as water through a radiator. As air passes across the fins, the heat is removed from the air within the cooler. This is called an air-to-air intercooler. Intercoolers can also be a water-to-air design, circulating engine coolant through the intercooler to aid in the removal of heat. Cooling the air makes it denser, effectively increasing the amount of oxygen that enters the cylinders on the intake stroke. This allows the engine to burn more fuel. For every 10°F (5.5°C) the air temperature is reduced, engine power is increased approximately 1 percent. Intercoolers are usually located in front of or beside the radiator and condenser to provide full airflow across the cooler **(Figure 13)**.

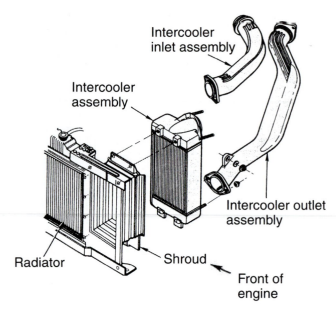

**Figure 13.**   A typical intercooler installation.

# Summary

- Airflow management is critical to the performance of an engine. The intake manifold is responsible for delivering the air/fuel mixture to the cylinders.
- The throttle body allows the driver to control the exact amount of air going into the engine. Throttle body styles will greatly differ among the various types of fuel injection systems.
- The air intake system is designed to deliver clean cool air to the engine. The MAF sensor is placed in a position so that it is allowed to measure all of the air flowing into the engine.
- The exhaust system is used to manage the flow of exhaust gases from the engine. The amount of backpressure within the exhaust system can greatly affect the performance of the engine.

- Forced induction uses a pump to pressurize the intake manifold in an engine to better fill the cylinders. The two most common methods of providing forced induction are the supercharger and the turbocharger.
- Supercharging uses a belt-driven pump to provide boost to the engine. Boost from a supercharger is available at all engine speeds. Supercharger boost is controlled by the use of a by-pass valve.
- A turbocharger is driven by the pressure and flow of exhaust. Turbo boost pressure is delayed until exhaust gas flow is sufficient to spin the turbine wheel. Boost pressure in the turbo is controlled by a wastegate.
- An intercooler can increase the engine's performance by cooling the inlet air. Air-to-air and water-to-air designs are two types of intercooler designs.

# Review Questions

1. Two technicians are discussing induction systems. Technician A says that intake manifolds can differ between different styles of fuel injection. Technician B says that obstruction in the induction system will affect the performance of the engine. Who is correct?
   A. Technician A
   B. Technician B
   C. Both A and B
   D. Neither A nor B

2. Technician A says that the PCM is able to control idle by-pass air using passages in the throttle body. Technician B says that if the passages for the idle by-pass air are clogged, the PCM will not be able make accurate adjustments. Who is correct?
   A. Technician A
   B. Technician B
   C. Both A and B
   D. Neither A nor B

3. Technician A says that a torn air filter seal could allow debris to contaminate the MAF sensor. Technician B says that a torn air duct between the MAF sensor and the throttle body could cause the MAF to read incorrectly. Who is correct?
   A. Technician A
   B. Technician B
   C. Both A and B
   D. Neither A nor B

4. All of the following statements about superchargers are true *except:*
   A. A belt drives the supercharger.
   B. Both pressure and vacuum are used to control the by-pass valve.
   C. A large leak in the by-pass valve diaphragm would cause the supercharger to create an excessive amount of boost pressure.
   D. A large leak in the by-pass valve diaphragm would cause the supercharger to produce little or no boost.

5. All of the following statements about turbochargers are true *except:*
   A. A severe restriction in the exhaust system will cause the turbine wheel to spin at a slower speed.
   B. If a wastegate valve were stuck in the closed position, it would cause the turbine wheel to spin at a slower speed.
   C. If a wastegate valve were stuck in the open position, it would cause the compressor wheel to spin at a slower speed.
   D. A restricted air filter could cause low boost pressures.

# Chapter 28

# Throttle Body Fuel Injection

## Introduction

Throttle body fuel injection (TBI) is a type of electronic fuel injection system. This relatively simple system is the forerunner to the more complex electronic fuel injection systems used today. The TBI systems served as a bridge between the carburetor age and modern electronic fuel injection. TBI systems share similarities with both the carburetor and direct port fuel injection. Like the carburetor, fuel and air are mixed in the throttle bores and the intake manifold. Like electronic fuel injection, the fuel is metered and controlled digitally by a fuel injector. These similarities with the carburetor allowed a relatively easy transition to fuel injection. TBI systems have served reliably for many years and millions of TBI vehicles are still on the road today. A thorough understanding of TBI systems will be a stepping-stone for interpreting operation of other electronic fuel injection systems that are covered in Chapter 29.

### THROTTLE BODY

As we discovered in Chapter 27, the throttle body allows the driver to control the amount of air that is allowed in the engine. In a TBI system, the throttle body does much more. The throttle body in a TBI system contains all of the control components of the fuel injection system in one unit. The throttle body as a single serviceable component contains the throttle bores, the throttle plates, and the control shaft with built-in throttle return spring, passages for fuel, vacuum, and idle air. These passages are shown in **Figure 1**. Additional components that are serviced separately are the injector(s), fuel pressure regulator, IAC valve, and the TPS.

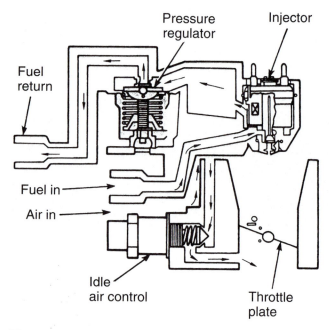

**Figure 1.** Operation of a common TBI unit.

The TBI unit has one throttle bore and throttle control plate for each fuel injector. A TBI unit will typically have one, two, or four fuel injectors, depending on engine size. Four-cylinder engines are equipped with one injector, six-cylinder engines typically have two injectors, and eight-cylinder engines can have either two or four injectors **(Figure 2)**.

### FUEL INJECTOR

The fuel injector(s) in a TBI system is located in the top of the throttle body, either above or slightly recessed in the

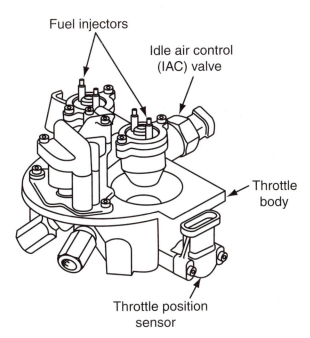

**Figure 2.** An external view of a common TBI unit.

throttle bore, located above the throttle plates. Because of the location of the fuel injector, all of the air that enters the engine has to pass by and around it. Because the air passes by the injector, the fuel can be easily mixed with the air by simply presenting an atomized mixture into the air stream. As we learned in Chapter 15, the fuel injector is an electronic solenoid that is used to control the flow of fuel **(Figure 3)**.

The fuel injector consists of an electrical coil and a moveable core built into a housing. The ends of the coil terminate at the top of the injector, where they are connected to the terminals used to control the injector. A plunger is

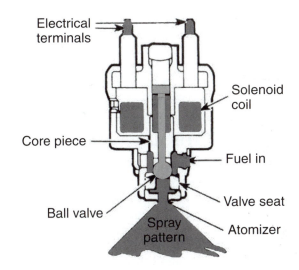

**Figure 3.** Cutaway view of a common TBI.

located within the center of the coil and acts as the solenoid core. The end of the plunger can be either tapered or ball-shaped. The shaped end of the core seals against a valve seat built into the injector body. When the injector is not energized, spring pressure holds the core on the valve seat, effectively cutting off fuel flow. Fuel typically enters the TBI injector at the bottom of the injector just above the valve seat. Before the fuel is allowed into the injector, it is passed through a very fine mesh screen to keep out any debris that might have passed through the fuel filter. Located at the bottom end of the injector below the seat is an insert called an atomizer. The atomizer has a small hole or series of holes that the fuel must pass through. The pressurized fuel is pushed through these holes and is broken into very small droplets as it exits the injector into the air stream. As the fuel exits the injector, it will have a distinct appearance called the spray pattern. The manufacturers may alter the shape of the pattern or the size of the droplets by modifying the atomizer. These patterns are altered based on the specific application in which the atomizer will be installed. The fuel pressure and the cleanliness of the injector itself also directly affect the spray pattern. Fuel injectors are rated by the amount of fuel that they are able to pass through the seat. This is called the injector flow rate. A typical measurement is the number of pounds of fuel that the injector is able to flow in 1 hour. This measurement is done at a specific and constant fuel pressure. When injectors are designed for a vehicle, the flow rate is matched to the available fuel pressure and the engine requirements. If fuel pressure is higher than what the injector is rated, the injector will deliver more fuel than is desired. Conversely, if the fuel pressure is lower, the injector will deliver less than the desired amount of fuel. Because of this, fuel pressure must be constantly adjusted for the specific engine demands.

## Fuel Pressure Regulator

For the fuel injector to operate properly, it must be supplied with a constant volume of clean fuel delivered within a specific pressure range. The fuel in a TBI system is provided by a high-volume electric pump and is regulated between 9 and 13 psi. The pump itself has the ability to build significantly more pressure than is needed in the system. For this reason, the pressure must be regulated. An in-line by-pass regulator is used to control the specific amount of fuel pressure. Fuel travels through the throttle body using internal passages cast within the throttle body itself. The throttle body has an inlet port, where filtered fuel is introduced from the tank, and an outlet port, where the return line leading back to the fuel tank is connected. The by-pass regulator is located in a position to control the amount of fuel returning to the tank.

The typical TBI system will use a pressure-sensitive diaphragm-type pressure regulator. The diaphragm is installed within a metal canister that is equipped with an inlet

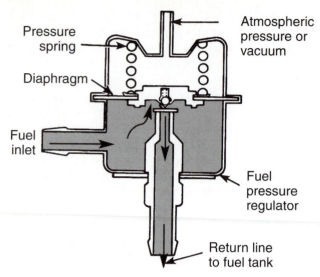

**Figure 4.** A common fuel pressure regulator. Note the fuel inlet, outlet, and external port on the top. This design is intended to have a vacuum line connected at the top. However, some applications might not require connection of a vacuum hose.

and outlet port. The diaphragm covers the top of the return line. A calibrated spring is located above the diaphragm and exerts a specific amount of pressure to keep the diaphragm from lifting. Any time that the fuel pump is running, fuel is allowed into the pressure regulator. The pressure acts upon the diaphragm. When pressure is sufficient to overcome spring pressure, the diaphragm lifts and fuel is allowed to flow into the return line. As fuel is consumed, the pressure drops and the spring closes the return line, blocking the flow of fuel back to the tank. This action allows the regulator to constantly control the fuel pressure available at the fuel injectors **(Figure 4)**.

*Interesting Fact*

*The design of the intake manifold affects the overall performance of the engine. Intake manifolds are tailored to supply a specific amount of air to an engine, based on a specific engine configuration. The requirements are based on the amount of airflow the engine needs as determined by cylinder head port flow and efficiency, the intake valve size, and the lift and duration specifications of the camshaft. Each different engine configuration needs a manifold with specific characteristics.*

## INTAKE MANIFOLD

The intake manifold is the component that is bolted to the cylinder head and provides passages for the air/fuel mixture to move from the throttle body to the intake valves. These passages are called intake runners. Intake runners are specifically designed to increase velocity and maintain consistent flow. Additionally, the intake runners are designed to promote some amount of turbulence to keep the fuel droplets suspended in the incoming air. The intake manifold used in a TBI system is designed so that the ends of the intake runners join in a common area called a plenum. The plenum serves as the inlet for the intake manifold. A mounting flange is provided at the inlet of the plenum and serves as the mounting point for the throttle body unit.

As the intake valves open, a low-pressure area is created within the entire plenum below the throttle plates of the throttle body. Each time a valve opens, a small pulse occurs within the plenum. These pulses can be seen when using a vacuum gauge connected to the manifold and provide the basis for vacuum testing as discussed in Chapter 10. The throttle plates separate the low-pressure area of the plenum from the atmospheric pressure outside of the engine. As the operator of the vehicle depresses the accelerator pedal, the throttle plates are opened and the atmospheric air rushes into the engine. As the air flows into the engine, fuel enters the throttle bores and is mixed with the incoming air charge. The air/fuel mixture then goes past the throttle plates and into the plenum of the intake manifold. The opening of the individual intake valves provides a low-pressure area within the plenum. Because the intake valves open at different times, each cylinder draws an air/fuel charge from the plenum as needed **(Figure 5)**.

There are many different styles of intake manifolds, and each is designed to fit a particular engine. Intake manifolds used with TBI systems are usually constructed from aluminum. As we learned in Chapter 27, the intake manifold also has other purposes and functions in addition to delivering air and fuel. Most intake manifolds have passages to allow for the recirculation of exhaust gases. On the V-type engine, the intake manifold might have passages to support the flow of coolant from one cylinder head into the other. The intake manifold can also serve as a mounting location for various accessories.

## IDLE AIR CONTROL

The ability to control idle speed is an important part of the throttle body's job. When the engine is idling, the engine receives just enough air and fuel to maintain a desired idle speed. There are two methods to providing this fuel and air: a base idle setting, called **minimum air rate**, and the PCM-controlled air bypass. When the throttle plates are completely closed, they are able to form a tight seal. The throttle blades have the ability to seal so tight that the engine does not receive enough air to run at all.

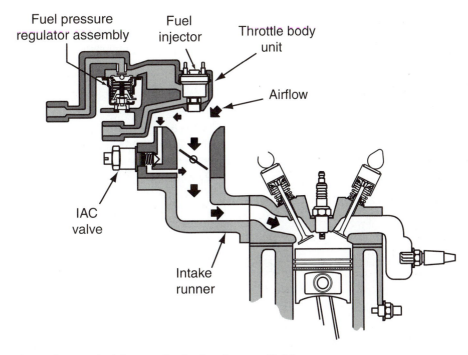

**Figure 5.**  Airflow through a typical four-cylinder intake manifold.

The minimum air rate establishes the minimum amount of air that should be allowed into the engine. This is accomplished by opening the throttle blades a very slight amount. This provides just enough airflow through the throttle blades to allow the engine to idle at the desired speed in the event that the PCM-controlled system failed. This setting generally provides enough air to allow the engine to idle at 400–600 rpm. The minimum air rate is generally set at the factory using specialized equipment and then sealed off to prevent tampering. This setting should not be altered under any circumstance unless directed by the manufacturer's service information. The amount of fuel delivered is controlled by the PCM and cannot be adjusted.

All modern fuel injection systems provide some method to allow for computerized idle speed control. The throttle body provides a passage that allows filtered air to bypass the throttle plates and enter the engine. The IAC valve controls the amount of air that is allowed through the by-pass passage; operation of the IAC is explained in Chapter 15. Depending on engine operating conditions, the IAC pintle can retract to increase engine idle speed or extend to slow engine speed down. The IAC pintle moves in very small steps called counts. These counts are the manner in which the PCM monitors IAC movement. Low numbers indicate IAC valve extension, whereas high numbers represent IAC retraction. The IAC valve and passage are illustrated in **Figure 6**.

The IAC counts can be viewed on most scan tools. The following are conditions in which the PCM makes adjustments to idle speed:

● *Engine temperature.* When an engine is cold, it requires more air to idle than it would at higher temperatures.

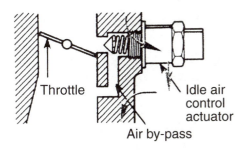

**Figure 6.**  IAC passages within the TBI unit.

As the engine warms up, less air is needed to keep the engine running and the IAC acts accordingly, extending the pintle to block off some of the by-pass air.

● *Engine load.* Anytime a load is placed on the engine, the idle speed must be increased. Common conditions that place load on the engine are placing the vehicle in gear and engaging the A/C compressor. Each of these actions places a significant strain on the engine and requires the engine speed to be increased. A change in engine rpm is very noticeable as the A/C compressor cycles. Engine speed might also need to be increased to compensate for system voltage fluctuations. These conditions will result in the PCM requesting that the IAC pintle retract a specific number of steps to allow air to pass through the by-pass passage.

● *Off idle operation.* As the throttle is opened up to allow for engine acceleration, the IAC will typically retract a proportionate amount. This accomplishes two distinctive things. First, it is a precautionary measure to make sure that enough airflow is available to the engine in the event that the throttle was to close suddenly. Second, it allows the maximum amount of airflow to be passed into the engine above what can be provided by the throttle plates alone.

- *Deceleration.* As the throttle blades are closed the IAC valve also closes. This allows the engine to return to the desired idle speed without engine stalling.

## BASIC FUEL CONTROL

The PCM has complete control of the fuel system. The PCM has ability to operate, adjust, and diagnose any system that it controls. The PCM program provides extensive diagnostic capabilities for those systems. For each system that it controls, the PCM has the ability to set a variety of DTCs.

Fuel control through the injectors is controlled by the PCM. A complete TBI control circuit is illustrated in **Figure 7**. The fuel injector usually receives a 12-volt ignition feed anytime that the ignition is in the start or run positions. The PCM controls the amount of time that the injector is energized using PWM to control the ground circuit of the injector. PWM is described in detail in Chapter 15. By using PWM, the PCM precisely controls the amount of injector "on" time. This is often displayed on the scan tool in milliseconds. Wider pulse widths result in the injector being held on longer and, as a result, more fuel being allowed into the engine. Short pulse widths have the opposite effect.

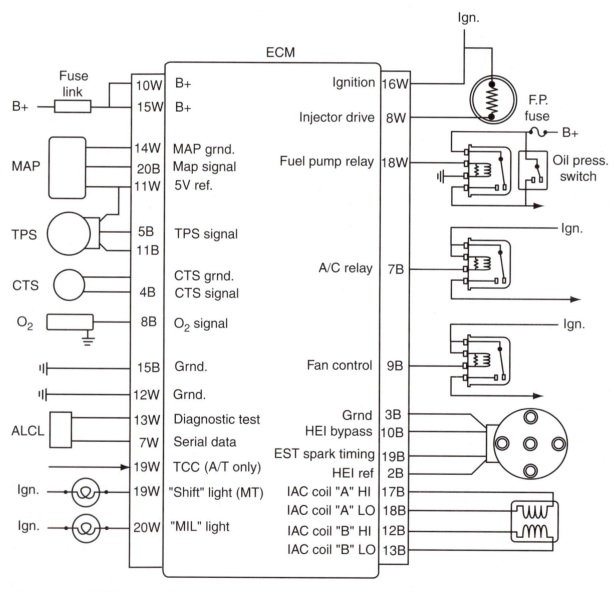

**Figure 7.** A typical TBI system wiring diagram with all of the inputs, outputs and auxiliary systems.

When the ignition is first turned to the on position, the PCM activates the fuel pump relay for approximately 2 seconds. This allows the system to build up fuel pressure to the throttle body. As the engine turns over, the distributor produces an rpm signal that is used by the PCM to determine that the engine is turning. When the PCM receives this signal, it will activate the fuel pump relay to establish constant fuel flow. This signal is also used to determine when to pulse the injectors. The PCM will continue to operate the fuel pump relay and pulse the injectors as long as an rpm signal is received. Once the engine has started, a basic fuel injector pulse width is established that is based on CTS, IAT, MAP, TPS, and BARO values.

Once the engine starts, it operates in open loop. Open loop is the state of engine operation when the engine fuel and timing controls are based on a limited number of sensor inputs that are strictly compared with a control map. Once the engine is running and idling in open loop, the PCM also takes MAP and rpm into consideration when making fuel and timing adjustments, but specifically ignores any information provided by the $O_2S$.

As the engine has been run a predetermined amount of time and reaches a specific temperature as determined by the manufacturer and the $O_2S$ is warm enough to function, the PCM will go into closed loop operation. In closed loop operation, the PCM has complete authority of fuel control and can make adjustments that are based on sensor readings and fuel map specifications. However, in closed loop, the PCM is allowed to deviate from the amount of fuel that is prescribed by the fuel control map to meet immediate needs.

When the $O_2S$ begins to function, it senses the amount of oxygen that is present in the exhaust system and communicates this information to the PCM. The PCM considers these signals and makes additional compensations to the injector pulse width to tailor the air/fuel ratio to the specific conditions in which the engine is operating. TBI systems can use heated and/or nonheated $O_2S$.

## SPECIAL MODES

There are several special modes of operation that the PCM has at its disposal to ensure reliable engine performance. These modes are enacted during very specific engine operating conditions. These examples are typical of most systems. You should familiarize yourself with the specific system that you are working on for specific modes.

### Clear Flood

If the engine should become flooded, a **clear flood mode** may be provided. Clear flood mode is activated when the PCM senses a TPS signal above a predetermined amount while cranking. When this occurs, the PCM turns the injector off or sends a pulse width to the injector that will produce a very lean air/fuel ratio. Once the TPS voltage drops back down or the engine starts, the PCM will return to normal fuel control.

### Acceleration

By monitoring the TPS and MAP signals, the PCM can sense rapid changes in engine load. When the PCM detects a high TPS signal and low vacuum, as indicated by the MAP, the PCM knows that the engine is being accelerated. In this situation, the PCM provides enough fuel to prevent engine hesitation and ensure smooth engine operation.

### Deceleration

When the PCM detects a rapid decrease in TPS and a sudden rise in vacuum, as indicated by the MAP, the PCM knows that engine is being decelerated. Under these conditions, the PCM has the ability to shorten injector pulse width and in some circumstances turn the injector off. This is done to prevent an excess amount of raw fuel from building up in the engine and exhaust systems. Excessive amounts of fuel could result in damaging backfire.

## *Summary*

- The TBI system has features of both port fuel injection and carburetion. Air and fuel are mixed within the throttle bores. The TBI contains all of the control components of the system.
- The injectors of a TBI system are located within the throttle bores. Injectors are electrically operated solenoids. Fuel flow is allowed when the injectors are energized.
- A TBI system typically needs 9–13 psi of fuel pressure. A fuel pressure regulator is located on the outlet side of the throttle body. The regulator controls fuel pressure by controlling the amount of fuel that is allowed to return to the fuel tank.

- The intake manifold provides passages for fuel and air to move into the cylinders. The opening of the intake valves creates a low-pressure area in the intake manifold. The air/fuel mixture is forced into the plenum of the manifold, where it is drawn into the cylinders as needed.
- Idle air control is controlled by allowing air to bypass the throttle plates. Minimum air adjustments should not be changed unless directed to do so by the manufacturer's service information. The IAC valve changes idle speeds based on engine temperature, load, and throttle plate position.

■ The injectors typically receive voltage whenever the key is in the run or on position. The PCM controls the injector pulse width by using PWM on the ground circuit. Longer pulse widths result in increased fuel delivery. Shorter pulse widths result in less fuel delivery.

■ The PCM uses a variety of sensors to control fuel delivery. Open loop is a fuel delivery mode that is based on fuel map specifications. Closed loop mode is also based on fuel map specifications but can be altered based on the current operating conditions of the engine.

## Review Questions

1. All of the following statements about TBI systems are correct *except:*
   A. Air and fuel are mixed together in the throttle bores.
   B. All of the control components of the TBI systems are contained or attached to the throttle body.
   C. The entire throttle body must be replaced if an injector fails.
   D. The throttle body has one injector located in each throttle bore.

2. Technician A says that an open circuit in the injector coil will keep the injector from opening. Technician B says that a clogged atomizer can cause a lean operating condition. Who is correct?
   A. Technician A
   B. Technician B
   C. Both A and B
   D. Neither A nor B

3. Technician A says that a broken fuel pressure regulator spring will cause fuel pressure to be lower than specifications. Technician B says that a restricted fuel return line can cause fuel pressure to be higher than specifications. Who is correct?
   A. Technician A
   B. Technician B
   C. Both A and B
   D. Neither A nor B

4. Technician A says that as engine load increases, the amount of by-pass air allowed into the engine also increases. Technician B says that an IAC valve that is stuck in the fully extended position will cause a high engine idle speed. Who is correct?
   A. Technician A
   B. Technician B
   C. Both A and B
   D. Neither A nor B

5. Technician A says that a lean $O_2S$ reading will result in injector pulse widths becoming longer. Technician B says that the PCM ignores the fuel control maps in closed loop operation. Who is correct?
   A. Technician A
   B. Technician B
   C. Both A and B
   D. Neither A nor B

# Chapter 29

# Port Fuel Injection Systems

## Introduction

A PFI system is an electronic fuel injection system in which each cylinder has its own individual injector. This system can also be referred to as a **multiport fuel injection (MPFI)** system. In this text, PFI and MPFI are interchangeable terms.

The injector is placed in a position to spray fuel directly into the cylinder head port directed at the back of the intake valve, making fuel available immediately when the valve opens **(Figure 1)**. There are many advantages to port fuel injection; among those advantages are better fuel distribution, better economy, and smoother running engines. Although there are many variations of port fuel injection, most them are similar in operation. In this chapter, we look at the most common configurations.

## THROTTLE BODY

Because no fuel flows through the throttle body of the PFI system, its construction is much more simplistic than a TBI system. The PFI throttle body typically has only one large throttle bore, although some have two small bores instead. A return spring is used to close the throttle plates of the throttle body. The throttle body still uses built-in passages for the IAC system to operate. Furthermore, the throttle body provides provisions for mounting the IAC valve, TP sensor, and MAF sensor.

Because the air moves through the throttle body at a very high velocity, it can sometimes become cold enough for the moisture within the air to freeze and allow ice crystals to form. For throttle body designs that have problems with icing, engine coolant passages are routed throughout the throttle body. These passages warm the throttle body

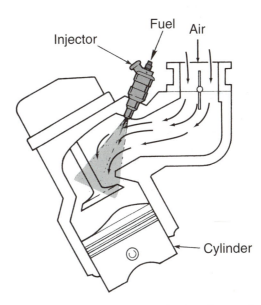

**Figure 1.** In a PFI system, the fuel is directed at the back of the intake valve.

to the point that icing does not occur **(Figure 2)**. The passages can receive engine coolant through external hoses, or the coolant can transfer directly through internal ports in the upper manifold.

Control strategies and operation of the IAC system are virtually the same as those found in TBI injection. As with the TBI units, minimum air rates are set at the factory and should not be adjusted in the field unless specifically directed.

## FUEL INJECTOR

The port fuel injectors are located in the intake manifold runners. Each injector nozzle protrudes through the

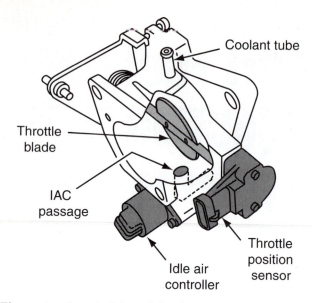

Figure 2. A typical throttle body used in a PFI system.

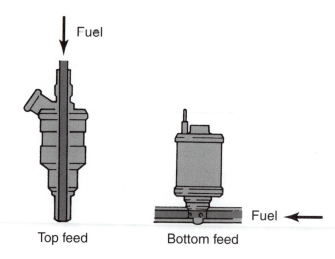

**Figure 4.** Examples of top-feed and bottom-feed injectors.

manifold with the tip being directed at the intake port of each cylinder. Electrical operation of the port fuel injector is virtually identical to that of the TBI systems. However, noticeable differences in design are obvious **(Figure 3)**.

The most noticeable of these differences is that the port fuel injector is an external injector, meaning that the injector is encased in a sealed housing. This sealed housing allows the injector to be exposed to the elements. Port fuel injectors can be of a bottom-feed design, in which fuel enters the injector just above the seat, or they can be of a

top-feed design **(Figure 4)**. With top-feed injectors, fuel enters at the top of the injector and flows down across the injector windings and down to the seat. The fuel in this type of injector serves to cool the injector coil and reduce operating noise of the injector. Each of these injectors is equipped with an inlet screen to trap any foreign particles that might have passed through the filter.

Port fuel injectors have lower fuel flow rates than do TBI systems but operate at higher system pressures. Because fuel is injected directly into the cylinder head port, it has little time to mix with the air before it enters the cylinder. Because of this, the fuel must be better atomized before it is introduced into the intake manifold. (Refer to Figure 1.) To accomplish this, the atomizers that are used in port fuel injectors are often smaller than those found in throttle body injectors and therefore require higher operating pressures to move the fuel. The higher pressures are utilized to not only push the proper amount of fuel out of the injector but also introduce it in a better-atomized state.

Because port fuel injectors are located outside of the intake manifold **(Figure 5)**, they have no built-in fuel delivery

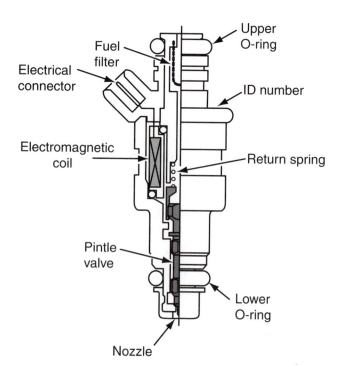

**Figure 3.** A cutaway view of a port fuel injector.

**Figure 5.** A typical port fuel injector location.

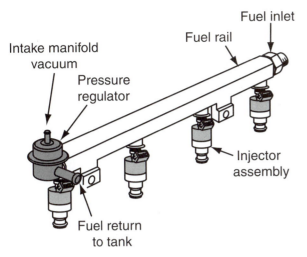

**Figure 7.** Typical fuel pressure regulator location.

**Figure 6.** A typical fuel rail assembly. Note the locations of the regulator and injectors.

mechanism like the throttle body injector and must have an external means of fuel delivery. A component called a **fuel rail** serves the purpose of delivering fuel to the injectors. The fuel rail is made from steel tubing or from an aluminum bar with holes drilled through the center for fuel flow. The inlet of each injector is equipped with an O-ring and protrudes into the fuel rail. The injector is held in place using a clip.

An in-line engine uses one straight fuel rail with an inlet on one end and an outlet on the other. The V-type engine typically uses two individual fuel rails that are attached with a balance tube **(Figure 6)**.

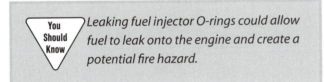

*Leaking fuel injector O-rings could allow fuel to leak onto the engine and create a potential fire hazard.*

## FUEL PRESSURE

A high-volume, high-pressure fuel pump delivers fuel from the fuel tank to the fuel rail. Before the fuel is allowed to enter the fuel rail, it is passed through a high-flow fuel filter to remove contaminants. The fuel pumps used in PFI systems are capable of producing significantly more fuel pressure than is actually desired and must be regulated. The required fuel pressure for a PFI system depends completely on the design of the fuel system. Some systems operate in a range of 28–32 psi, whereas another might operate from 56 to 62 psi. It is imperative that you consult the factory service manual for the specifications of the system in which you are working.

Fuel pressure in the PFI system is controlled by allowing unused fuel to return to the fuel tank. A by-pass fuel

pressure regulator is located on the fuel rail to control the return of fuel to the tank **(Figure 7)**. Operation of the fuel pressure regulator is similar to that of the one found in the TBI unit. The typical PFI system will use a pressure-sensitive diaphragm-type pressure regulator. The diaphragm is installed within a metal canister that is equipped with an inlet and outlet port on one side and a vacuum port on the other. The diaphragm covers the top of the return line and a calibrated spring exerts a specific amount of pressure to keep the diaphragm from lifting. The spring side of the diaphragm is exposed to engine vacuum. Under normal conditions when engine vacuum is high, fuel demands are typically low. The vacuum acts upon the diaphragm and allows fuel to return to the tank, thus lowering the fuel pressure. When vacuum is low, the fuel demand is typically higher, resulting in longer injector pulse widths. This causes a drop in the fuel pressure in the fuel rail. To compensate, the spring pressure overcomes the low vacuum signal, and the diaphragm is again seated against the return pipe, causing fuel pressure to rise **(Figure 8)**. If the vacuum line

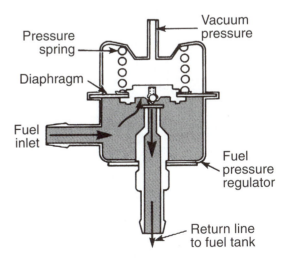

**Figure 8.** A cutaway view of a fuel pressure regulator.

comes loose from the regulator, the only control within the system is that provided by the spring pressure in the regulator. In this situation, fuel pressure will be higher than desired but will be low enough to prevent damage to the pump. These actions allow the regulator to constantly control the fuel pressure available at the fuel injectors.

## RETURNLESS FUEL SYSTEMS

The returnless fuel system continues to gain popularity among manufacturers. Returnless systems are able to control the fuel pressure and volume available to the injector without the aid of an auxiliary return line from the fuel rail back to the fuel tank. There are two distinct designs of

returnless fuel systems: the regulator controlled and the electronic controlled.

Because of its simplicity, the regulator-controlled system is the most popular form of returnless system at this time. In this system, a fuel pressure regulator is combined with the fuel filter in one unit. The filter/regulator assembly is often located either in the modular sending unit along with the pump or a short distance from the fuel tank **(Figure 9)**. The regulator used in this system uses a pressurized chamber with an internal diaphragm. The diaphragm is used to block a pressure release port. Fuel is allowed to flow into the inlet of the regulator to fill the chamber and directly to the outlet port. A spring forces the diaphragm to seal against a seat. When fuel demand is low, the pressure

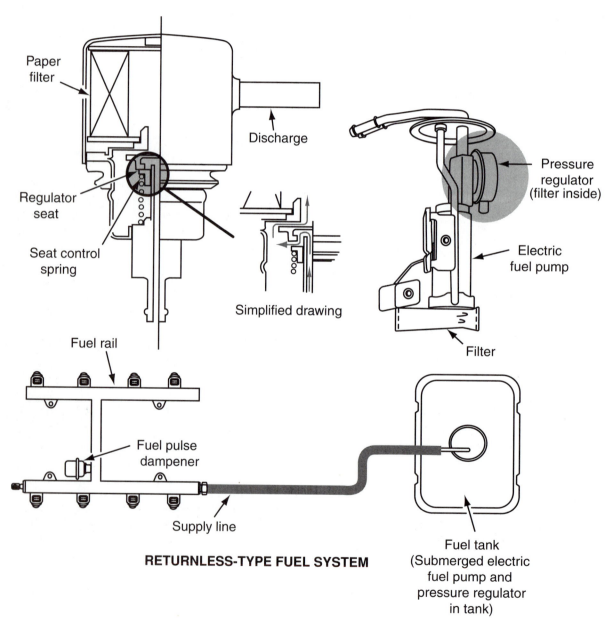

**Figure 9.** A fuel pressure regulator used in a returnless system.

within the chamber builds up and exceeds the pressure that is applied by the diaphragm spring. This forces the diaphragm to open and allow fuel to flow back into the fuel tank, thus lowering system pressure. When fuel demand is high, the pressure within the chamber drops, and the spring will seat the diaphragm and allow pressure to again build up. Under most conditions, the fuel pressure is such that the diaphragm is in a partially opened state.

The electronic returnless system is more complicated but provides for precise electronic control of the fuel system pressure. The ability to regulate pressure in this system is controlled by varying the fuel pump speed through PWM. A fuel pressure sensor is located on the fuel rail that sends feedback about fuel pressure to the PCM. The PCM then sends commands to a separate fuel pump control module near the fuel tank that controls the speed of the pump using PWM.

## INTAKE MANIFOLDS

The intake manifold for the PFI system is significantly different from that of other fuel delivery methods. When the manifold is forced to deliver both air and fuel into the cylinders, serious compromises to airflow must be taken to ensure that the fuel stays suspended within the air/fuel mixture. The first thing to understand is that no fuel is flowing in the manifold. Because there is no fuel flow in the manifold, there are no design concerns for keeping the fuel suspended within the air. This allows designers to focus on designing manifolds that will efficiently deliver air into the cylinders. Some PFI manifolds have an appearance of a TBI manifold with relocated injectors. These manifolds work well; however, they do not take full advantage of airflow possibilities into the engine.

We will look at the most prevalent design that is found today. This design consists of a lower manifold and a separate upper manifold, often called a plenum **(Figure 10)**. The

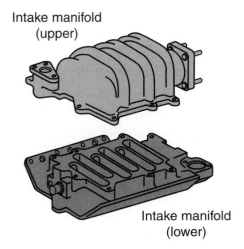

Intake manifold (upper)

Intake manifold (lower)

**Figure 10.** Components of an upper and lower manifold assembly.

lower manifold is bolted to the cylinder head and is responsible for delivering air into the engine. The lower manifold provides the mounting base for the upper manifold and contains the coolant passages, if they exist. For the most part, when the air reaches the lower manifold, it is traveling at a high velocity and has a straight shot to the cylinders.

The most common upper manifold design in use today is separate from the lower manifold. However, some one-piece units do exist. Once air passes through the throttle body, it enters the upper manifold area of the manifold. Connected within the upper manifold, each cylinder has an individual runner for air delivery to the cylinder head port. It is the intent of the designers to make all of these runners an equal length. By doing this, air velocity and volume among all cylinders can be equalized. These runners are engineered to optimize the velocity and airflow volume.

As each intake valve is opened, air is drawn from this common area into the individual runners. As the air travels, velocity is increased. Just as the air is about to enter the cylinder head port, a mist of fuel from the injector is picked up by the column of air and rushed around the intake valve into the cylinder.

## Variable Induction

Many of the intake systems in use today use adjustable runners. By adjusting the effective length of the intake runners, the power curve of the engine can be modified for current engine conditions. These upper manifold systems usually have two runners per cylinder: low-speed runners and high-speed runners. When engine speed is low, the long runners produce a smaller volume of air but at a higher velocity. This helps the engine to produce more low-speed torque. When the engine is operating at higher rpm, natural velocity is higher and the long runners become a restriction to airflow. Because of this, short tubes are used for high-speed operation. Although short tubes do not produce high velocity, they deliver more volume.

A shutter valve located inside the upper manifold separates the runners. The PCM controls the valve using either an electric or a vacuum actuator. The valve is opened or closed in response to engine operating conditions. Every major vehicle manufacturer produces some variation of this system. Because of this, it is important to research the particular system in which you are working **(Figure 11)**.

## PORT FUEL INJECTION SYSTEMS

As most other systems within the vehicle, the fuel systems have evolved as well. There are several styles and variations of PFI in use today. We examine standard PFI, sequential fuel injection (SFI), **central port injection (CPI)**, and direct fuel injection.

If you were to visually inspect a PFI system and an SFI system, it would be difficult to tell the difference. Although

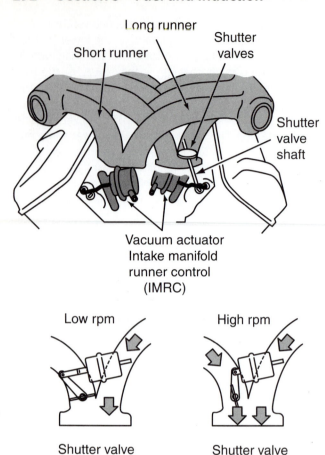

Long runner
Short runner
Shutter valves
Shutter valve shaft

Vacuum actuator
Intake manifold
runner control
(IMRC)

Low rpm

High rpm

Shutter valve
closed

Shutter valve
open

**Figure 11.** A representation of a variable induction manifold.

they look the same, their operation is much different. The difference between the two systems is in how the injectors are activated. Injectors in a PFI system are group fired. Group firing is a method of pulsing the injectors in pairs or groups **(Figure 12)**. For example, on a V-type engine, all of the injectors residing on the same bank of the engine are

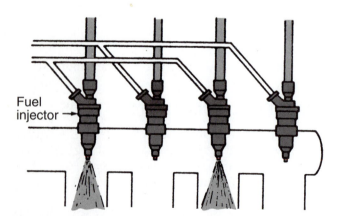

Fuel injector →

**Figure 12.** Group-fired fuel injection.

pulsed at the exact same time. An in-line engine pairs the injector for every other cylinder together. Not only are the injectors fired at the same time, all grouped injectors are spliced into the same PCM control circuit. In this system, each group will fire alternately one time per crankshaft revolution. This is the same method used to pulse throttle body injectors. The advantages of this system are that fuel is delivered right to the back of the intake valve, providing better fuel control. The main disadvantage is that fuel is forced to stand at the back of the intake valve, until the valve is opened.

The SFI systems were an upgrade from the popular PFI systems. The PCM used in the SFI systems have individual control circuits for each injector. By controlling each injector individually, each one can be opened in the proper relationship to the opening of the intake valve. This allows the fuel to be sprayed into the air stream for better fuel atomization. This method further prevents fuel from standing at the back of the valve, where it collects heat from the cylinder head and makes it more difficult to introduce into the air stream.

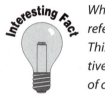

**Interesting Fact**

*When first introduced, CPI was often referred to as "poor man's port fuel." This nickname was based on the relative expense compared with the cost of conventional PFI and SFI systems.*

The SFI systems were made possible by the advances in computer and ignition system technology. The ignition system advances include the development of high sample rate crank and cam sensors. These sensors allow the PCM to monitor the crankshaft rotation in precise increments. The camshaft is also monitored; it can be monitored in specific intervals or simply as a measure of relative position of the piston in the number 1 cylinder. However, by monitoring both the cam and crankshafts, the PCM is able to calculate the exact timing needed in which to pulse the injector.

## Central Port Injection

The CPI system is a version of PFI that uses a single injector to provide fuel to all of the cylinders. The first variation of this system used one fuel injector that was located in the center of the intake manifold. The injector feeds one port for each cylinder. Each cylinder is equipped with a plastic tube that connects to the injector at one end and a **poppet nozzle** in the other end. The poppet nozzle consists of a ball seat and a ball. A spring is placed in such a manner as to force the ball against the seat. The nozzle is located in the same relative position as a port fuel injector would be. When the injector is opened, fuel flows into each of the plastic tubes. When fuel pressure within the tube

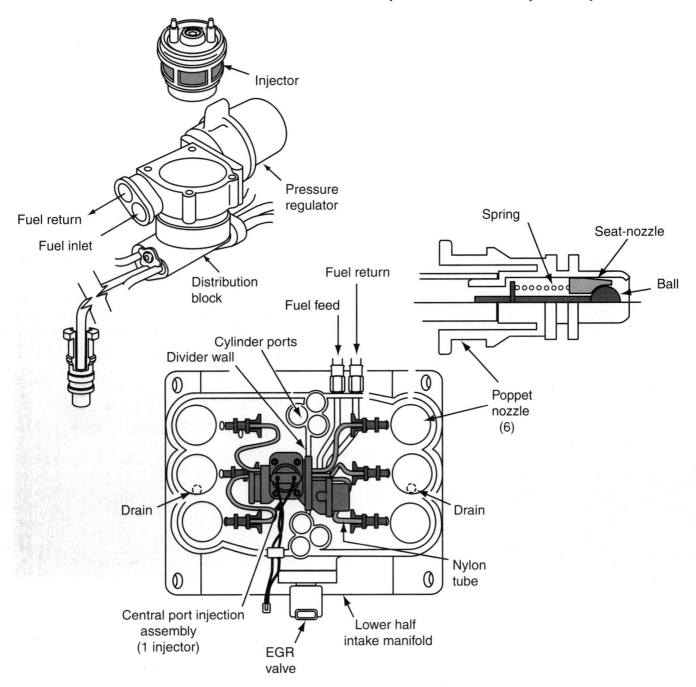

**Figure 13.**   The components of a CPI system. A CSFI system is very similar to this.

reaches a pressure strong enough to overcome spring pressure and move the ball, typically 37–43 psi, fuel will then enter the intake port. Because the opening of the nozzle depends entirely on fuel pressure, any amount under the minimum will not move the ball and open the nozzle. All of the components of this system are contained beneath the upper manifold and cannot be accessed without removal of the upper manifold **(Figure 13)**.

Later versions of the CPI system, called **central sequential fuel injection (CSFI)**, use a small electronic injector

located opposite the poppet nozzle. These individual injectors take the place of the single injector. The use of individual injectors allows this system to operate sequentially. Injectors in both of these systems are required only to fill the tube with fuel. All fuel atomization takes place at the ball valve.

## Gasoline Direct Injection

The latest development in fuel injection is that of **gasoline direct injection (GDI)**. The GDI system is an emerging technology that has seen usage on some vehicle models in Japan and Europe for several years but has yet to emerge on a wide basis in North America. The GDI system works on the premise of injecting fuel directly into the cylinder.

Unlike other injection systems, in which fuel delivery into the cylinder depends on the operation of the camshaft to deliver fuel on the intake stroke, the GDI systems are able to inject fuel into the cylinder during the compression stroke.

Given the ability to deliver fuel on the compression stroke, the timing of the fuel delivery can be closely coordinated to that of the ignition timing. By closely controlling these relationships, leaner fuel mixtures can be used, resulting in better fuel economy and lower hydrocarbon emissions.

Because a great deal of pressure exists in a cylinder on the compression stroke, fuel pressures required to operate a GDI system can exceed 1700 psi. To produce these very high operating pressures, a mechanically driven high-pressure fuel pump must be used.

*Interesting Fact*

*Gasoline direct fuel injection systems are very similar to those systems found on diesel engines.*

In these systems, a fuel pump located in the fuel tank will be charged with delivering fuel to the high-pressure pump. From there, fuel is delivered to the electronically controlled fuel injectors. The remaining PCM inputs and outputs remain similar to those of a typical SFI system.

## BASIC FUEL CONTROL

The PCM has complete control of the fuel system. The PCM has the ability to operate, adjust, and diagnose any system that it controls. The PCM program provides extensive diagnostic capabilities for those systems. For each system that the PCM controls, it has the ability to set a variety of DTCs.

As we begin to look at fuel control, the first thing that must be determined is the amount of load that is placed on the engine. This can be determined by knowing the amount of air that is flowing into the engine. Two methods exist for determining airflow: the speed density method and the MAF system.

The speed density method is a calculation that is based on information from the TP sensor and the MAP, rpm, ECT, and IAT signals. By calculating these parameters, the PCM can determine the relative amount of air that is flowing into the engine. This allows the PCM to make proper air/fuel calculations. Vehicles that use the speed density system do not require the use of a MAF sensor.

By using the MAF system, the PCM has the ability to measure the amount of air that is flowing into the engine. Systems using this method do not require any additional adjustments to determine density. The MAF system provides the most accurate measurement.

Fuel control through the injectors is controlled by the PCM. The fuel injectors are supplied with 12V in two different ways. The first is a direct feed from the ignition whenever the key is in the start or run position. The second method uses a relay that is energized only when the PCM detects an rpm signal.

The PCM controls the amount of time that the injector is energized using PWM of the injector ground circuit. PWM is described in detail in Chapter 15. By using PWM, the PCM can precisely control the amount of injector "on" time. This is often displayed on the scan tool in milliseconds. Wider pulse widths result in the injector being held on longer and, as a result, more fuel being allowed into the engine. Short pulse widths have the opposite effect. In control of SFI systems, the PCM uses individual circuits for each fuel injector. This allows each injector to be opened individually. The standard PFI uses the group-fire method of control **(Figure 14)**. The amount of fuel that is needed by the engine is decided by the PCM. A typical SFI control system is shown in **Figure 15**.

When the ignition is first turned to the on position, the PCM activates the fuel pump relay for approximately 1–2 seconds. This allows the system to build up fuel pressure to the fuel rail. As the engine turns over, the PCM receives signals from the CKP and the CMP sensors that indicate that the engine is turning. As soon as the PCM notes a CKP signal indicating that the engine is turning, the PCM will activate the fuel pump relay, and constant fuel flow will be established. Once this occurs, the PCM monitors the CKP and CMP inputs to make decisions on when to pulse the injectors and in what sequence. The PCM continually monitors these inputs to open the proper injectors at the proper time. The fuel pump and injectors will continue to operate as long as a CKP signal is received.

Once the engine is started, a basic fuel injector pulse width is established based almost entirely on CTS, IAT, MAP, TPS, and BARO values. Once the engine starts, it operates in open loop.

Open loop is the state of engine operation when the engine fuel and timing controls are based on a limited number of sensor inputs that are strictly compared with a control map. Once the engine is running and idling in open loop, the PCM also takes MAF (if used) and rpm into consideration when making fuel and timing adjustments but specifically ignores any information provided by all of the $O_2Ss$ or $HO_2Ss$.

As the engine has been run a predetermined amount of time and reaches a specific temperature as determined

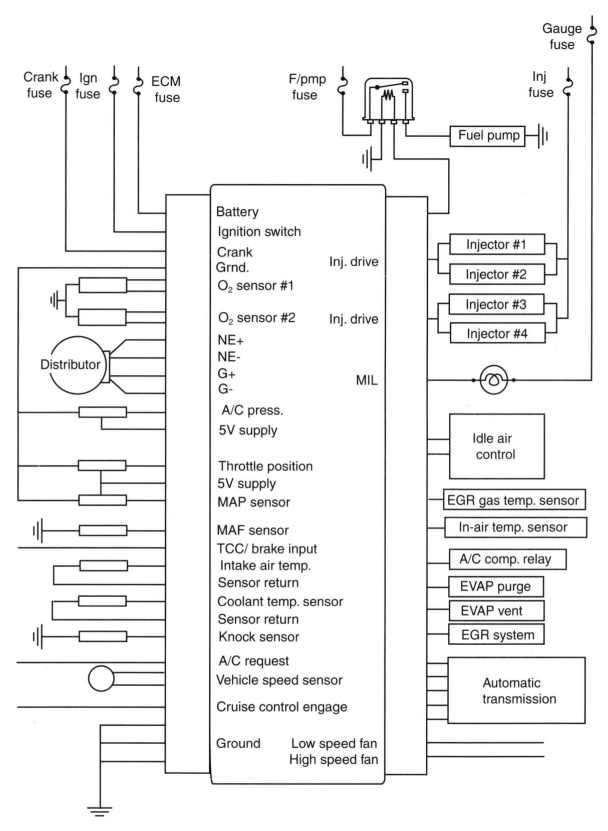

**Figure 14.**  This diagram represents a typical group-fire management system.

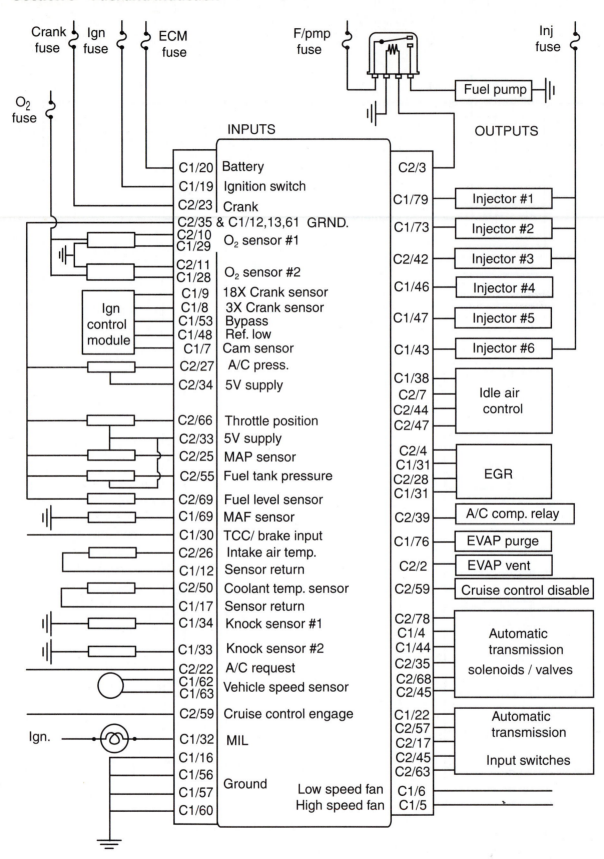

**Figure 15.**  This diagram represents a typical sequential engine management system.

by the manufacturer and the HO₂S is warm enough to function, the PCM will go into closed loop operation. OBDII-equipped vehicles use HO₂Ss to help the sensors reach operating temperature sooner.

In closed loop operation, the PCM has complete authority of fuel control and can make adjustments that are based on sensor readings and fuel map specifications. However, in closed loop, the basic fuel map serves only as a basic guide from which fuel delivery is determined. When operating in closed loop, the PCM is allowed to deviate from the amount of fuel that is prescribed by the fuel control map to meet the immediate need of the engine based on sensor readings.

When the HO₂S begins to function, it detects the amount of oxygen that is present in the exhaust gases. A signal is then sent to the PCM about the fuel mixture. The PCM will adjust injector pulse width to continually keep the $O_2S$ in transition from rich to lean. A typical PFI system may use one to four $O_2Ss$ or $HO_2Ss$, depending on the specific application. Those systems that use multiple $HO_2Ss$ have a different injector pulse width for each bank of cylinders.

## Fuel Trim

Fuel trim applies to both TBI and PFI systems. Because it has gained more notoriety as a diagnostic tool and is monitored by OBDII, we will study it as a PFI concept; however, the information does apply to TBI vehicles.

Because no standard fuel calibration will be accurate in all situations and conditions, the manufacturers developed a method to allow the PCM to change the fuel map calibrations to some degree to make the engine as efficient as possible. As we learned in Chapter 13, fuel trim is the adaptive strategy used by the PCM to monitor short-term and long-term fuel management. This chapter gives us the opportunity to study that a little more in depth.

*General Motors first made fuel trim data available to the technician. This information was displayed on a scan tool as integrator and block learn. Integrator was equivalent to short-term fuel trim and block learn was equivalent to long-term fuel trim. These designations remained in use until OBDII pull-ahead models appeared on the market.*

As noted, once the system enters closed loop, the system is able to make adjustments to air/fuel ratio that deviate from the control map. These changes are tracked by the PCM program and are stored in the PCM. This information can be viewed on the scan tool and is expressed as a percent. Under ideal conditions, both long- and short-term fuel trim values would register 0 percent.

As the $O_2S$ detects that the fuel mixture is rich or lean and fuel must be added or removed, the PCM makes note of this and registers the amount of deviation as a "+" or "−" percentage. When toggling of the HO₂S above and below the 0.450-volt threshold has slowed, the short-term fuel trim will react by adding or subtracting fuel while monitoring the HO₂S response to restore toggling of the sensor to the desired frequency. As short-term fuel changes begin to increase, the long-term fuel trim numbers move in the same direction as the short-term. A positive fuel trim value indicates that the HO₂S is reading a lean condition and the PCM is compensating by adding fuel. A negative reading indicates that the HO₂S is reading a rich condition and the PCM is subtracting fuel.

*Fuel trim values can be a valuable diagnostic tool. They allow you to assess what changes to fuel calibration have been made recently.*

Long-term fuel trim is actually a semipermanent adjustment to the fuel map strategy. This means that as the PCM processes data, it substitutes the long-term fuel trim adjustments for what is located in the fuel map. Long-term fuel trim adjustments will remain as the substitute until they themselves have been substituted or the PCM memory has been reset. Because the fuel map has been effectively changed, the short-term corrections are reversed and fuel trim starts to move back toward zero. This allows the HO₂S to continue to toggle normally. Both long- and short-term fuel trim have specific limits as to the amount of adjustment that is allowed. When the short- and long-term fuel trim exceed these limits in either direction, the MIL will be illuminated and a DTC will be set.

## SPECIAL MODES

There are several special modes of operation that the PCM has at its disposal to ensure reliable engine performance. These modes are enacted during very specific engine operating conditions. These examples are typical of most systems. You should familiarize yourself with the specific modes of the specific system in which you are working.

## Clear Flood

If the engine should become flooded, a clear flood mode might be provided. Clear flood mode is activated when the

PCM senses a TPS signal above a predetermined amount while cranking. When this occurs, the PCM sends a pulse width to the injector that will produce a very lean air/fuel ratio. Once the TPS voltage drops back down or the engine starts, the PCM will return to normal fuel control.

## Acceleration

By monitoring the TPS and MAP signals, the PCM can sense rapid changes in engine load. When the PCM detects a high TPS signal and low vacuum, as indicated by the MAP, the PCM knows that the engine is being accelerated. In this situation, the PCM provides enough fuel to prevent engine hesitation and ensure smooth engine operation.

## Deceleration

When the PCM detects a rapid decrease in TPS and a sudden rise in vacuum, as indicated by the MAP, the PCM knows that the engine is being decelerated. Under these conditions, the PCM has the ability to shorten injector pulse width and in some circumstances turn the injector off. This is done to prevent an excess amount of raw fuel from building up in the engine and exhaust systems. Excessive amounts of fuel could result in damaging backfire.

## Accessory

In addition, of the fuel management system, the technician must remember that the PCM also maintains control of many of the emissions system functions, transmission functions, and accessory functions. It is important to understand the capabilities of the system in which you are working. A malfunction in the control of any one of these circuits can result in a driveability concern and a customer complaint. Because these systems are controlled by the PCM, many diagnostic tests of these systems can result in testing of some engine management components.

# Summary

- PFI systems have one injector for each individual cylinder. Fuel is directed at the back of the intake valve.
- The throttle body of the PFI system controls only the amount of air that enters the engine. The PFI throttle body is equipped to prevent icing conditions.
- PFI fuel injectors are mounted in the intake manifold runners. These injectors use an external fuel rail to provide fuel flow.
- Fuel pressure is controlled with a by-pass regulator. The by-pass regulator has a vacuum line to change fuel pressure in relation to engine load. Electronic returnless fuel systems control fuel pressure without the aid of a fuel pressure regulator.
- Intake manifold systems used on today's vehicles typically consist of an upper and lower manifold. The upper manifold is tuned to maximize airflow characteristics.

Many upper manifolds are equipped with tunable intake runners designed to further maximize airflow.
- The PFI and MPFI injectors are group-fired. SFI systems fire the fuel injectors in relation to individual intake valve opening points. Advances in ignition system and computer technology allowed for the development of SFI systems.
- CPI and CSFI are styles of port fuel injection that use poppet nozzles to inject fuel into the ports. The GDI system is emerging technology that uses extremely high fuel pressures to inject fuel directly into the cylinder.
- The PCM uses various input sensors to control the pulse width of the injectors. Fuel trim is the computer's ability to monitor and adjust fuel system calibrations. The PCM provides several special operational modes that are used under very specific conditions.

# Review Questions

1. Technician A says that a short in one bank-fired port fuel injector will affect only that particular cylinder. Technician B says that PFI systems provide better fuel control than do TBI systems. Who is correct?
    A. Technician A
    B. Technician B
    C. Both A and B
    D. Neither A nor B

2. Technician A says that air moving through the throttle body can sometimes cause an icing condition on the throttle blades. Technician B says that the throttle body contains passages in which air is allowed to bypass the throttle plates. Who is correct?
    A. Technician A
    B. Technician B
    C. Both A and B
    D. Neither A nor B

3. All of the statements about port fuel injectors are true *except:*
   A. The PFI fuel injectors must have an external fuel supply.
   B. Injector tips extend into the manifold runners.
   C. A clogged inlet screen will restrict the flow into the injector.
   D. A clogged injector will cause an increase in system fuel pressure.

4. Technician A says that a ruptured fuel pressure regulator diaphragm could allow raw fuel to enter the engine through the vacuum hose. Technician B says that a broken pressure regulator spring would cause fuel pressure to rise above specifications. Who is correct?
   A. Technician A
   B. Technician B
   C. Both A and B
   D. Neither A nor B

5. Which of the following statements about intake manifolds is correct?
   A. The PFI intake manifolds are designed to minimize air/fuel separation.
   B. The length of the intake runners has little if any effect on an engine's power output.
   C. Variable induction systems typically use the long runners during high-speed operation.
   D. Variable induction systems typically use short runners during high-speed operation.

6. Technician A says that port fuel injectors are timed to the opening events of the intake valves. Technician B says that a single sequential fuel injector will open only once per crankshaft revolution. Who is correct?
   A. Technician A
   B. Technician B
   C. Both A and B
   D. Neither A nor B

7. Technician A says that a slight decrease in performance of the fuel pump for a CPI system can cause an engine no-start condition. Technician B says that the CSFI system fires the injectors on the combustion stroke. Who is correct?
   A. Technician A
   B. Technician B
   C. Both A and B
   D. Neither A nor B

8. Technician A says that the fuel pump relay will be energized for approximately 2 seconds when the ignition is first turned to the on position. Technician B says that the PCM will energize the fuel pump relay as long as a CKP signal is received. Who is correct?
   A. Technician A
   B. Technician B
   C. Both A and B
   D. Neither A nor B

9. Technician A says that a defective $HO_2S$ heater could extend the amount of time required for the system to go into closed loop fuel control. Technician B says that a cooling system thermostat that is stuck in an open position could keep the system from going into open loop. Who is correct?
   A. Technician A
   B. Technician B
   C. Both A and B
   D. Neither A nor B

10. Technician A says that a long-term fuel trim changes according to short-term fuel trim. Technician B says that a fuel injector leaking fuel into the intake port of the cylinder head could cause the fuel trim to register a "−" number. Who is correct?
    A. Technician A
    B. Technician B
    C. Both A and B
    D. Neither A nor B

# Chapter 30

# Basic Fuel System Testing and Diagnosis

## Introduction

As an engine performance technician, a major portion of your time will be spent in the diagnosis of fuel system concerns. Problems within the fuel system can contribute to a wide array of customer concerns. However, many problems caused by the fuel system might not even manifest themselves as driveability concerns but might only cause the illumination of the MIL. These problems are often intermittent in nature and can be difficult to diagnose. In this chapter, we focus on some specific fuel system tests and the

equipment used to perform these tests. As with any diagnostic procedure, the technician should perform a thorough visual inspection before any in-depth diagnosis takes place.

> **You Should Know** *Gasoline in the fuel system is under high pressure. Fuel system pressure should be relieved before performing any fuel system service. Refer to the specific vehicle service manual for the proper method of relieving fuel system pressure.*

### FUEL SYSTEM DIAGNOSIS

Driveability symptoms that are associated with fuel system problems are lack of power, stalling, no-start conditions, and in some cases misfiring. Anytime that you begin diagnosis of a driveability concern, you should take a fuel pressure reading. This is especially critical if the problem is intermittent. To perform a fuel pressure test, you will need a special piece of equipment called a fuel pressure gauge. A fuel pressure gauge can be purchased in the form of a kit that will contain a gauge specially designed to withstand exposure to gasoline, a hose to connect to the fuel system, and variety of adapters to connect the equipment to the vehicle **(Figure 1)**. The gauge should also be equipped with an auxiliary valve used to bleed air out of the hose and to gather fuel samples.

When selecting a gauge, you must make sure that it is capable of reading the pressures that you will encounter. For example, if you primarily work on PFI-equipped vehi-

> **You Should Know** *Extreme caution must be exercised when working with gasoline. Gasoline is an extremely volatile chemical and should be handled with great care. The spillage or leakage of gasoline is commonplace when servicing fuel systems. Every effort should be made to minimize leaks and spills. There should be no open flames, sparks, or heat sources in the vicinity of any area in which fuel system service is being performed. A fire extinguisher should be within arm's reach anytime a fuel system is being serviced. Safety glasses should be worn at all times, and measures should be taken to limit skin exposure to gasoline. Technicians should be aware of the dangers static electricity could pose when working with gasoline. Failure to follow all safety rules can cause serious burns and/or death.*

**Figure 1.**   A typical fuel pressure test kit with various adapters.

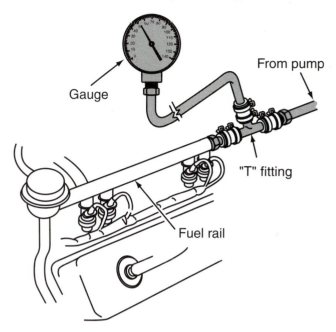

**Figure 2.**   Installation of a fuel pressure gauge using the T method.

cles, you need to be sure that the gauge is calibrated to read up to at least 80–100 psi. But if you perform a large amount of TBI diagnosis, you need to be sure that the gauge you are using will accurately read fuel pressures below 20 psi. Ideally, you should have one gauge of each calibration available.

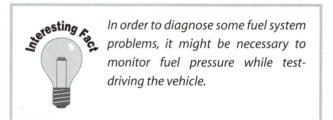

*In order to diagnose some fuel system problems, it might be necessary to monitor fuel pressure while test-driving the vehicle.*

To check fuel pressure, you must tap into the fuel line in a location between the fuel filter and the fuel pressure regulator. If a pressure reading is taken after the regulator, you will be measuring the pressure in the return line. In many cases, the fuel rail or supply line will have a **Schrader valve** available specifically for checking fuel pressure. A Schrader valve is a threaded fitting with a spring-loaded valve inside, similar to what is found in a tire valve stem. The fuel pressure gauge has an internal piece called a depressor that, when attached to the fitting, pushes the valve open and allows fuel to enter the gauge. Schrader valves used for fuel system testing can be one of several sizes. If a Schrader valve exists, it will typically be located on the fuel rail, on the throttle body near the regulator, or in the supply line **(Figure 2)**. If a valve is not available, you will have to open the fuel supply line in order install the gauge. In this type of installation, the gauge needs to be "T-ed" into the systems, forming a connection that resembles the letter "T." Special adapters that attach to the vehicle's factory fuel line

connection must be used. There are a variety of different styles of connections used **(Figure 3)**. Many of these will require special tools to disconnect the lines.

Once the fuel supply lines have been disconnected, the supply line is then routed to the fuel gauge. Another adapter will attach the gauge to the fuel rail side of the fuel system. (Refer to Figure 2.) With the gauge connected in this manner, the vehicle can be started and operated while fuel pressure is observed. **Figure 4**, **Figure 5** and **Figure 6** show the steps to installing a fuel pressure gauge using a Schrader valve.

## FUEL PRESSURE TESTING

Before testing and analyzing fuel pressure, you will need to have the specifications available for the particular vehicle that you are testing. Fuel pressure specifications are often listed as "key-on–engine-off " and "engine running." Fuel pressures are given as a range, for example, 9–13 psi. A range is used to account for variations among pumps, gauges, pump voltage levels, and pump wear. The fuel pressures that you observe on the gauge should fall well within the range of the specifications.

Once the gauge is connected and the air has been bled from the gauge, turn the key to the on position. The pump should run 1–2 seconds. This should be long enough to build up fuel pressure. Observe the reading on the gauge and write it down. This number represents the engine off pressure. Now start the vehicle. The pressure observed will represent the engine running specification. The specifications with the engine off will typically be higher on a PFI vehicle. This is because there is no vacuum acting on the

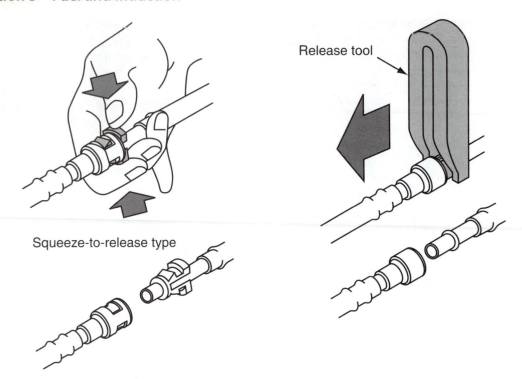

Release tool

Squeeze-to-release type

**Figure 3.** Common fuel hose connection styles and disassembly methods.

**Figure 4.** A typical Schrader valve location.

**Figure 5.** Installation of the fuel pressure gauge.

**Figure 6.** Observing fuel pressure readings.

fuel pressure regulator. When the engine is started and vacuum builds, it acts on the regulator diaphragm and fuel pressure drops slightly.

## INTERPRETING RESULTS

Common fuel system symptoms and causes are low fuel pressure, no fuel pressure, and high fuel pressure.

### Low Fuel Pressure

Low pressure will cause lean air/fuel mixtures that can surface as either an engine no-start condition or a loss of engine power. Further indications can be positive fuel trim numbers and MIL illumination. Low fuel pressure can be caused by low fuel level, a faulty fuel pump, a blocked supply line, a severely clogged fuel filter, or a fuel pressure regulator that is stuck in the open position.

## No Fuel Pressure

Conditions in which readings of zero are indicated will cause a no-start condition. These problems are caused by a faulty fuel pump or a fault in the fuel pump control circuit. If the vehicle does run and no pressure is indicated, check for proper connection of the equipment.

> **You Should Know** *The accuracy of fuel level gauge readings can change over time. Because of this, it is always wise to add approximately a gallon of fuel to the fuel tank anytime fuel pressure readings are very low or are zero. This is especially true if the gauge reads one-quarter tank or less; however, this is not a bad practice to adopt with all low fuel pressure concerns.*

## High Fuel Pressure

High fuel pressure readings will cause the engine to run rich. This can be indicated by negative fuel trim numbers, illumination of the MIL, and the possibility of black exhaust smoke. Typical causes of high fuel pressure are a no-vacuum at the fuel pressure regulator, a faulty fuel pressure regulator, or a restriction in the return line.

## Fuel Trim

Fuel trim numbers can be a valuable tool in the diagnosis of the fuel system. This data is especially helpful in the diagnosis of intermittent concerns. When you encounter an intermittent driveability symptom that you believe is fuel system related, pay particular attention to the fuel trim data. Viewing the long-term data will give you particular insight into how the fuel system has been performing over a long period. If high negative numbers are observed or a visible positive correction is taking place, a low or intermittently low pressure problem or a fuel injector restriction might be indicated.

## Dead Head Pressure

Another useful test in determining the maximum capacity of the fuel pump is the dead head pressure test. The dead head pressure test is performed with the same gauge connection of a normal fuel pressure test, with the exception that the fuel return line is blocked for this test. By doing this, the pump is forced to produce its maximum amount of fuel pressure. Although there are no specifications available for maximum fuel pressure, pressures will typically double that of the normal engine on readings. The dead head pressure test is also a useful test to perform when an intermittent fuel pressure concern is suspected.

> **You Should Know** *Measuring maximum fuel pressure in this manner places a tremendous strain on the fuel pump. Because of this, the return line should never remain blocked more than 2 seconds. If it does, damage to the fuel pump can occur.*

What often occurs is that the capacity of the fuel pump motor or the pump itself can become diminished, and consistent fuel pressures might not occur. Under these variable conditions, fuel pressures can be adequate under some operating conditions but not others. A low dead head pressure reading would indicate that the pump might not be capable of producing adequate pressure and volume in all circumstances. Using this information coupled with pressure and volume information, an accurate diagnosis can be made.

## FUEL PUMP VOLUME

Fuel pump volume is another test that is used to test fuel pump performance. For the engine to run, the fuel pump must be able to deliver an adequate amount of fuel at very specific pressures. Loss of pump volume can result from a clogged fuel filter, a restricted or crimped fuel inlet line, or a faulty fuel pump. As with maximum fuel pressure, there are typically no specifications available for fuel pump volume. Because of this, fuel pump volume tests must be assessed using a great deal of experience. Fuel pump volume can be tested using the bleed valve from the fuel pressure gauge. Note that fuel volume test results can vary among specific fuel gauge units.

One method of fuel pump volume testing is:

1. With the fuel gauge connected as described above, allow the engine to idle and insert the extension hose from the bleed valve into an approved container.
2. Open the bleed valve, allowing the fuel to drain into the container. The engine rpm might dip slightly when the valve is first opened but should recover very quickly.
3. If the engine idles rough or dies, a lack of fuel pump volume is indicated.
4. While performing the test, pay special attention to the amount of air that is present within the fuel stream. An excessive amount of air in the fuel stream is an indication of pump aeration and further indicates a fuel pump problem.

A second method of volume testing is described below:

1. With the fuel gauge connected as described earlier, start the engine and allow it to idle.

2. Disconnect the vacuum hose from the fuel pressure regulator. The fuel pressure should rise slightly.
3. While observing the gauge, quickly snap the throttle to WOT. The pressure should remain relatively steady. If fuel pressure drops more than 4 psi, a lack of fuel volume might be indicated.

## Fuel Filter Testing

To positively check fuel filter restriction, the filter must be removed from the vehicle. Once the filter is removed, empty the fuel from the inlet end while watching for any dirt or debris that might come out. Large amounts of debris might indicate restriction. Once the fuel has been drained, blow through the filter from the inlet end. A clean filter should have no obstruction to airflow. If any effort is required to force air through the filter, the filter should be replaced. As a rule, the fuel filter in most applications should be replaced every 30,000 mi.

## BASIC FUEL PUMP CIRCUIT DIAGNOSIS

If no fuel pressure is available, a pump failure or electrical circuit failure might exist. The first test to perform is to have an assistant turn the ignition to the run position while you listen near the fuel tank. If the pump can be heard running, an electrical problem does not exist. If the pump cannot be heard, test the fuel pump fuse. Once the integrity of the fuse is verified, locate the fuel pump relay and listen to see if it clicks when the ignition is turned on. If it does click, the control portion of the relay circuit is in working order. If a clicking is not heard, refer to specific vehicle service information for diagnosing the fuel pump relay and control circuit.

**Interesting Fact** *Lightly tapping the bottom of the fuel tank with a rubber mallet might cause an inoperative fuel pump to begin operating. If this occurs, the fuel pump should be replaced.*

At this point, with a test lamp to ground, back probe the fuel pump relay on the voltage supply wire that leads to the fuel pump. Again turn the ignition to the on position. The test lamp should illuminate. If it does illuminate, an electrical problem is not the likely cause of the concern. If the relay clicks but the lamp does not illuminate, the relay is the likely cause of the concern.

## PRESSURE LEAKDOWN

When the engine is turned off, the pressure should hold steady for several minutes. It is normal for pressure to naturally bleed off at a slow rate when the pump is off. In situations in which pressure drops immediately, a problem is indicated. Rapid loss of pressure can indicate a leaking fuel injector, faulty fuel pressure regulator, or defective fuel pump check valve. To test for a rapid fuel pressure loss, use this short series of tests; **Figure 7** illustrates the fuel system test points:

1. With the fuel gauge connected as described earlier, turn the ignition on to allow the system to build pressure.
2. Block the fuel supply line between the pump and the fuel rail or TBI.
3. If pressure holds, a faulty fuel pump check valve is indicated. If pressure continues to bleed down, go to step 4.
4. Repeat step 1. Block the return line.
5. If pressure holds, a leaking fuel pressure regulator is indicated.
6. If pressure continues to bleed off, a leaking fuel injector is indicated.

**You Should Know** *If the fuel supply and return lines are made from rubber, they can be temporarily pinched off using special pliers with smooth jaws. On vehicles with hard nylon or steel lines, special adapters with in-line valves must be used.*

## LEAKING INJECTORS

If a leaking fuel injector is suspected, it must be diagnosed and serviced. Failure to do so will fill the cylinder with an excessive amount of fuel. When this occurs, gasoline will leak past the piston rings and dilute the engine oil, resulting in increased cylinder wear and possible bearing damage. In cases where severe leaking exists, the cylinder in question could hydra lock. This could cause severe damage, such as a bent connecting rod, to internal engine components.

**You Should Know** *Gasoline might leak from the fuel injectors. Extreme caution should be exercised. The ignition system should be disconnected and all heat sources removed from the immediate area. A fire extinguisher should be within arm's reach during any fuel system service.*

When a leaking fuel injector is suspected, the injector can be isolated by performing an injector balance test or

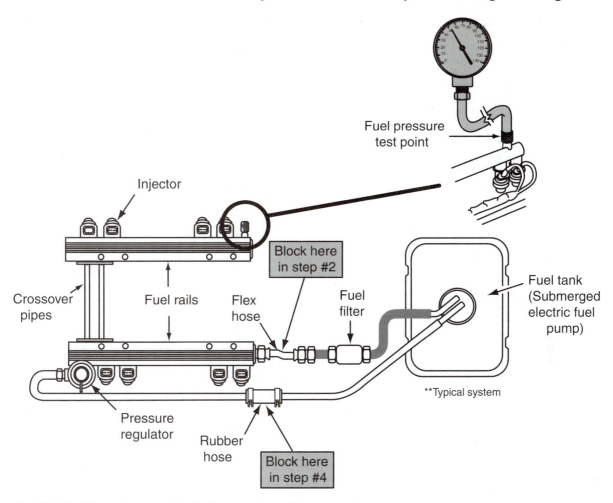

Fuel pressure test point

Injector

Crossover pipes

Fuel rails

Flex hose

Block here in step #2

Fuel filter

Fuel tank (Submerged electric fuel pump)

Pressure regulator

Rubber hose

Block here in step #4

**Typical system

**Figure 7.** Block off locations used to isolate engine off pressure drop.

by observing the injector tips with the fuel rail removed. A visual inspection can be performed by removing the fuel rail and injectors from the intake manifold. Disable the ignition system following the procedure outlined in the service manual. Loosen the fuel rail to expose the tips of the injectors **(Figure 8)**. Turn the key to the on position without

starting the engine to build up fuel pressure. If an injector is leaking, it should be visible at this time **(Figure 9)**. Due to the location of TBI injectors, no disassembly is required to observe the injector tip. Leaking injectors must be serviced by either replacement or cleaning. Refer to injector cleaning later in this chapter.

**Figure 8.** Removal of the fuel rail is required to observe fuel injector tips.

**Figure 9.** An illustration of a leaking fuel injector.

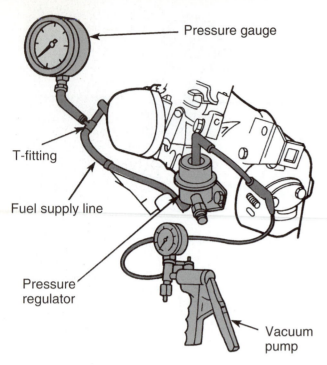

**Figure 10.** Testing a fuel pressure regulator using a vacuum pump.

## FUEL PRESSURE REGULATOR DIAGNOSIS

A faulty fuel pressure regulator can cause both high and low fuel pressure readings. Testing of the fuel pressure regulator should also be performed with the fuel pressure gauge connected. To test the operation of the regulator, quickly snap the throttle open. The gauge should momentarily rise to account for the loss of vacuum. When vacuum rises, the pressure will again settle back near the first reading. Disconnecting the vacuum hose from the regulator will cause the pressure to again rise. The pressure should go back down when the regulator is again connected. When performing this test, look for the presence of fuel in the vacuum line. If fuel is present, the regulator diaphragm is leaking and should be replaced. If further diagnosis of the regulator is desired, a hand-held vacuum pump can be connected in place of the vacuum line (**Figure 10**). With the engine running, apply approximately 15 inches of vacuum. This should cause the fuel pressure to drop. If the regulator holds vacuum but pressure does not change, the regulator is defective. A regulator unable to hold a vacuum would also indicate a faulty regulator.

## INJECTOR TESTING

Malfunctioning fuel injectors can be a source of many different driveability concerns, lean conditions, rich conditions, and single-cylinder misfires. Clogging of either the inlet screen or the nozzles usually causes lean conditions in injectors. Rich conditions can be caused by internal damage

to the injector or debris stopping the injector needle from seating, resulting in a leak. Either of these conditions will often lead to single-cylinder misfire. In most cases, an injector problem can be narrowed down to one individual cylinder. This can be accomplished by performing a power balance test or by observing misfire data with a scan tool.

The first step in diagnosing an injector problem is to verify that the injector is being supplied with both power and ground. This can be accomplished in two different ways. The simplest is to listen to the injector coil with a stethoscope (**Figure 11**). You should be able to hear the injector click. If not, a check of the injector harness is needed. This can be accomplished with a **noid light**. A noid light is a special test lamp that is plugged directly into the injector harness (**Figure 12**). If the light illuminates while the engine is cranking, further inspection of the injector is needed. If, however, the noid light fails to illuminate, further diagnosis of the control circuit is needed. Remember that

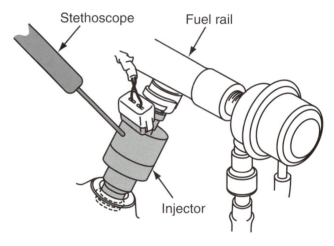

**Figure 11.** A stethoscope can be used to test the operation of a fuel injector.

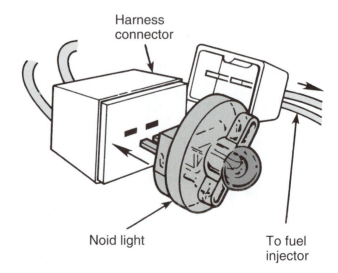

**Figure 12.** A noid light is used to check the operation of the injector control circuit.

the injectors should be supplied with 12V when the engine is cranking over. The ground is controlled using PWM from the PCM. The PCM provides a ground path when it receives an rpm signal.

Once correct injector control has been verified, further injector diagnosis can proceed. There are many tests that can identify a fuel injector problem. For the scope of this text, we look at two methods that will identify the majority of the injector failures encountered by technicians. The resistance test is the first and simplest test to be performed. Simply unplug the injector connector and place the leads of an ohmmeter across the injector terminals **(Figure 13)**.

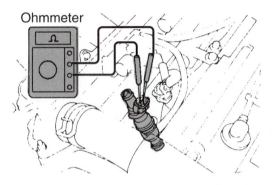

**Figure 13.** An ohmmeter is connected directly to the injector terminals to test coil resistance.

The resistance measured can be compared with that found in the vehicle service manual. Many technicians might simply compare the readings of the suspect injector with those of several others on the engine. This test operates on the premise that all of the other injectors are in good working order. If the resistance is incorrect, the injector must be replaced.

Once resistance has been verified, a balance test is in order. A balance test measures the amount of pressure drop provided by the injector. Testing the pressure drop allows us to test the flow of fuel through the injectors. A lower-than-expected drop could indicate a restricted injector tip or restricted inlet screen. A higher-than-expected pressure drop can be caused by debris holding the injector open when it should be closed. Every style of injector will have a specific amount of desired pressure drop.

An injector pulse tester is used to simulate injector control from the PCM. The pulse tester is a small box that will pulse the injector for a specific amount of time. The box has two sets of leads; one set connects to the injector, and the other set provides a connection to the positive and negative leads of the battery. To test the injector, the technician depresses the momentary switch on the box. This activates the tester and pulses the injector in a manner similar to what the PCM normally would. When testing multiple injectors, it might be helpful to make a chart to keep track of all results **(Figure 14)**. The steps to performing a balance test are as follows:

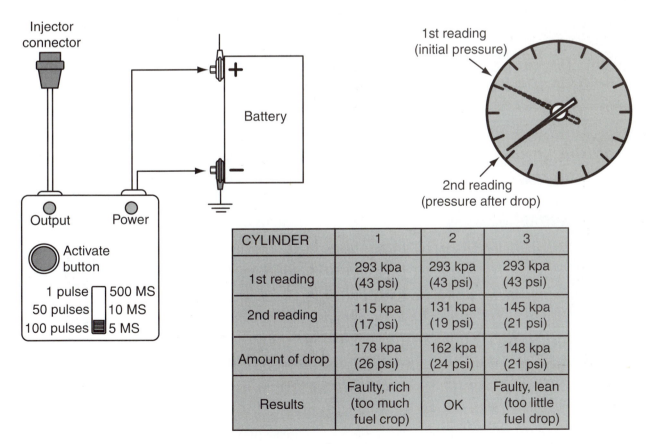

| CYLINDER | 1 | 2 | 3 |
|---|---|---|---|
| 1st reading | 293 kpa (43 psi) | 293 kpa (43 psi) | 293 kpa (43 psi) |
| 2nd reading | 115 kpa (17 psi) | 131 kpa (19 psi) | 145 kpa (21 psi) |
| Amount of drop | 178 kpa (26 psi) | 162 kpa (24 psi) | 148 kpa (21 psi) |
| Results | Faulty, rich (too much fuel crop) | OK | Faulty, lean (too little fuel drop) |

**Figure 14.** Using a chart to help organize injector test information.

1. Connect a fuel pressure gauge as described earlier in this chapter.
2. Connect the pulse tester to the injector being tested.
3. Turn the ignition key to the on position to build up fuel pressure. Record this reading.
4. Depress the button on the pulse tester. The fuel pressure on the gauge will drop. Record this second reading.
5. Subtract the second reading from the first. This is the amount of pressure drop for this injector.
6. If specifications are not available for pressure drop, perform steps 1–5 on the remaining injectors and calculate the average amount of drop. Compare the injector in question to the calculated average.
7. If all pressure drops are within 10 percent of each other, the injector flow should be considered acceptable. If flow is significantly different on the suspect injector, that injector must be serviced. In some instances, cleaning the injector might correct injectors that have excessive flow or restricted flow. (Refer to **Figure 15**, **Figure 16**, **Figure 17**, **Figure 18**, and **Figure 19** for the steps to performing an injector balance test.)
8. Repeat steps on additional injectors as needed.

**Figure 15.**   Connect the fuel pressure gauge to the fuel system.

**Figure 16.**   Disconnect the injector to be tested.

**Figure 17.**   Connect the pulse tester to the injector and battery, observing polarity.

**Figure 18.**   Cycle the ignition to build adequate test pressure. Note the reading.

**Figure 19.**   Depress and release the switch of the pulse tester. Observe the new pressure reading.

## FUEL INJECTOR CLEANING

Fuel injector cleaning is a procedure in which strong chemicals are circulated through the fuel rail or throttle body to remove deposits from the injectors. In many cases, cleaning can restore the performance of an injector. Cleaning can be used to correct a lean condition or rich condition. However, in some instances, injector deposits might be too difficult to remove and the injector(s) must be replaced. A secondary benefit that is often as important as cleaning the injectors is that the cleaner is often absorbed into the carbon deposits on the back of the valves, pistons, and combustion chambers. This aids in the removal of carbon deposits from these areas.

To perform this procedure, special equipment must be used. Fuel system cleaners that are added into the fuel tank are not strong enough to remove tough injector deposits. However, these additives can be useful as a maintenance item to use every few thousand miles to discourage the buildup of debris.

The most common type of equipment connects to the fuel system in the same manner that a fuel pressure gauge would be connected and uses cleaner in either an aerosol can or liquid that is pressurized with compressed air in a special vessel. The chemical is introduced into the fuel system at or about the same pressure that is needed for fuel system operation. Once the equipment is connected, the vehicle is then started and allowed to run using only the cleaning chemical. Many different fuel injection cleaning systems exist, and specific equipment directions should be observed. A common style of injector cleaning system is shown in **Figure 20**.

**Figure 20.** A commonly used injector cleaning unit.

**You Should Know** *It is critical that all steps provided by the equipment manufacturer be followed without modification. Failure to do so can result in personal injury and/or equipment and vehicle damage.*

General injector cleaning steps are as follows:
1. Disable the fuel pump. This can be accomplished by removing the fuel pump fuse or relay.
2. Block the fuel return line to prevent cleaning chemicals from entering the fuel tank.
3. Attach cleaning equipment per instructions. This is usually done in the same location as the fuel pressure gauge is attached.
4. Set the equipment to the proper operating pressure.
5. Start and run the engine. Hold the engine speed at a fast idle, approximately 2000–2500 rpm.
6. Allow the engine to run until all of the cleaning solution has been used.
7. Restore the engine to operational condition.
8. If the cleaning procedure was performed to correct a problem that you suspect was caused by debris, recheck the suspect injector at this time.

**You Should Know** *Because the chemicals used to clean fuel injectors are highly corrosive, some manufacturers might advise against cleaning fuel injectors. This is because the chemicals can damage the protective coating on the fuel injector coils.*

## FUEL QUALITY DIAGNOSIS

Driveability problems caused by poor fuel quality are often difficult to locate. Driveability concerns associated with contaminated fuel or excessive alcohol can often appear as intermittent stalling, chugging, or surging and detonation. Because these concerns are also common with other systems, it is recommended, if a driveability concern is not readily evident, to perform the following fuel quality tests early in your diagnosis.

The first step in analyzing fuel quality concerns is to obtain a sample of the fuel. This can be accomplished by using the bleeder valve on the fuel pressure gauge. Drain the sample into a clear container. The fuel should be readily transparent and present a clear to a golden color. If the fuel sample has an excessive amount of floating trash and dirt particles or appears cloudy, the fuel tank should be drained and cleaned out.

Another contributor to fuel-induced driveability concerns is alcohol. Although alcohol is a common gasoline additive, concentration by volume should be 10 percent or less. When concentrations exceed 10 percent, driveability concerns such as those mentioned at the beginning of this section are common. Testing alcohol is a relatively simple task to perform.

1. Obtain a 100-ml container with 1-ml graduations.
2. Add 90 ml of the suspect fuel to the cylinder.
3. Add 10 ml of water to the container.
4. Install a sealed cap and shake the mixture vigorously for 15 seconds.
5. Relieve any pressure and shake it again.
6. Remove the cap and let stand approximately 5 minutes.
7. The alcohol and water will mix together and fall to the bottom of the container. This will form visible layers in the container.
8. If the volume of water at the bottom of the container now exceeds 10 ml, excessive amounts of water or alcohol are present in the fuel. For example, if the water volume were now 20 ml, this would indicate that at least 20 percent of the sample is alcohol or water. This indicates that the fuel tank should be drained, cleaned, and refilled. It should be noted that all of the alcohol will not be drawn out of the fuel using this procedure, making it possible that alcohol content can be higher than indicated.

# Summary

- Fuel system problems are often recognized as a loss of power, an engine no-start condition, or misfiring. A fuel pressure reading should be taken anytime driveability diagnosis is performed.
- The fuel pressure gauge must be selected for the range of pressure that will be measured. Fuel pressure must be checked at a location between the fuel pressure regulator and the fuel filter.
- Fuel pressure measurements are obtained with the key-on–engine-off and with engine running.
- Low fuel pressure can be caused by a weak fuel pump, a blocked supply line, a severely clogged fuel filter, or a defective fuel pressure regulator. Higher-than-normal fuel pressure can be caused by a restriction in the return line or a defective fuel pressure regulator.
- Maximum fuel pressure can be obtained by momentarily blocking the fuel return hose.
- Fuel pressure should hold steady for several minutes after the fuel pump has turned off. Rapid loss of fuel pressure can indicate a leaking fuel injector, faulty pressure regulator, or defective fuel pump check valve.
- A faulty fuel pressure regulator can be the cause of both high and low fuel pressure readings. When vacuum is applied, the fuel pressure should drop. When vacuum is removed, the pressure will rise. If fuel is present in the vacuum hose connected to the regulator, the regulator should be replaced.
- If the engine stalls with the fuel pressure gauge bleed valve open, low fuel pump volume might be indicated.
- Leaking fuel injectors can eventually lead to serious engine damage such as excessive cylinder wear or damaged engine bearings. A leaking fuel injector can be identified by performing an injector balance test.
- The first step in injector diagnosis is to verify the operation of the control circuit. An injector balance test measures the pressure drop through the injector. Cleaning the injectors involves the use of strong chemicals to remove injector deposits.
- Driveability problems caused by contaminated fuel are rare. However, when they do surface, they are difficult to identify.

# Review Questions

1. All of the following statements concerning fuel systems are true *except:*
   A. Fuel system problems can cause a variety of customer concerns.
   B. A fuel pressure reading should be taken anytime a driveability diagnosis is performed.
   C. A thorough visual inspection should always be the first diagnostic step.
   D. Fuel pressure readings should be taken only to verify a fuel system concern.

2. Technician A says that one fuel pressure gauge will test all fuel injection systems. Technician B says that fuel pressure readings can be taken at any location between the fuel filter and the fuel pressure regulator. Who is correct?
   A. Technician A
   B. Technician B
   C. Both A and B
   D. Neither A nor B

3. Technician A says that fuel pressures taken with the engine running should be higher than those with the engine off. Technician B says that maximum fuel pump specifications should be lower when a vacuum is applied to the fuel pressure regulator. Who is correct?
   A. Technician A
   B. Technician B
   C. Both A and B
   D. Neither A nor B
4. Technician A says that a fuel pressure regulator that is stuck in the open position will cause the fuel pressure to be higher than normal. Technician B says that a fuel pressure regulator that is stuck closed will cause fuel pressure to be higher than normal. Who is correct?
   A. Technician A
   B. Technician B
   C. Both A and B
   D. Neither A nor B
5. All of the following statements about fuel pressure are correct except:
   A. Fuel pressure should hold steady for several minutes after the fuel pump stops running.
   B. A restricted fuel filter can cause fuel pressure to drop once the fuel pump stops running.
   C. An external leak at the fuel gauge connection will cause fuel pressure to drop once the fuel pump stops running.
   D. A dirty fuel injector can cause the fuel pressure to bleed down once the fuel pump has turned off.
6. Technician A says that a leaking fuel injector can cause the engine to run rich. Technician B says that a leaking fuel injector can be identified measuring the coil resistance. Who is correct?
   A. Technician A
   B. Technician B
   C. Both A and B
   D. Neither A nor B
7. Technician A says that blocking off the fuel return hose can help isolate an internal leak caused by the fuel pressure regulator. Technician B says that fuel leaking into the regulator vacuum hose can cause a lean air/fuel ratio. Who is correct?
   A. Technician A
   B. Technician B
   C. Both A and B
   D. Neither A nor B
8. Technician A says that a dirty fuel injector can cause a lean condition by restricting fuel flow. Technician B says that a dirty fuel injector can cause a rich condition by not allowing the injector needle to seat. Who is correct?
   A. Technician A
   B. Technician B
   C. Both A and B
   D. Neither A nor B
9. Technician A says that an injector with excessive pressure drop might indicate that the injector is sticking open. Technician B says that a single injector with lower-than-normal pressure drop might be due to a restriction between the fuel pump and the fuel rail. Who is correct?
   A. Technician A
   B. Technician B
   C. Both A and B
   D. Neither A nor B
10. All of the following statements about fuel quality are correct except:
   A. The fuel should be clean with no visible dirt or debris present.
   B. Water and alcohol will mix together.
   C. Contaminated fuel can mimic other fuel system concerns.
   D. Alcohol is a fuel system contaminant.

# Chapter 31

# Induction and Exhaust System Diagnosis

## Introduction

In Chapter 30, you learned how to perform several fuel system tests. With Chapter 31, we add to your diagnostic arsenal of testing procedures. By the conclusion of this chapter, you will also have a good foundation on how and when to apply the knowledge that you have acquired. In addition to these tests we look at some common service procedures that apply to almost any fuel-injected engine. As with any specific service procedures, you should always refer to manufacturer's service information for particular details.

## AIR INDUCTION SYSTEM

Whenever we begin diagnosis of the air induction or fuel systems, we should begin with a thorough visual inspection. This inspection should begin by examining the air cleaner housing and ducts to ensure that they are properly sealed. Clamps attaching the ducts should be tight, and ducts should be clear of any obstructions or kinks. The air filter is the only line of defense to keep dirt out of the engine. We inspect the cleanliness of the filter and the condition of the seal. You should further examine the area between the pleats for any trapped dirt and oil. If a filter is dirty, it should be replaced **(Figure 1)**.

 The throttle blades have very sharp edges. Extreme care should be taken when working around the throttle plates to avoid cuts and abrasions.

Torn seal

**Figure 1.** Inspect the filter element for excessive dirt and debris. Inspect the seal for tears or other damage. Closely examine the area between the pleats for dirt and debris.

The inspection should expand to include the fuel system and vacuum lines. Look for any loose connections, chafed wires, and loose or broken hoses. We should also perform at least part of the inspection with the engine running. We do this so we can listen for any unusual noises or vacuum leaks. Chapter 11 outlines various methods for locating both internal and external vacuum leaks. Any problems located during the visual inspection should be repaired before continuing with the diagnosis.

## THROTTLE BODY SERVICE

As a vehicle is driven, exhaust gases are allowed to enter the intake manifold to mix with fresh air and fuel in

**Figure 2.** Carbon buildup is common after only a few thousand miles of driving. Carbon buildup affects many engine operational characteristics.

order to reduce emissions. The introduction of exhaust gases forms layers of soot on the inner surfaces of the intake manifold and especially the passages within the throttle body. As the vehicle is driven over several thousand miles, the soot can become such a problem that minimum air rates are affected, including the ability of the IAC system to function properly. This can often manifest itself as engine stalling, idle surge, or high IAC counts (**Figure 2**).

Because the throttle plate opening is preset and should not be changed, periodic cleaning of the throttle body is required to maintain desired idle characteristics. This should entail cleaning the IAC passages and the backside of the throttle plates. This task can be accomplished using a brass brush and a commercially available aerosol throttle body cleaner or top engine cleaner (**Figure 3**).

Severe carbon buildup can require throttle body removal. Because TBI throttle bodies have fuel flowing across the throttle plates, carbon buildup on the plates is rarely a problem. However, it is still common for IAC and vacuum passages to become restricted.

> **You Should Know** *Throttle blades are often coated with a special coating that resists carbon buildup. Some manufacturers advise against using specific chemicals or cleaning altogether. These throttle bodies are usually labeled as such. If cleaning is prohibited, consult the specific service manual for service procedures.*

## IDLE AIR CONTROL SYSTEM DIAGNOSIS

The IAC system is a frequent cause of driveability concerns such as stalling or engine surge. These problems are often the result of carbon buildup on the valve and clogged passages. By comparing current IAC counts to an expected range found in the service manual, the technician can develop an idea of what might be occurring within the system (**Figure 4**).

When airflow is restricted, idle rpm will be lower. The PCM reacts by retracting the IAC valve to allow more airflow. This can be seen on a scan tool as high IAC counts. If IAC counts are higher than normal, blockage of the IAC passages might be indicated. Note that IAC counts will also be higher when the engine is under a load, such as when the A/C compressor is engaged.

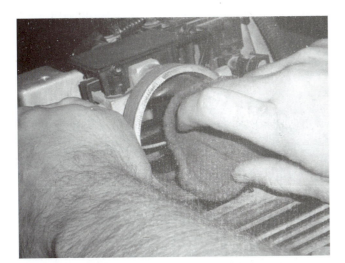

**Figure 3.** A throttle body can usually be cleaned while still on the vehicle.

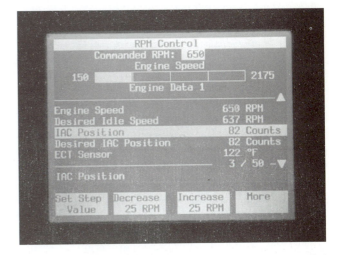

**Figure 4.** Observing the IAC position can be useful in diagnosing some engine stalling conditions.

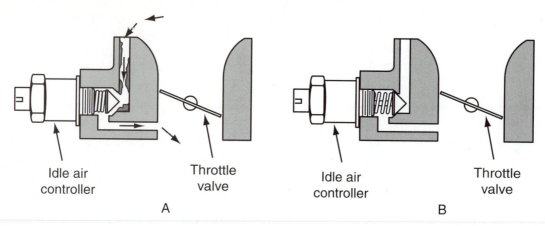

Idle air
controller

Throttle
valve

A

Idle air
controller

Throttle
valve

B

**Figure 5.**   (A) The PCM will retract the IAC valve pintle to compensate for excessive carbon buildup. (B) The PCM will extend the IAC pintle to compensate for vacuum leaks.

Low IAC counts indicate that unmetered air is entering the engine. This is often caused by a vacuum leak. When a vacuum leak is present, the engine receives more than enough air to idle. The PCM reacts by trying to close off as much air as possible. This can be seen as very low IAC counts; in some cases, IAC counts might be zero. Both of these examples are illustrated in **Figure 5**.

> **You Should Know** *As a point of reference, install a scan tool to monitor the engine data. Set it to monitor IAC counts. Change the engine load by turning on the A/C or moving the transmission from drive to park. Adding load to the engine should cause the counts to increase. Next, create a small vacuum leak, and again note the change to the IAC counts. This simple experiment will give you an idea of how the PCM responds to various load conditions. You will be surprised at how the slightest change in load will affect IAC.*

Most scan tools have a special function test that allows the technician to fully extend or retract the valve. If the IAC valve can be fully extended and retracted with the scan tool, a control circuit problem is not likely. However, if the valve does not extend and retract as it is commanded, the valve could still be faulty. To further isolate the problem, plug a known good IAC valve with the correct connector into the vehicle harness. Using the scan tool, command the IAC to extend and retract. If the valve operates in concert with the commands, the control circuit is working. If the valve still does not move, a control circuit failure is indicated and further diagnosis must be performed.

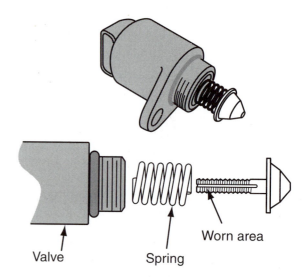

Valve

Spring

Worn area

**Figure 6.**   Mechanical wear of the IAC valve will adversely affect the operation of the valve.

A common problem with IAC valves is that the delicate threads on the pintle that cause it to extend and retract become worn **(Figure 6)**. This allows the pintle to vibrate out to the fully extended position. When this occurs, the engine stalls when the throttle is closed. Because the PCM has not issued a command to extend the pintle, it believes that the valve is in the proper position and the IAC counts reflect the last position that was commanded. When the vehicle is returned to idle, the PCM expects the IAC to provide adequate airflow, which it will not be able to do. This causes the engine to stall. When this occurs, the engine can be restarted by slightly opening the throttle. If the IAC counts were viewed after this occurrence, they would be higher than normal. This is an indication that the PCM is trying to compensate by opening the valve further.

# EXHAUST SYSTEM INSPECTION

The exhaust system is relatively simple. Many of the problems caused by the exhaust system can be detected by means of a visual inspection. Most exhaust system concerns deal with noises; these come in the form of exhaust leaks, rattles, or vibrations.

*Most technicians will save various sensors and actuators for circuit testing purposes. These are typically parts that are no longer able to properly perform on the vehicle but remain in a condition that allows them to test the operation of various control circuits. Although this should not be the extent of a technician's diagnostic capability, it can be a quick, easy, and accurate way to verify the operation of some control circuits that would otherwise be difficult to test.*

## Exhaust Leaks

Because exhaust system components are located beneath shields or other components, exhaust gas leaks are often difficult to detect. A proven method for detecting exhaust leaks is performed by introducing smoke into the engine **(Figure 7)**. A widely accepted method for many years involved introducing small amounts of transmission fluid into the intake manifold while the engine is running.

**Figure 7.**   A smoke machine is useful for locating many different types of engine leaks.

When transmission fluid burns, it produces a bright white smoke. The smoke can easily be seen escaping from any leak in the exhaust system. Because of the difficulty required to introduce the fluid into the intake manifold on PFI-equipped vehicles and the risk of contamination to $O_2Ss$ and the catalytic converter, this method is rarely used today.

A new method that has gained acceptance in recent years is the use of a smoke machine. Smoke machines are used to force smoke throughout the engine. The smoke machine produces the same effect as transmission fluid. However, the smoke machine is used with the engine off. This makes testing much easier, alleviating the difficulty of working around the hot exhaust components. The smoke machine has many other uses, such as finding **Evaporative emissions control system (EVAP system)** leaks, oil leaks, and vacuum leaks.

## Noise and Vibration

Rattling within the exhaust system is often caused by loose heat shields or broken system hangers. In some cases, the material inside the catalytic converter or muffler will loosen and cause a rattle within that is noticeable to the customer. These noises are often heard while driving the vehicle. A thorough visual inspection while jarring the exhaust system will often locate these problems.

Vibrations within the exhaust system will usually occur at a very specific engine speed or load condition. As the exhaust pulses enter the exhaust system, they set up natural vibrations. Vibrations can be transferred to the passenger compartment if the components of the exhaust system are misaligned or touch the frame or body. Vibrations can also be transferred if an engine or exhaust system mount is broken. To properly diagnose these types of problems, it might be necessary to have an assistant duplicate the vibration while you are observing where the components might be touching. Pay particular attention to any damage that might be evident.

## BACKPRESSURE TESTING

Backpressure is a result of a restriction in the exhaust system. Every internal combustion engine has some amount of exhaust system backpressure; in fact, engines and systems are designed to use some amount of exhaust backpressure. However, if backpressure should become too great, engine performance will suffer. Excessive backpressure will manifest itself as a severe loss of engine power.

Two pieces of equipment are available for diagnosing engine exhaust backpressure conditions: a vacuum gauge and a back pressure gauge. The vacuum gauge is the simplest piece of equipment to attach. The steps for testing backpressure with a vacuum gauge are as follows:

1. The vacuum gauge is attached to a manifold vacuum source.
2. Start the engine and record the vacuum reading at idle.

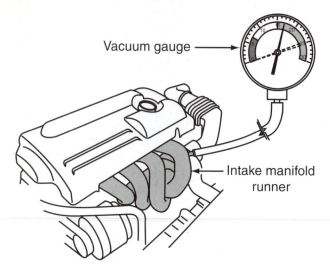

**Figure 8.** The vacuum gauge is a useful tool for locating a restriction in the exhaust system.

**Figure 9.** A backpressure gauge can be used to locate an exhaust restriction.

3. Steadily accelerate the engine to approximately 2500 rpm and maintain this speed.
4. Observe the gauge reading.

A normally operating engine should obtain a reading equal to that recorded at idle and remain steady. A vehicle with a severe exhaust restriction will allow the vacuum to steadily fall, and a noticeable decrease in engine rpm might be noticed. In cases of severe restriction, vacuum might fall to near zero **(Figure 8)**. Restriction can be isolated to a specific section of the exhaust by disconnecting individual sections of the exhaust one at a time and rerunning the test.

> **You Should Know** *Exhaust restrictions are often a result of a severely overheated catalytic converter. When this occurs, the converter material can melt and form a solid block, causing a restriction. In some cases, the material might break up and pass into the muffler, causing a restriction in the muffler. If the converter shell is found to be empty, the muffler should be inspected.*

The second method of testing for excessive backpressure is by using a backpressure gauge. A backpressure gauge is a special pressure gauge that is attached in place of the $O_2S$ **(Figure 9)**. Because the gauge attaches in place of the $O_2S$, it can be moved between pre- and postcatalyst $O_2Ss$ to specifically check the amount of backpressure caused by the catalytic converter. The steps to using a backpressure gauge are as follows:

1. Remove one of the precatalyst $O_2Ss$.
2. Install the backpressure gauge in place of the sensor.
3. Start the engine and accelerate the engine to 2500 rpm.
4. Backpressure should not exceed 2–3 psi.

If pressure exceeds this reading, move the gauge to the postcatalyst $O_2S$ fitting. If pressure is below 2–3 psi, the converter is the cause of the restriction. If pressure is above 2–3 psi at this point, the restriction is located downstream of the converter and must be diagnosed by component removal.

An infrared thermometer is also useful for locating exhaust restrictions. Temperature among exhaust system components should steadily and gradually decrease near the end of the exhaust. If the exhaust system is restricted, a large temperature differential can be detected on either side of the restriction.

## TESTING THE FORCED INDUCTION SYSTEM

Forced induction systems typically provide many years of trouble-free service. When an internal failure of the supercharger or turbocharger occurs, replacement of the complete unit is the most economical repair. Because of this, our diagnosis centers on very basic testing.

The diagnosis of a supercharger or turbocharger should begin with a visual inspection. This includes inspection of the air filter, ducts, hoses, and connectors. When proceeding through the basic diagnostic, listen for any unusual noises. Supercharger operation will be relatively quiet, whereas the turbo has some amount of inherent whine as speed increases.

Boost pressure can be measured with a gauge connected to the manifold. When measuring boost pressure on a supercharged engine, the measurement must be taken between the supercharger outlet and the cylinder head. This can be difficult on some designs. If the measurement is taken near the inlet, a vacuum will be measured. Improper boost pressures can often be traced to the by-pass valve operation. To ensure proper diagnosis of a supercharger and by-pass system, refer to the vehicle service manual.

*Interesting Fact*

*Low power concerns caused by low fuel pressure and a restricted exhaust can feel almost identical when the vehicle is driven. A scan tool will help determine which system is causing the problem. When test driving the vehicle, observe the O$_2$S under heavy acceleration. If the O$_2$S voltage is low, a fuel system failure is likely. If the O$_2$S voltage is high, a restricted exhaust might be the cause.*

Because the turbo uses ducts to connect to a normal manifold, a pressure reading can be taken anywhere on the intake manifold or in the boost control hose on the wastegate actuator. Like the supercharger, improper boost pressures might be linked to operation of the wastegate. Before performing an in-depth diagnosis, make sure that the wastegate and the linkage connecting the wastegate actuator to the wastegate have the ability to move freely. Refer to the service manual for specific diagnosis of the turbocharger and wastegate system.

## DRIVEABILITY SYMPTOMS

There are a variety of driveability symptoms that the engine performance technician is exposed to. Many of these are related to the fuel, air induction, and exhaust systems. Problems with these systems often have a definite signature. Problems associated with these systems are often recognized by the fact that they occur over time. What this means is that in most cases fuel, air induction, and exhaust system problems give some indication, such as a surge or a loss of power, that a problem exists. An ignition problem or electrical problem will often appear without warning and disappear just as rapidly. For example, both an ignition system failure and a fuel system failure can cause the engine to stall. The ignition problem is likely to stall the engine with no prior warning, whereas the fuel system failure will typically cause the engine to surge or lose power before stalling. This is information that is usually obtained from the customer. Technicians use this information in the process of narrowing down a concern to a specific system. Here are some examples of common fuel system–related customer concerns:

- *No-start condition.* No-start conditions can be caused by lack of fuel pressure, a severely restricted exhaust, a defective IAC system, or an injector control malfunction.
- *Lack of power.* Lack of power is usually identified with low fuel pressure and/or volume or a restricted exhaust.
- *Stalling.* Stalling can be caused by a fuel pressure concern or an IAC malfunction.
- *Hesitation.* Hesitation is very often caused by low fuel pressure, clogged injectors, or a restricted fuel filter.
- *Surging.* Surging is typically caused by a vacuum leak, low fuel pressure, a restricted fuel filter, or an incorrectly operating IAC system.

# *Summary*

- All diagnosis should begin with a visual inspection. Problems located during the visual inspection should be repaired.
- Carbon buildup in the manifold and throttle body will affect operation of the throttle body and IAC system.
- Carbon buildup can result in stalling, surging, or high IAC counts. Carbon can be removed from the throttle body by using a brass brush and a throttle body cleaner.
- Many IAC problems are a result of carbon buildup. High IAC counts might indicate a restriction in the IAC passages. Low IAC counts might indicate a vacuum leak.

- Many exhaust system problems are recognized as noises. Most exhaust system concerns can be diagnosed through visual inspection.
- Backpressure is the result of a restriction in the exhaust system. Excessive backpressure will greatly affect engine performance.
- The vacuum gauge and backpressure gauge are used to detect exhaust system restrictions. A damaged catalytic converter is often a source of excessive exhaust system backpressure.

# Review Questions

1. Technician A says that a visual inspection can locate the cause of many driveability concerns without extensive diagnosis. Technician B says that a leaking air induction system duct should be repaired as part of the visual inspection. Who is correct?
   A. Technician A
   B. Technician B
   C. Both A and B
   D. Neither A nor B

2. Technician A says that excessive carbon buildup in the throttle body can cause an engine to stall during deceleration. Technician B says that excessive carbon buildup in the IAC passages will cause lower-than-normal IAC counts. Who is correct?
   A. Technician A
   B. Technician B
   C. Both A and B
   D. Neither A nor B

3. Technician A says that low IAC counts can be caused by a vacuum leak. Technician B says that all throttle bodies can be cleaned with a brass brush and throttle body cleaner. Who is correct?
   A. Technician A
   B. Technician B
   C. Both A and B
   D. Neither A nor B

4. All of the following problems concerning exhaust system diagnosis are true *except:*
   A. Many exhaust system problems can be identified through a thorough visual inspection.
   B. Some technicians use transmission fluid to detect exhaust leaks.
   C. An overheated catalytic converter can cause an exhaust system rattle.
   D. A loose heat shield can be removed to stop an exhaust rattle.

5. Technician A says that an overheated catalytic converter might require that both the converter and muffler be replaced. Technician B says that a leaking fuel injector can eventually lead to a restricted exhaust. Who is correct?
   A. Technician A
   B. Technician B
   C. Both A and B
   D. Neither A nor B

# Section 6

## Emissions Control

**Interesting Fact**

*A 1969 Mustang equipped with a 302-cubic-inch engine produces more than 100 times more tailpipe emissions than a 2003 model equipped with a similar-sized engine.*

## SECTION OBJECTIVES

After you have read, studied, and practiced the contents of this section, you should be able to:

- Recognize the importance of air quality issues and how the combustion process affects air quality.
- List the major exhaust and evaporative pollutants and their effects on the environment.
- Discuss how oxides of nitrogen are formed and how exhaust gas recirculation lowers the production of NOx.
- List the various causes of evaporative emissions.
- Explain how a PCV system reduces crankcase vapors.
- Explain how the evaporative emissions system controls the release of gasoline vapors.
- Explain how exhaust aftertreatment can reduce harmful tailpipe emissions.
- Explain the difference between a reducing catalyst and an oxidizing catalyst.
- Discuss how a secondary AIR system affects exhaust emissions and list the major system components.
- List the symptoms of failures in the exhaust gas recirculation system.
- Explain how to test the PCV system.
- Discuss test procedures used in locating faults in the evaporative emissions system.
- Explain how to test a catalytic converter for proper operation.
- List steps necessary to verify that the AIR system is functioning correctly.
- Defend the need for service publications when conducting emissions system diagnostics.
- List the gases measured by a four-gas and a five-gas analyzer.
- Explain how a scan tool can be used to diagnose emissions-related concerns.
- Discuss the various monitors used by the OBDII system that are used to diagnose emissions system failures.
- List the various emissions tests recognized by the Environmental Protection Agency.
- List causes for tailpipe emissions failures and explain the relationship among the various gases.

# Chapter 32

# Introduction to Air Quality and Emissions Control Systems

## Introduction

Dating as far back as the early 1960s, air pollution was becoming a serious concern. Particularly in California, **air pollution** was an ever-increasing problem. In 1963, Congress passed the first Clean Air Act. An amendment passed in 1965 added motor vehicles to the list of pollution sources.

As a result of the Clean Air Act, automotive manufacturers began a program of vehicle emission controls to meet regulations, first in California in 1966 and nationally in 1968. Congress established the Environmental Protection Agency (EPA) in 1970 to enforce the policies set forth in the Clean Air Act.

Since that time, the Clean Air Act has been amended several times to include strict exhaust and evaporative emissions limits on passenger cars and light trucks and, most recently, diesel-powered trucks. As air quality issues evolve, emissions control devices will likely be included on all fossil fuel–powered engines, from snowmobiles to lawnmowers.

## AIR QUALITY CONCERNS

The concern over air quality has existed since the 1940s, when a relationship between vehicle emissions and air quality was suggested by research in California. However, motor vehicles are not the only source of air pollution. **Fossil fuel–burning** power plants, industrial processing plants, commercial and residential heating systems, and refuse disposal facilities all contribute to air pollution. **Smog** is a term that was coined over 35 years ago to mean a mixture of smoke and fog. Over the years, it has come to be known as a noxious mixture of air pollutants. This mixture can often be seen as a haze in the air.

Fog formation was particularly offensive in many European cities when burning coal was the primary fuel. Because of the shift to oil and natural gas and the addition of exhaust emission controls, this type of air pollution has been substantially reduced.

Smog was initially identified in Los Angeles as a brown cloud hanging over the city. Unfortunately, many cities around the world, including New York and Tokyo, are affected by smog. Smog results from photochemical and chemical reactions occurring in the atmosphere. Sunlight aggravates the problem because it causes the formation of ozone.

## A BRIEF HISTORY OF EMISSIONS CONTROL

Emissions control was first introduced on California vehicles in 1961. Before any state or federal legislation was introduced, vehicle manufacturers began installing a positive crankcase ventilation, or PCV, valve to recycle crankcase HC vapors into the engine's intake system. By 1963, automotive manufacturers extended the use of a PCV valve nationally.

> **Interesting Fact**
>
> The term "London fog" has nothing to do with atmospheric fog caused by tiny droplets of water vapor suspended in the air. Instead, it is related to air pollution caused from smoke and gases arising from burning **hydrocarbon (HC)** fuels such as coal. Poor air circulation and cool, damp conditions aggravate the problem.

In 1970, California vehicles incorporated an evaporative emissions control system designed to contain unburned HC vapors. These vapors were previously allowed to escape from the carburetor and fuel tank. Evaporative controls became standard equipment on 1971 models across the nation.

During the 1960s and 1970s, improved control of spark advance helped reduce tailpipe emissions, while improving fuel economy. Electronic ignition was on the horizon in the 1960s but was not standard equipment until the early 1970s. Electronic control of spark advance was still many years away.

Carburetor preheat systems helped reduce tailpipe emissions by permitting more consistent fuel metering, including the use of leaner fuel mixtures. A heat stove fastened around the exhaust manifold supplied a source of warm air to the engine's air intake system. This warm air provided a more stable air density under a variety of ambient temperatures, resulting in better control of air/fuel mixtures.

In an effort to combat the formation of **oxides of nitrogen (NOx)** produced by lean air/fuel ratios, an EGR valve was used in many applications. The EGR system continues to be an effective method in reducing the formation of NOx in today's engine technologies.

By 1975, many vehicles included a catalytic converter designed to treat the engine exhaust. This exhaust aftertreatment provided significant reductions to hydrocarbon (HC), carbon monoxide (CO), and oxides of nitrogen (NOx) emissions.

In 1975, Chrysler introduced the first "lean burn" engine technology on a production engine. Using electronic control of certain engine parameters, the first generation of electronic engine control was born.

Since the mid-1970s, the electronics industry has evolved tremendously. As with all areas affected by developments in the electronics field, substantial changes have occurred since the early days of engine management. Modern engine designs pollute less, provide superior fuel economy, and offer excellent driveability compared with early designs using a carburetor. Through the 1980s, 1990s, and into the millennium, the use of electronic engine control is perhaps the single most important factor in the quest for cleaner-burning vehicle technologies.

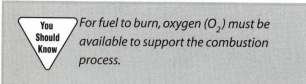

> **You Should Know**
> *For fuel to burn, oxygen ($O_2$) must be available to support the combustion process.*

## COMBUSTION PROCESS

To better understand how vehicle exhaust affects air quality, it is necessary to understand how the combustion process occurs. The combustion process should be thought

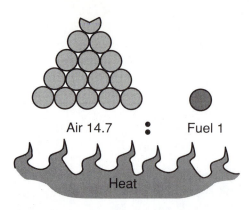

**Figure 1.** Combustion is a reaction between fuel and oxygen when activated by heat.

of as a reaction between fuel and oxygen ($O_2$) that is activated when sufficient heat is supplied (**Figure 1**).

More specifically, however, the amount of fuel and oxygen must be present in an appropriate ratio, depending upon the fuel used. Common fuels include gasoline, methanol, ethanol, compressed natural gas, and liquid petroleum gas (LPG). Each fuel is a unique combination of hydrogen and carbon atoms. For this reason, they are referred to as hydrocarbon fuels.

In the case of gasoline, an air/fuel ratio of 14.7:1 by weight is considered balanced. This is also called the stoichiometric point of combustion. Air/fuel ratios less than 14.7:1 are considered rich, in that more gasoline is present for a given amount of $O_2$. If the air/fuel ratio is greater than 14.7:1, the fuel mixture is classified as being lean.

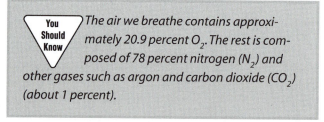

> **You Should Know**
> *The air we breathe contains approximately 20.9 percent $O_2$. The rest is composed of 78 percent nitrogen ($N_2$) and other gases such as argon and carbon dioxide ($CO_2$) (about 1 percent).*

To develop useful power from the engine, the air/fuel charge must fill each cylinder at the proper time determined by the camshaft. The piston must compress the air/fuel charge, an important step in the combustion process. Using the energy stored in the ignition coil, the spark plug provides the heat necessary to ignite the compressed air/fuel mixture (**Figure 2**).

## Chemical Nature of Gasoline

Gasoline is not composed exclusively of one type of hydrocarbon (HC) molecule. Instead, gasoline is a combination of many complex HC molecules, which form a chemical structure with hydrogen atoms around the outside of the molecule and carbon atoms in the center (**Figure 3**).

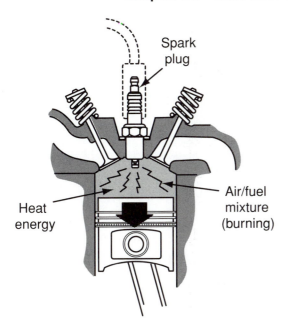

**Figure 2.** Compressed air-fuel charge is ignited by the spark plug. This changes energy stored in fuel into heat energy.

Heptane

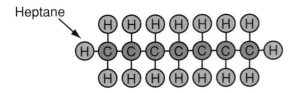

**Figure 3.** Gasoline is a combination of many complex hydrocarbon molecules.

This HC molecule is heptane, or $C_7H_{16}$, a common HC molecule found in gasoline. Other HC molecules present in gasoline are isooctane ($C_8H_{18}$) and benzene ($C_6H_6$).

Combustion is a chemical process that produces heat, light, and new molecules **(Figure 4)**. Because HCs will be combined with $O_2$ and nitrogen ($N_2$), these will be the key factors in the combustion process. Using a special chemical formula, the chemical reaction that occurs during combustion can be determined. In the case of gasoline ($C_8H_{18}$), the formula can be written as:

$$O_2 + N_2 + C_8H_{18} = CO_2 + N_2 + H_2O + O_2 + heat$$

From the equation, it is evident the products of the combustion process are carbon dioxide ($CO_2$), water ($H_2O$), a very small amount of $O_2$, and heat. During the reaction, gasoline and oxygen combine, forming the new molecules. Ideally, all of the HC molecules are completely oxidized by the $O_2$. Gases left behind from the combustion process travel out of the combustion chamber and cylinders into the exhaust manifold. The gases are discharged into the tailpipe, where they become part of the atmosphere.

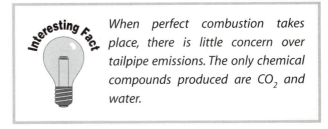

*When perfect combustion takes place, there is little concern over tailpipe emissions. The only chemical compounds produced are $CO_2$ and water.*

It is interesting to note that $N_2$ does not enter into the reaction at low temperatures. The presence of $N_2$ in the

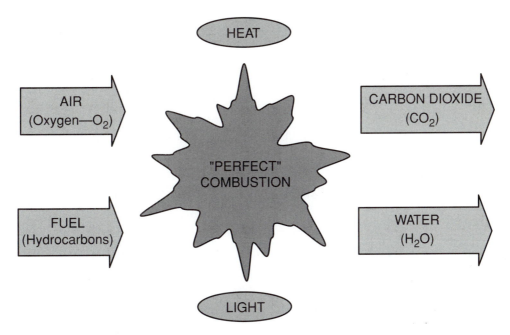

**Figure 4.** Combustion produces heat, light, and new molecules.

atmosphere is responsible for NOx emissions only when combustion temperatures exceed approximately 2500°F.

Although it appears any amount of fuel and $O_2$ can be combined together to produce combustion, this is not the case. To utilize all of the $O_2$ and fuel in the combustion chamber requires precise amounts of each. Considering that air contains about 21 percent $O_2$, the ideal or stoichiometric air/fuel ratio is 14.7:1, by weight. This translates into 14.7 pounds of air to completely burn 1 pound of fuel—in this case, gasoline. When perfect combustion takes place, no gasoline remains unburned at the end of the reaction. Likewise, no $O_2$ is left in the exhaust gases after combustion.

*Although $CO_2$ is not classified as a noxious pollutant, it is considered to be a **greenhouse gas**, which can contribute to global warming.*

## HOW RICH AIR/FUEL RATIOS AFFECT COMBUSTION

When a rich air/fuel mixture burns, some of the HC molecules do not properly enter into the reaction. Recall that a rich air/fuel mixture has more fuel present than needed for perfect combustion. Because of the rich air/fuel ratio, sufficient $O_2$ is not available for proper combustion. Two additional products of combustion result: unburned HC molecules and CO, a deadly gas.

Because of the rich air/fuel mixture, some HC molecules will *not* enter into the combustion process. When this occurs, unburned HC molecules will be present in the exhaust gas. The quantity of unburned fuel remaining after the combustion process depends upon how rich the air/fuel mixture is.

Additionally, some HC molecules will react with *less* oxygen. Those that react with less oxygen cause the production of CO.

The new chemical reaction contains other products not present when perfect combustion takes place. The chemical equation for a rich air/fuel mixture is:

$$O_2 + N_2 + C_8H_{18} = CO_2 + H_2O + CO + C_8H_{18} + N_2 + heat$$

Although the reaction still produces water and $CO_2$, the combustion process also produces CO and unburned gasoline ($C_8H_{18}$).

## HOW LEAN AIR/FUEL RATIOS AFFECT COMBUSTION

Lean air/fuel mixtures occur when there is a greater amount of $O_2$ present than required to burn all of the fuel present in the combustion chamber. Although there is little danger of any CO being produced, there is a threat of a condition known as "lean misfire."

When air/fuel ratios are excessively lean, it is possible for the combustion process to terminate, or quench, before all of the fuel has been burned in the reaction. In this case, a cylinder misfire occurs, leaving excessive amounts of HC molecules to pass into the exhaust stream, where they are discharged into the atmosphere.

Lean air/fuel ratios can also cause combustion chamber temperatures to increase significantly. This leads to the production of NOx. Although lean air/fuel ratios are often desirable, a means of controlling the production of NOx is necessary. This is usually in the form of an exhaust gas recirculation (EGR) system. EGR systems are discussed in Chapters 33 and 36.

# VEHICLE EXHAUST EMISSIONS

The term "exhaust emissions" refers to the discharge of combustion gases and particles into the atmosphere and is sometimes referred to as "tailpipe emissions" because the tailpipe is the point of discharge. Certain gases produced from the combustion process—such as water vapor—have no negative impact on the environment. Others, such as unburned HCs and CO, are dangerous to humans, animals, and vegetation. In this case, methods that reduce or eliminate toxic gases and particles are important in terms of protecting the environment.

*Interesting Fact*
*Tailpipe emissions are measured using a gas analyzer. Most analyzers measure five distinct gases: HC, CO, $CO_2$, $O_2$, and NOx.*

## Unburned Hydrocarbons

Gasoline and other fuels are based upon HC molecules. When perfect combustion occurs, no HC molecules should remain after the reaction. However, engines do not always operate under ideal circumstances. Variations in engine operating temperature, engine load, and other factors contribute to unburned HCs in engine exhaust. By minimizing rich air/fuel mixtures whenever possible, unburned HCs in the exhaust stream can be minimized.

Fast engine warmup also reduces the formation of unburned HCs by helping to vaporize and burn the fuel more completely. Accordingly, cold engines produce significantly higher tailpipe pollutants than engines running at normal operating temperatures.

The use of a catalytic converter can oxidize the HC molecules into water and $CO_2$, much like the combustion process itself. This reduces HC tailpipe emissions to low levels. Unfortunately, the catalytic converter is not effective until it reaches operating temperature, so it is not especially effective at reducing tailpipe emissions on a cold engine.

Most people tend to equate unburned HC emissions with those emanating from the tailpipe exclusively. However, a substantial amount of HCs can escape from the fuel system if evaporative controls are not used. A sealed fuel system prevents gasoline vapors from escaping into the atmosphere. Evaporative emissions controls are discussed in Chapter 34.

## Effects of Hydrocarbons on the Environment

Unburned HCs cause respiratory problems when inhaled. Small amounts inhaled over long time periods can cause a variety of serious health concerns, including cancer.

When unburned HCs react chemically with certain molecules in the air, smog is produced. In many instances, sunlight plays an important role in smog formation. Photochemical reactions are complex and often depend upon the amount of HCs present.

## Effects of Carbon Monoxide on the Environment

Carbon monoxide is a deadly, poisonous gas that is absorbed into the body's blood supply when inhaled. It prevents blood from absorbing oxygen, causing death in as little as several minutes, depending upon the concentration. Early symptoms include dizziness and headache.

To minimize the amount of carbon monoxide (CO) produced during the combustion process, more combustion air must be made available for the amount of fuel provided. Provided adequate oxygen is present, each carbon molecule will combine with two oxygen molecules, minimizing the amount of CO produced. Good control over air/fuel mixtures is essential in preventing excessive CO production. It is also important to use combustion chamber designs that prevent quenching the flame front. This helps reduce the formation of CO by allowing sufficient time for all carbon to oxidize into $CO_2$.

*Interesting Fact*

*Total CO emissions peaked in 1970, have declined through the ensuing years, and are approaching 1940 levels.*

The catalytic converter can also oxidize any residual CO present in the engine's exhaust stream into $CO_2$. A properly functioning catalytic converter is essential in reducing CO production.

In certain localities, depending upon population density and atmospheric conditions, the use of oxygenated gasoline is required. Oxygenated gasoline contains additional $O_2$ molecules to help reduce the production of CO. Fuels containing oxygenates are often used in urban areas during the winter months to reduce CO in the atmosphere. Low ambient temperatures can cause increased CO levels because engines are operating at lower temperatures for longer periods of time. When engine temperature is low, the engine requires more fuel. Accordingly, efficiency is reduced, and CO production increases.

## Effects of Oxides of Nitrogen on the Environment

By itself, $N_2$ does not readily combine with other molecules. To do so requires high energy levels. Because $N_2$ is present in combustion air, the potential for creating nitrogen oxide compounds exists if combustion chamber temperatures escalate past about 2500°F. These extreme temperatures provide the energy levels necessary for $N_2$ to react with $O_2$ molecules present during combustion.

Nitrogen oxides are commonly called "NOx," where "x" is a number representing a specific type of nitrogen oxide. Nitrogen monoxide, or NO, accounts for about 90 percent of NOx. The remaining 10 percent is $NO_2$, or nitrogen dioxide.

The chemical formula for combustion using gasoline as a fuel can be modified to include NOx and a residual amount of CO and unburned HCs. It then follows that:

$$O_2 + N_2 + C_8H_{18} = CO_2 + H_2O + CO + C_8H_{18} + NOx + O_2$$

That nitrogen oxides are produced is sufficient cause for concern.

When NOx react with HCs in the presence of sunlight, ozone is produced. Ozone is a serious health concern because it causes reduced pulmonary function, respiratory problems, and eye irritation.

Because the formation of ozone occurs slowly—over several hours—areas having little or no air movement are at the highest risk. Los Angeles is located within the California Basin, an area that sees little air movement. Accordingly, there are frequent ozone warnings in and around Los Angeles. The combination of high ambient temperatures, a stagnant air mass, and sunlight provides a catalyst for ozone formation in areas where high amounts of NOx and HCs exist.

### Connection Between NOx and Acid Rain

When NOx combine with gases in the atmosphere to produce nitric acid, acid rain is formed. Acid rain possesses an acidic pH compared with that of ordinary rainwater, which has a neutral pH. Because of the acidic content, acid rain damages soil and vegetation, including forests. It is also responsible for killing fish and other water life in lakes, rivers, and streams.

Controlling NOx production consists of two basic methods:

- EGR (exhaust gas recirculation)
- Exhaust aftertreatment using a catalyst

*Passenger cars and trucks contribute only about 30 percent of NOx to the atmosphere in states east of the Mississippi River. The remaining comes from power plants and other industrial processes.*

EGR recycles a portion of spent, or burned, exhaust gases into the intake system, where they displace an amount of the air/fuel charge. This reduces combustion chamber temperatures, which decreases the formation of NOx. EGR flow occurs mainly at part-throttle cruising conditions when NOx production is highest.

The use of a reducing catalytic converter decreases NOx tailpipe emissions. When used in conjunction with an oxidizing catalyst, effective control of CO, HCs, and NOx can be accomplished. More information relating to catalytic converter technology is presented in Chapter 35.

## EFFECTS OF SULFUR AND OTHER COMPOUNDS ON THE ENVIRONMENT

Although sulfur levels in gasoline have decreased in recent years, the production of sulfur dioxide during the combustion process remains a concern. Sulfur dioxide, like NOx, contributes to acid rain. This occurs in the form of sulfuric acid.

The sulfur content in gasoline can also cause hydrogen sulfide to be produced during the combustion process. It has a pungent, rotten egg odor. This chemical is highly toxic and can cause death if inhaled in moderate to strong concentrations. Fortunately, it can be smelled at low concentrations.

The combustion process takes place at high temperatures. Because of the energy levels involved by these high temperatures, many new chemicals can be formed from fuel and combustion air impurities. Some of these chemicals, such as ammonia and aldehydes, can have hazardous health and environmental implications. Scientists continue to study the effects combustion chemicals have on the environment and its inhabitants.

# Summary

- In an effort to control air pollution, the first Clean Air Act was passed in 1963.
- Congress established the EPA in 1970 to enforce policies set forth in the Clean Air Act.
- Besides pollution generated by motor vehicles, fossil fuel–burning power plants and industrial processing plants contribute to air pollution.
- Smog is a noxious mixture of air pollutants and is often seen as a haze in the atmosphere.
- The first pollution control device on a motor vehicle was the PCV valve. The PCV system was designed to control the release of crankcase vapors.

- Combustion within a cylinder occurs when an appropriate amount of air and fuel are activated by heat.
- When perfect combustion occurs, only $CO_2$ and $H_2O$ are produced.
- Rich air/fuel mixtures can cause an increase in CO and HC emissions.
- Carbon monoxide is a deadly, poisonous gas.
- Lean air/fuel ratios increase combustion temperatures. This results in an increase in NOx emissions.
- When HCs and NOx react in the presence of sunlight, ozone is produced. Ozone is a serious health hazard because it causes respiratory problems and eye irritation.

# Review Questions

1. Technician A says that all air pollution comes from motor vehicles. Technician B says that industrial processing and power plants cause air pollution. Who is correct?
   - A. Technician A
   - B. Technician B
   - C. Both A and B
   - D. Neither A nor B

2. Technician A says that the Environmental Protection Agency was formed in 1970 to enforce policies set forth in the Clean Air Act. Technician B says that the first pollution control device used on automobiles was the PCV valve. Who is correct?
   - A. Technician A
   - B. Technician B
   - C. Both A and B
   - D. Neither A nor B

3. Technician A says that ozone is formed when HCs react with acid rain. Technician B says that NOx contribute to the formation of acid rain. Who is correct?
   A. Technician A
   B. Technician B
   C. Both A and B
   D. Neither A nor B

4. Technician A says that a rich fuel mixture contains more $O_2$ than does a lean mixture. Technician B says that lean air/fuel mixtures cause the formation of NOx. Who is correct?
   A. Technician A
   B. Technician B
   C. Both A and B
   D. Neither A nor B

5. Technician A says that air from a breather vent is added to the combustion chamber to reduce cylinder temperatures. Technician B says that the PCV system is designed to increase crankcase pressures. Who is correct?
   A. Technician A
   B. Technician B
   C. Both A and B
   D. Neither A nor B

6. Technician A says that a PCV system helps reduce crankcase evaporative vapors from escaping into the atmosphere. Technician B says that crankcase vapors contain HC vapors. Who is correct?
   A. Technician A
   B. Technician B
   C. Both A and B
   D. Neither A nor B

7. Technician A says that a catalytic converter helps eliminate HCs from tailpipe emissions. Technician B says that CO is safe to breathe. Who is correct?
   A. Technician A
   B. Technician B
   C. Both A and B
   D. Neither A nor B

8. Technician A says that an air/fuel ratio that causes all of the fuel to react completely with all of the $O_2$ during combustion is called a lean air/fuel mixture. Technician B says that when all of the fuel reacts with all of the $O_2$ during combustion, the air/fuel ratio is called stoichiometric. Who is correct?
   A. Technician A
   B. Technician B
   C. Both A and B
   D. Neither A nor B

9. Technician A says that when more fuel is supplied to a cylinder than necessary for combustion, the air/fuel mixture is lean. Technician B says that when more fuel is supplied to a cylinder than necessary for combustion, the air/fuel mixture is rich. Who is correct?
   A. Technician A
   B. Technician B
   C. Both A and B
   D. Neither A nor B

10. Technician A says that an EGR valve helps reduce combustion chamber temperatures, reducing NOx formation. Technician B says that NOx are produced when combustion chamber temperatures exceed 2500°F. Who is correct?
    A. Technician A
    B. Technician B
    C. Both A and B
    D. Neither A nor B

# Chapter 33

# Precombustion Emissions Control

## Introduction

Precombustion emissions control includes components and engine design improvements that reduce harmful tailpipe emissions before they enter the exhaust stream. Improved combustion chamber and intake manifold designs, along with precise spark timing, help maintain optimum air/fuel ratios, minimizing hydrocarbon (HC) and carbon monoxide (CO) emissions. Recycling exhaust gases reduces combustion chamber temperatures, which, in turn, decreases NOx emissions.

Electronic engine control provides the ability to optimize various engine operating characteristics. This enables the engine to operate very efficiently. The driver experiences sensitive throttle response and maximum fuel economy, yet vehicle tailpipe emissions remain low, protecting the environment.

### CHANGES IN ENGINE DESIGN

Engine efficiency has improved significantly over the past three decades. In 1971, a dramatic reduction in engine horsepower swept the nation. This occurred because vehicle manufacturers were responding to recently enacted legislation, calling for reduced tailpipe emissions. The death of the muscle car seemed imminent. The latest trend focused on fuel economy and reduced emissions, rather than raw horsepower.

Slowly, however, horsepower began to increase during the 1980s. Manufacturers began using a combination of improved intake and cylinder head designs, along with electronic controls. These changes provided increased power while keeping emissions in check. As the electronics industry continues to evolve, motorists can expect improved performance, reduced fuel consumption, and lower emissions.

## Modified Combustion Chamber Designs

The use of compact, effective combustion chamber designs helps reduce the **quench area**. The quench area is the area within the chamber that causes the flame front to extinguish. This causes the combustion reaction to terminate before oxidation of the fuel can fully complete. The result is elevated HC and CO tailpipe emissions.

The hemispherical cylinder head design is effective in promoting efficient combustion (**Figure 1**). In this design,

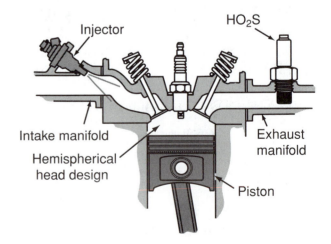

**Figure 1.** The hemispherical cylinder head is effective in promoting efficient combustion.

the spark plug is centrally located between the intake and exhaust valves. Wedge-shaped and other complex combustion chamber designs can allow HCs to hide from the combustion process, increasing HC emissions. Simple chamber designs minimize HC hiding.

The use of combustion chamber designs that promote controlled swirling of the air/fuel charge helps to ensure reliable flame front propagation, while minimizing spark knock. Leaner air/fuel mixtures are often used with swirl chamber designs because the probability of source ignition and flame propagation is improved. Whenever the combustion reaction is able to fully complete, HC and CO emissions output is substantially reduced.

## Improved Intake Manifold Designs

Efficient combustion depends upon equal distribution of combustion air to each cylinder. Extensive use of plastics has resulted in lightweight, precision designs that offer smooth runner surfaces, providing reduced airflow resistance.

Tuned intake runners provide improved volumetric efficiency, yet offer excellent performance over a wide range of engine speeds. In many designs, electronic control of manifold runners provides the balance necessary over wide engine speed variations. The system provides long-runner performance at low engine speeds, offering high torque **(Figure 2)**. When maximum power at high speeds is required, short runners bypass the long runners, providing an unrestricted path for combustion air.

**Figure 2.** Electronic control of intake manifold runner length provides high torque at low speeds, while offering excellent performance at high speeds.

## Improved Engine Temperature Control

Compared with engines of the muscle car era, modern engines operate at higher temperatures. This provides an advantage in terms of reducing HC and CO emissions.

Using PCM-controlled electric cooling fans, temperature management is much more precise. This results in more stable engine temperatures, increasing engine efficiency and reducing harmful emissions. Increased engine temperatures are not without drawbacks. Elevated temperatures can increase oxides of nitrogen (NOx) output, requiring additional emissions control components in the form of exhaust gas recirculation (EGR).

## SPARK TIMING

Significant reductions in NOx emissions are possible by reducing spark advance. This occurs because peak combustion temperatures are reduced. However, retarding the spark advance increases exhaust temperature and increases the amount of unburned fuel in the exhaust. Additionally, fuel economy and driveability suffer.

When a vehicle is fitted with a three-way catalytic converter, increased spark advance is possible, while maintaining low NOx emissions. Computer-controlled spark advance is carefully tailored to engine speed and load, resulting in improved fuel economy and driveability.

## EXHAUST GAS RECIRCULATION

The NOx are formed by high combustion chamber temperatures. NOx output can be reduced by:
- Reducing the compression ratio.
- Enriching the air/fuel charge entering the combustion chamber.
- Recycling cooler exhaust gases back into the combustion chamber.

Of the three alternatives, the most efficient way to lower NOx emissions is by recycling a small amount (5–10 percent) of the exhaust gases back into the combustion chamber. Reducing the compression ratio can lead to inefficient combustion and elevated HC emissions. Enriching the air/fuel mixture hurts fuel economy and increases HC and CO emissions.

By recirculating a portion of cooler exhaust gases back into the combustion chamber, the production of NOx is reduced. The exhaust gases reduce combustion chamber temperatures because they contain little or no $O_2$ to react with the combustion process. In this respect, exhaust gases are considered inert. An EGR system is shown in **Figure 3**. EGR systems began appearing on vehicles in 1971. The heart of the EGR system is the EGR valve **(Figure 4)**.

The introduction of EGR flow into the intake manifold has an effect on air/fuel mixtures. This occurs because the

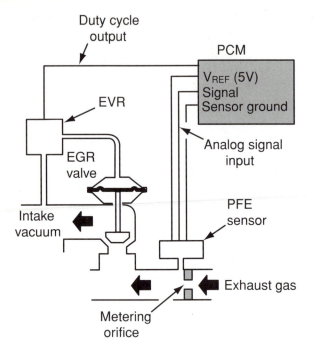

**Figure 3.** An EGR system.

**Figure 4.** Typical EGR valve.

inert exhaust gases displace the incoming combustion air so less $O_2$ is available for combustion. When EGR flow occurs, the PCM must change fuel injector pulse width to maintain an appropriate air/fuel ratio. Otherwise, the air/fuel ratio will be excessively rich because of the reduced $O_2$ available. Monitoring EGR flow is essential and is discussed shortly.

Another effect of adding EGR relates to ignition timing. Adding an inert gas into the combustion chamber slows the combustion process. Additional spark advance is necessary to provide sufficient time for the air/fuel charge to completely burn.

Because of the lowered peak combustion temperatures, EGR can effectively reduce spark knock. An engine

that gradually shows signs of engine ping most likely has an inoperative EGR system. Clogged passages or an inoperative EGR valve are often to blame.

## ENGINE OPERATING CONDITIONS

Because EGR flow can easily disturb the delicate balance between combustion air and fuel, several engine operating conditions must be met to ensure the engine will perform satisfactorily.

### Engine Temperature

Because EGR flow lowers combustion chamber temperatures, the PCM prevents EGR flow until coolant temperature reaches approximately 130°F. Allowing EGR flow when engine temperatures are low will affect driveability, causing poor engine performance. Furthermore, adding EGR flow makes little sense when engine temperatures are low. Formation of NOx is a concern only when combustion chamber temperatures exceed 2500°F.

### Throttle Position

EGR flow should occur only when the throttle is opened past idle. This parameter is measured using the signal from the TPS. At closed throttle, EGR flow is disabled.

### Vehicle Operating Characteristics

It is desirable to reduce high combustion chamber temperatures. This normally occurs during part-throttle cruise conditions.

By monitoring engine load and vehicle speed, the PCM can establish the current driving conditions. An appropriate amount of EGR flow is then calculated. The PCM commands the EGR valve to open, allowing EGR to flow.

Feedback signals to the PCM ensure the requested amount of EGR flow has occurred. These signals originate from either an **EGR Valve Position (EVP) sensor** or pressure transducer **(Figure 5)**.

A signal from the park-neutral switch ensures that the vehicle is in gear. Often, a signal from the **torque converter clutch (TCC)** is used to determine if the TCC is in lockup before allowing EGR to flow. Many EGR strategies are used, depending upon the application.

## EXHAUST GAS RECIRCULATION STRATEGIES

Engine operational characteristics are different, depending upon driving situations. From idle to WOT, air/fuel requirements and combustion chamber temperatures vary widely. The EGR strategies must be carefully tailored to each engine operating condition to provide the best driveability possible, while maintaining low NOx tailpipe emissions.

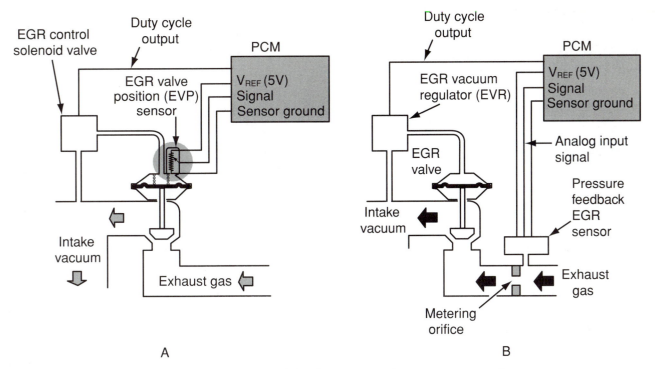

**Figure 5.**   (A) An EGR system using EVP sensor. (B) EGR system using pressure transducer.

## Exhaust Gas Recirculation Operation at Idle

At idle, very little air and fuel are supplied to the combustion chambers. Accordingly, no EGR flow should occur because virtually any dilution of the intake charge will severely affect the intake airflow, causing the engine to misfire or stall. Therefore, the PCM prevents EGR flow at idle.

The formation of NOx is minimal at idle. Engine load is very low, which results in low combustion chamber temperatures.

> **You Should Know** *A sticky EGR valve can cause engine idling concerns and throttle tip-in problems. If the valve has excessive carbon deposits, it should be replaced. A new gasket should be installed whenever the valve is removed from the engine.*

## Exhaust Gas Recirculation Operation at Wide-Open Throttle

To prevent loss of power under high-demand conditions, the PCM prevents EGR flow at WOT. At WOT, maximum power occurs when the combustion chamber is supplied with the largest air/fuel charge possible. Therefore, dilution of the mixture from EGR must be avoided.

## Part-Throttle Cruise

At cruising speeds where engine speed is moderate and engine load is low, EGR flow is at a maximum. There is sufficient airflow—and therefore fuel delivery—into the cylinders to prevent substantial dilution of the air/fuel charge.

Filling the cylinder with as much as 10 percent exhaust gases improves fuel economy and reduces NOx emissions. Combustion temperatures decrease by several hundred degrees during maximum EGR flow.

Because combustion is slowed by the addition of the inert exhaust gases, spark advance is increased by the PCM. As a general rule, ignition timing is advanced between 0.5 and 1 degree for each percent increase in EGR flow.

If engine load increases, EGR flow is reduced to increase power. This occurs until EGR flow eventually stops when WOT is reached or manifold vacuum decreases to a value sufficiently low enough to prevent keeping the EGR valve open (vacuum-operated valves only). Some vehicles, most notably light and medium duty trucks, employ a vacuum storage canister to provide a stable source of vacuum to vacuum-operated accessories such as the EGR valve. In this case, EGR flow will continue as long as the vehicle is not operated at WOT and still meets the PCM's requirements for allowing EGR operation.

Engines equipped with electronically operated EGR valves are not limited by engine vacuum, although the strategies used are similar to vacuum-operated EGR systems.

## Deceleration

During deceleration, EGR flow is disabled because no fuel is dispensed to the combustion chambers under this condition. No combustion takes place under vehicle deceleration. Combustion chamber temperatures decrease from the pumping action of the pistons drawing in ambient air.

## TYPES OF EGR VALVES

EGR valves can be of the conventional, vacuum-operated type or the digital, electronic variety. Because of the high operating temperatures involved, all EGR valves are made of cast iron and steel.

## Conventional EGR Valve

The conventional EGR valve uses a vacuum diaphragm and return spring to control EGR flow. A valve shaft is attached to the diaphragm and contains a pintle on the end of the shaft, which seats against a seat, closing off the valve. As vacuum causes the diaphragm to lift the pintle off its seat, exhaust gases are allowed to flow back into the combustion chamber **(Figure 6)**.

## Controlling the Conventional EGR Valve

Using one or more vacuum solenoids, the PCM controls the amount of EGR flow. Some solenoids are normally open; others are of the normally closed variety. If a solenoid

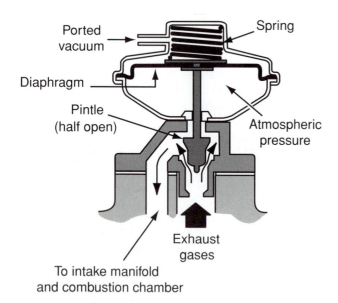

**Figure 6.** Diaphragm causes pintle to lift from seat, causing exhaust gases to flow back into the intake manifold, where they enter the combustion chambers.

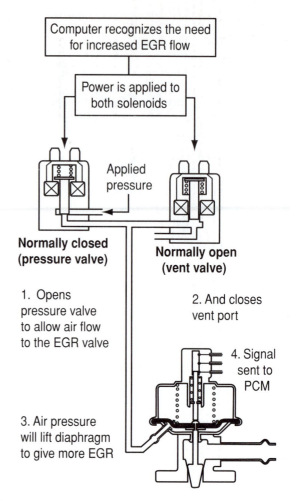

**Figure 7.** Dual-solenoid EGR system.

fails, the system is designed to fail with the EGR valve closed. A typical EGR system using two solenoids and a vent is shown in **Figure 7**.

By monitoring EGR valve position, the PCM knows how far the pintle is raised from its seat. From this information, the PCM calculates an EGR flow rate, adjusting fuel and spark advance accordingly.

## Monitoring EGR Flow

Several types of EGR flow monitoring devices are used, depending upon system specifics. In each case, a representation of EGR flow is provided to the PCM. Because a feedback signal from the EGR system is provided to the PCM, it is considered to be of the closed-loop variety. Changes in EGR flow are reflected in either valve position or pressure measurements as described next.

### EGR Valve Position Sensor

Several methods exist for monitoring EGR flow. The simplest uses an EGR valve position (EVP) sensor mounted

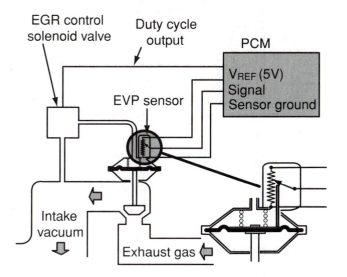

**Figure 8.**   Monitoring EGR flow using an EVP sensor.

on top of the valve **(Figure 8)**. The EVP sensor is a potentiometer. It behaves similarly to a throttle position sensor (TPS), except it operates in a linear mode, rather than a rotary mode.

As the EGR pintle moves in response to the vacuum signal applied to it, a signal is developed at the wiper terminal of the sensor **(Figure 9)**. The signal's amplitude depends on how far the valve is open. The more the pintle is lifted off its seat, the greater the DC voltage at the sensor. When the valve is closed completely, a small voltage is present at the sensor for system diagnostic purposes.

## Pressure Feedback EGR Sensor

The **pressure feedback EGR (PFE)** sensor monitors exhaust system backpressure to determine EGR flow. The PFE sensor is a ceramic capacitive pressure transducer. The PFE sensor does not measure actual EGR flow. Instead, it measures the drop in exhaust system pressure when the EGR valve opens **(Figure 10)**. A greater EGR valve opening causes a greater drop in exhaust system pressure.

The PFE sensor voltage varies from about 3.5V (engine idling, no EGR flow) to less than 0.5V (maximum EGR flow). Using a scan tool, it is possible to check for a restricted exhaust system by verifying the value of the PFE PID. If the value exceeds approximately 3.5V, there is a possibility of an exhaust restriction.

## Differential Pressure Feedback EGR Sensor

The **differential pressure feedback EGR (DPFE) sensor** operates similar to the PFE sensor, except it measures a change in pressure **(Figure 11)**. This is commonly called a differential pressure. Instead of measuring actual exhaust system pressure, the DPFE measures the pressure drop across a metering orifice. Pressure drop measurements provide a more accurate representation of EGR flow.

The signal produced by the sensor is less than that produced by a PFE sensor. Because the DPFE sensor measures a pressure drop, its signal cannot be used to check for exhaust system restrictions.

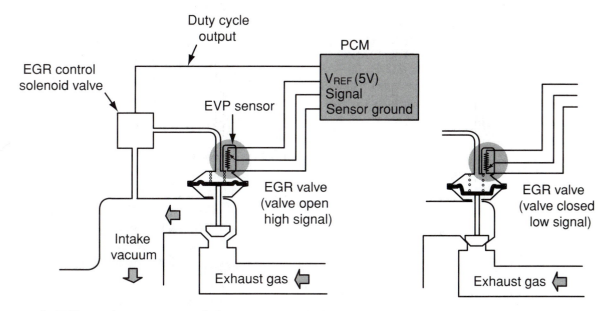

**Figure 9.**   As EGR pintle moves, signal changes across EVP sensor.

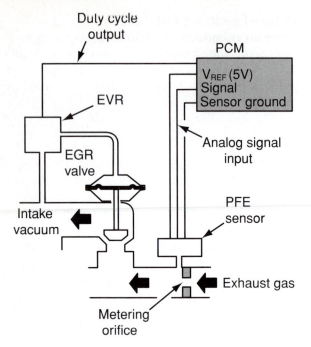

**Figure 10.** A PFE sensor measures drop in exhaust pressure when EGR valve opens.

## DIGITAL EGR VALVE

The digital EGR valve uses one or more solenoids mounted on the valve itself **(Figure 12)**. The solenoid(s) control the valve position without the need for a vacuum source. This reduces the dependency on manifold vacuum and increases the response time of the valve considerably. Signals from the PCM control the operation of solenoids within the valve.

Later model digital EGR valves use a stepper motor instead of a solenoid. The stepper motor provides accurate EGR flow adjustments, controlled by the PCM. EGR valves equipped with a stepper motor are called linear EGR valves.

## Controlling the Digital EGR Valve

When the PCM commands one or more solenoids to open, its armature opens the valve, causing EGR flow into the intake manifold. If the EGR valve uses more than one solenoid, the PCM can adjust EGR flow by activating one, two, or all three solenoids to achieve the proper amount of EGR flow **(Figure 13)**.

In the case of the digital linear EGR valve, a stepper motor controls pintle position in much the same way an idle air control (IAC) motor operates **(Figure 14)**. A signal from the PCM tells the stepper motor how far to shuttle the valve in either direction. Because the system makes use of a digital stepper motor, EGR flow can be regulated precisely.

## Monitoring the Digital EGR Valve

A feedback signal from the EGR valve provides the PCM with position data. The sensor is commonly called a **pintle position sensor (PPS)** because it provides data relating to the position of the valve pintle. The PCM can calculate EGR flow rates based on the position of the pintle.

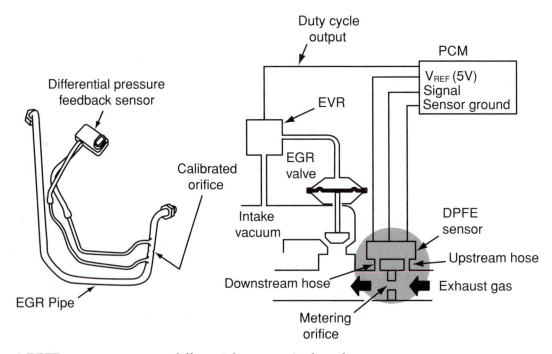

**Figure 11.** A DPFE sensor measures a differential pressure in the exhaust stream.

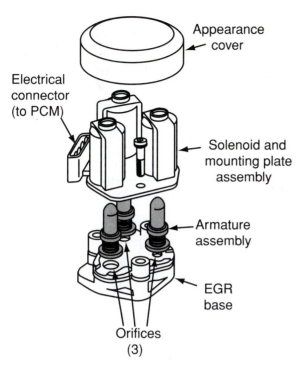

**Figure 12.** Digital EGR valve uses one or more solenoids mounted on the valve itself.

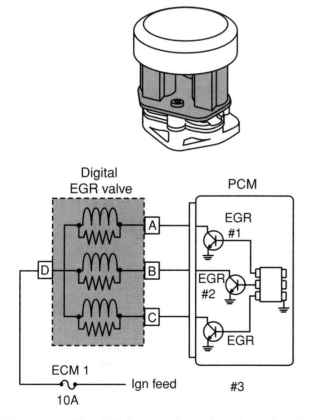

**Figure 13.** The PCM controls each solenoid independently, allowing one or more solenoids to activate, regulating EGR flow.

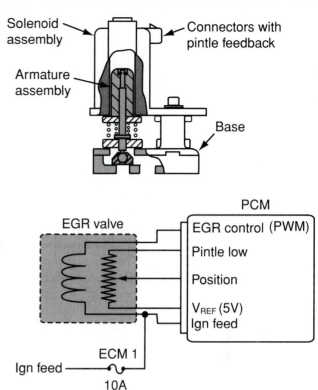

**Figure 14.** Digital linear EGR valve uses a stepper motor to control pintle position.

Because the signal provided by the PPS is of the feedback variety, the digital EGR system is also considered to be a closed loop system. If the PCM issues a command to increase EGR flow, the feedback signal tells the PCM if the requested flow adjustment occurred by checking the PPS signal.

## DIAGNOSING EGR SYSTEM FAULTS USING A SCAN TOOL

The PCM constantly monitors the EGR sensor activity and is programmed to set a diagnostic trouble code (DTC) if a fault occurs. For a given set of conditions, if an appropriate EGR flow rate is not received by the PCM, the system will set a DTC and illuminate the check engine lamp. Always retrieve DTCs before engaging in any type of EGR system repair.

## MEETING NOx EMISSIONS REQUIREMENTS WITHOUT AN EGR VALVE

In some cases, a specific engine family can meet NOx emissions requirements without the use of an EGR system. This is accomplished by using a camshaft that provides a large amount of overlap. Depending on many engine design factors, increasing valve overlap can provide an effect similar to using EGR, which reduces combustion temperatures. By allowing inert exhaust gases to remain in the combustion chamber after combustion has occurred, peak temperatures are decreased, causing a reduction in the amount of NOx formation.

It is important to realize that most engines marketed for sale in California and some Northeast states are equipped with an EGR valve. This is necessary to meet stringent emissions requirements in states where smog is a serious problem.

## Summary

- Precombustion emissions control includes components and engine design improvements that reduce harmful tailpipe emissions before they enter the exhaust stream.
- By combining improved intake manifold and cylinder head designs, manufacturers have reduced tailpipe emissions, while enhancing performance.
- By carefully balancing spark advance to engine demand, tailpipe emissions of NOx can be controlled without sacrificing driveability or fuel economy.
- Because NOx are formed when combustion chamber temperatures exceed 2500°F, the use of EGR significantly reduces NOx output by lowering combustion temperatures.
- No EGR flow is permitted at idle or WOT.
- The EGR flow is greatest at part-throttle cruise conditions.
- The EGR valves can be of the conventional (vacuum-operated) or digital variety.
- Monitoring EGR flow is important to ensure an appropriate amount of fuel is delivered to the engine under varying EGR flow rates.
- Certain EGR faults can be diagnosed using a scan tool.

## Review Questions

1. Technician A says that increased spark advance can cause an increase in NOx emissions. Technician B says that advancing ignition timing increases combustion chamber temperatures. Who is correct?
   A. Technician A
   B. Technician B
   C. Both A and B
   D. Neither A nor B

2. Technician A says that adding EGR flow to an engine causes some of the intake air charge to be displaced. Technician B says that to compensate for the smaller intake air charge, the amount of fuel delivered to the engine must be reduced to prevent a rich air/fuel mixture. Who is correct?
   A. Technician A
   B. Technician B
   C. Both A and B
   D. Neither A nor B

3. Technician A says that no EGR flow is permitted at part-throttle cruising conditions. Technician B says that no EGR flow is permitted at idle. Who is correct?
   A. Technician A
   B. Technician B
   C. Both A and B
   D. Neither A nor B

4. Technician A says that EGR valves are operated by vacuum. Technician B says that not all engines use an EGR valve. Who is correct?
   A. Technician A
   B. Technician B
   C. Both A and B
   D. Neither A nor B

5. Technician A says that adding EGR flow to an engine results in an increase of power. Technician B says that EGR flow is controlled by the air pump. Who is correct?
   A. Technician A
   B. Technician B
   C. Both A and B
   D. Neither A nor B

6. Technician A says that a DPFE sensor is used to measure EGR flow by sampling a pressure change in the exhaust stream. Technician B says that the $HO_2$ sensor is often used to measure EGR flow. Who is correct?
   A. Technician A
   B. Technician B
   C. Both A and B
   D. Neither A nor B

7. Technician A says to always use a new gasket when installing an EGR valve. Technician B says that cleaning the old gasket is sufficient. Who is correct?
    A. Technician A
    B. Technician B
    C. Both A and B
    D. Neither A nor B
8. Technician A says that some EGR valves are controlled by vacuum solenoids. Technician B says that most EGR valves are made of high-temperature plastic. Who is correct?
    A. Technician A
    B. Technician B
    C. Both A and B
    D. Neither A nor B
9. Technician A says that some EGR valves use a sensor mounted over the valve to indicate pintle position. Technician B says that an EGR valve with heavy deposits should be replaced. Who is correct?
    A. Technician A
    B. Technician B
    C. Both A and B
    D. Neither A nor B
10. Technician A says that a stuck-closed EGR valve can cause an engine to idle poorly. Technician B says that an inoperative EGR valve can cause spark knock to occur. Who is correct?
    A. Technician A
    B. Technician B
    C. Both A and B
    D. Neither A nor B

# Chapter 34

# Evaporative Emissions Control

## *Introduction*

Evaporative emissions include hydrocarbon (HC) vapors created during the combustion process and from evaporation of gasoline from fuel system components, most notably the fuel tank. Without crankcase and fuel system evaporative controls, a significant amount of unburned HCs will be emitted from the vehicle in the form of evaporative emissions.

In addition to the air quality issues surrounding the release of unburned HCs, a safety issue exists because gasoline vapors are extremely flammable. The containment of unburned HCs is important in terms of both air quality and vehicle safety. Evaporative emissions control systems are effective in reducing the release of unburned HCs to the atmosphere.

## CRANKCASE EVAPORATIVE EMISSIONS

The positive crankcase vent (PCV) system removes condensation, unburned HCs, and other harmful gases from the crankcase. It was introduced in California in 1961 and became standard equipment across the nation in 1963.

Before PCV systems were installed on passenger cars and light trucks, the crankcase was vented using a road draft tube. The road draft tube pulled unburned HCs and other vapors out of the crankcase by the natural suction effect that occurs while the vehicle is driven on the highway. Unfortunately, the road draft tube did little to curb evaporative emission from the engine's crankcase.

## How Crankcase Vapors Are Formed

The engine is an air pump. When it is operating, a certain amount of the air/fuel charge and exhaust gases escape past the piston rings. These gases make their way into the crankcase, where they can dilute the oil, causing serious lubrication concerns. Because of the combustion process, the vapors are often corrosive.

## Keeping the Crankcase Clean Inside and Out

The PCV system helps rid the crankcase of any condensation, unburned HCs, and other corrosive vapors by recycling them back into the intake manifold. Here, the vapors are burned as part of the combustion process. Because the vapors are consumed in the combustion process, no harmful crankcase vapors are released to the atmosphere.

Crankcase gases are often corrosive, causing deterioration of engine oil. In the case of water vapor, sludge can accumulate from the mixing of oil and water. Premature failure of internal engine components can occur if condensation and corrosive vapors are not removed from the crankcase. The PCV system helps to keep the oil free of moisture, gasoline, and other damaging chemicals.

## SYSTEM COMPONENTS

A typical PCV system contains the following components:
- A PCV valve
- A PCV hose
- An intake manifold vacuum port

A PCV system is shown in **Figure 1**. A hose or tube connects the PCV valve to the intake manifold port. The air cleaner assembly provides a source of make-up air to the crankcase in addition to blowby gases. This permits scavenging the crankcase of vapors by providing an adequate supply of air to maintain an appropriate flow rate through the crankcase.

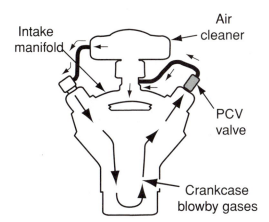

**Figure 1.** The PCV system.

## PCV Valve Function

The PCV valve is the heart of the PCV system. It is a check valve serving two functions:
- It regulates the rate of crankcase airflow to the intake manifold under varying operating conditions.
- It provides protection against crankcase vapors igniting.

A typical PCV valve is shown in **Figure 2**. One side of the valve is connected to the intake manifold. The other portion

**Figure 2.** Typical PCV valve.

of the valve is exposed to engine crankcase pressure. The plunger within the valve is spring-loaded to help control valve flow rates. The spring tends to open the valve by pushing the plunger away from the intake manifold side of the valve.

*Interesting Fact*

*Replacing the PCV valve with an incorrect part can cause the engine to idle poorly or not at all. The internal spring calibration affects how much airflow occurs at idle. This is important because the PCV valve supplies 25–30 percent of the engine's idle air requirements.*

## PCV Valve Operation: Idle and Part-Throttle Cruise

Under light engine loads such as part-throttle cruise and idle, intake manifold vacuum pulls the valve toward the closed position **(Figure 3)**. Under these conditions, PCV flow is low (between 0.2 and 3 cfm).

## PCV Valve Operation: Acceleration and High Engine Loads

As intake manifold vacuum decreases in response to high engine loads, the spring inside the valve causes the plunger to open further, increasing the flow rate **(Figure 4)**. Under conditions of heavy acceleration or high engine loads, PCV flow increases dramatically (between 3 and 8 cfm).

## Induction Backfire

If the engine backfires, pressure within the intake manifold increases dramatically, causing the valve plunger to

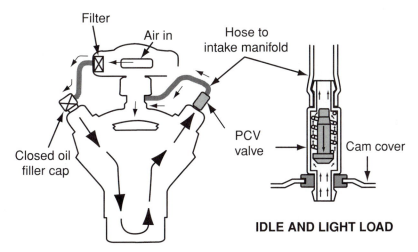

**IDLE AND LIGHT LOAD**

**Figure 3.** PCV valve operation at idle and light engine loads.

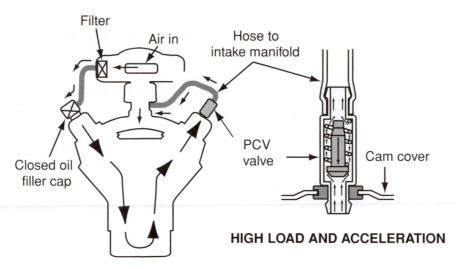

**Figure 4.** PCV valve operation at high engine loads and acceleration.

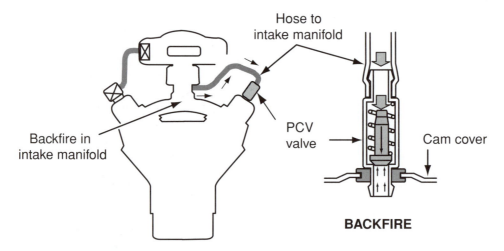

**Figure 5.** The PCV valve closes in the event of engine backfire.

seat on the crankcase side of the valve **(Figure 5)**. This prevents an explosion from occurring, should the crankcase vapors ignite.

## ORIFICE-TYPE PCV SYSTEMS

Some engines use a calibrated orifice instead of a PCV valve. Crankcase vapors are drawn into the intake manifold through the orifice, where they are burned during the combustion process.

## SEPARATOR PCV SYSTEMS

Some engine applications use an oil separator instead of a PCV valve. In this system, fresh air enters a vent hose at the throttle body and travels to the crankcase. Using an oil separator to minimize any chance of oil consumption, crankcase vapors return to the throttle body through a second vent hose **(Figure 6)**. Depending upon engine operating conditions, ports in the throttle body divert the crankcase vapors to the intake manifold at appropriate flow rates.

## FUEL SYSTEM EVAPORATIVE EMISSIONS

Approximately 20 percent of the total HC emissions produced by a vehicle are caused by fuel evaporation. The HC vapors are present in the fuel tank and fuel system components. As the ambient temperature increases, gasoline evaporation within the fuel tank also increases. If allowed to vent into the atmosphere, the vapors could become a significant source of HC emitted by the vehicle.

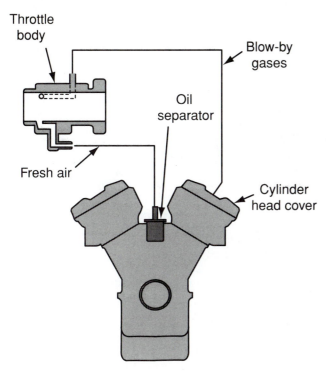

**Figure 6.** Oil separator used in place of PCV valve.

## PURPOSE OF THE EVAPORATIVE EMISSIONS CONTROL SYSTEM

The purpose of the evaporative (EVAP) emissions control system is to contain any gasoline (HC) vapors before they can escape into the atmosphere. The evaporative emissions control system is commonly called the EVAP system.

## COMPONENTS OF THE EVAP SYSTEM

A typical EVAP system **(Figure 7)** consists of the following components:
- A carbon storage canister
- Vapor lines
- Canister purge solenoid valve
- Fuel tank and filler neck
- Fuel tank cap
- Rollover/vent valve check valve (depending upon application)

At the heart of the EVAP system is the carbon canister. The canister is a temporary storage container for HC vapors that would otherwise vent into the atmosphere. Typical canister mounting locations are in the engine compartment

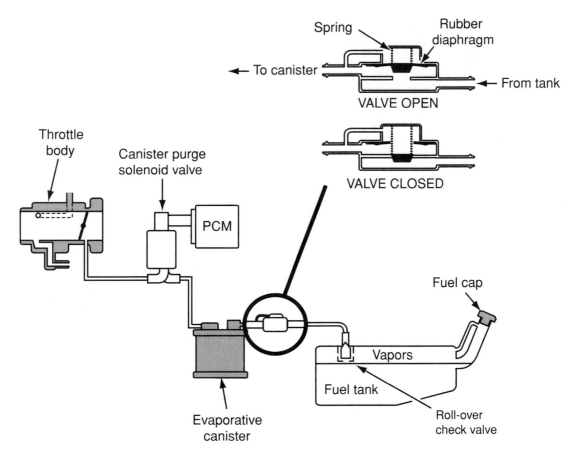

**Figure 7.** Typical EVAP system.

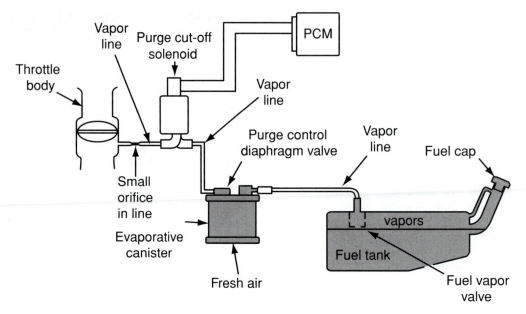

**Figure 8.**   Vapor lines connect the canister to the fuel tank and throttle body.

or near the fuel tank. Some vehicles use two canisters, depending upon the number of fuel tanks present and the volume of fuel stored.

Carbon is chosen because of its ability to attract HC vapors. The carbon granules can attach HC vapors of approximately ⅓ of their own weight by **adsorption**. Adsorption is a process whereby HC molecules are lightly attracted to the carbon surface, making it easy to remove them by pulling fresh air though the canister.

The effective surface of the activated carbon (charcoal) is extremely large. A carbon canister that is filled with 625 g of carbon has a surface area equal to about 165 football fields. Because of this enormous storage capacity, a large quantity of HC vapors can be stored in the canister for later burning during the combustion process.

Hoses connect the carbon canister to the fuel tank and throttle body. These hoses are commonly called vapor lines **(Figure 8)**.

## EVAP SYSTEM OPERATION: ENGINE OFF

When the engine is not running, the carbon canister temporarily absorbs vapors generated by the fuel tank or fuel system components. Gasoline in the fuel tank continues to evaporate, causing pressure in the tank to increase. The gasoline vapors are directed through the rollover/vent valve to the vapor lines and finally into the canister **(Figure 9)**. Here, the vapors are stored until they can be **purged** into the engine during the combustion process.

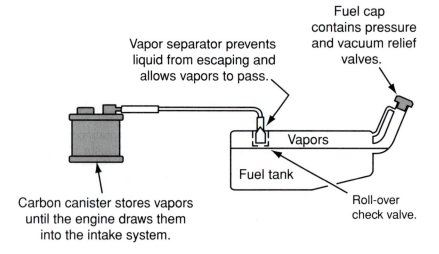

**Figure 9.**   Vapor lines carry gasoline vapors from the fuel tank to the carbon canister.

## EVAP SYSTEM OPERATION: ENGINE RUNNING

After the engine is started and all conditions are satisfied, the PCM energizes the **canister purge (CANP) solenoid (Figure 10)**. By activating the CANP solenoid, ported manifold vacuum draws the stored vapors into the engine, where they are burned along with fuel supplied by the injectors.

Because the purged vapors behave as ordinary HC molecules, the PCM must reduce fuel injector pulse width to compensate for the additional fuel supplied by the carbon canister.

Depending upon the application, a number of conditions must be satisfied before the PCM will activate the EVAP system, allowing purging to occur. These include:

- Engine coolant temperature must be above a minimum value.
- Engine run time must be beyond a minimum value.
- The engine must be operating in a closed-loop mode.

Some engine management systems allow canister purging only if the vehicle is in motion above a minimum speed. However, some systems also purge the canister while the vehicle is idling. Always consult the service manual for details regarding specific operational strategies.

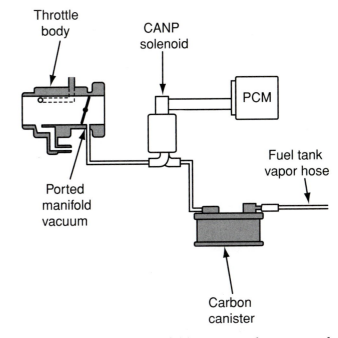

**Figure 10.** Ported manifold vacuum draws stored vapors from the canister into the engine, where they are burned, along with fuel supplied by the injectors.

# Summary

- Controlling evaporative emissions from the crankcase is accomplished using a PCV system.
- The PCV system recycles unburned HC vapors, condensation, and other harmful chemicals into the combustion chambers, where they are burned in the combustion process.
- The PCV valve regulates crankcase airflow and protects the engine in the event an induction backfire occurs.
- Some PCV systems do not use a PCV valve. Instead, a calibrated orifice or oil separator is used to regulate airflow and prevent oil consumption from occurring.

- Evaporative emissions in the form of unburned HCs account for about 20 percent of the total HC emissions produced by a vehicle.
- As ambient temperature increases, the amount of gasoline vapors in the fuel tank increases. These vapors are routed to a temporary storage container called the "carbon canister."
- Using one or more CANP solenoids, the stored gasoline vapors are directed into the intake manifold. The vapors are combined with gasoline supplied by the fuel injectors and burned in the combustion process.

# Review Questions

1. Technician A says that crankcase evaporative emissions are controlled using an EGR valve. Technician B says that the PCV system recycles crankcase vapors into the combustion chambers, where they are burned in the combustion process. Who is correct?
   A. Technician A
   B. Technician B
   C. Both A and B
   D. Neither A nor B

2. Technician A says that the crankcase vent protects the engine in the event an induction backfire occurs. Technician B says that the PCV valve protects the engine in the event an induction backfire occurs. Who is correct?
   A. Technician A
   B. Technician B
   C. Both A and B
   D. Neither A nor B

3. Technician A says that the PCV valve should be cleaned periodically. Technician B says that the valve should be replaced at intervals recommended by the manufacturer. Who is correct?
   A. Technician A
   B. Technician B
   C. Both A and B
   D. Neither A nor B

4. Technician A says that PCV flow is maximum at idle. Technician B says that PCV flow increases when engine load is high. Who is correct?
   A. Technician A
   B. Technician B
   C. Both A and B
   D. Neither A nor B

5. Technician A says that some PCV systems use an oil separator instead of a PCV valve. Technician B says that some PCV systems use a calibrated orifice instead of a PCV valve. Who is correct?
   A. Technician A
   B. Technician B
   C. Both A and B
   D. Neither A nor B

6. Technician A says that as ambient temperature increases, gasoline evaporation in the fuel tank decreases. Technician B says that unburned HCs in the form of evaporative emissions account for approximately 20 percent of the total HCs emitted from a motor vehicle. Who is correct?
   A. Technician A
   B. Technician B
   C. Both A and B
   D. Neither A nor B

7. Technician A says that the fuel filler cap is a part of the EVAP emissions system. Technician B says that the carbon canister stores HC vapors until they can be burned during the normal combustion process. Who is correct?
   A. Technician A
   B. Technician B
   C. Both A and B
   D. Neither A nor B

8. Technician A says that vapor hoses contain only air. Technician B says that most EVAP systems use one or more solenoids to direct HC vapors to the intake manifold. Who is correct?
   A. Technician A
   B. Technician B
   C. Both A and B
   D. Neither A nor B

9. Technician A says that the engine temperature must reach a minimum value before purging of the EVAP system can occur. Technician B says that the EVAP system purges only when the engine is cold. Who is correct?
   A. Technician A
   B. Technician B
   C. Both A and B
   D. Neither A nor B

10. Technician A says that the PCM must reduce fuel injector pulse width whenever the EVAP system purges the canister of HC vapors. Technician B says that the PCM increases fuel injector pulse width whenever EVAP purging occurs to prevent the engine from stalling. Who is correct?
    A. Technician A
    B. Technician B
    C. Both A and B
    D. Neither A nor B

# Chapter 35

# Postcombustion Emissions Control

## Introduction

Postcombustion emissions controls are used to clean up exhaust emissions left behind after the combustion process. Effective combustion should produce minimal harmful emissions. However, variations in engine operating conditions and shortcomings in the combustion process can produce undesirable combustion products.

The production of harmful exhaust emissions requires exhaust aftertreatment to meet federal emission standards. Aftertreatment consists of the use of a catalytic converter and possibly an **air injection reaction (AIR) system**.

## EXHAUST AFTERTREATMENT

Exhaust aftertreatment consists of using one or more catalytic converters to reduce tailpipe emissions. Catalytic converters were first installed on vehicles in the mid-1970s. By 1977, most passenger vehicles and light trucks were fitted with a catalytic converter.

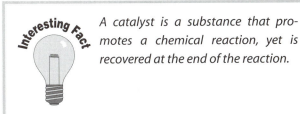

*A catalyst is a substance that promotes a chemical reaction, yet is recovered at the end of the reaction.*

The catalytic converter is located between the exhaust manifold and muffler **(Figure 1)**. Depending upon the

**Figure 1.** Typical catalytic converter location.

exhaust system design, multiple catalytic converters may be used. In the case of V-type engine designs employing dual exhaust pipes, a separate catalytic converter is often used for each cylinder bank.

## CATALYST ELEMENTS

The catalytic converter uses **noble metals**, such as **palladium (Pd)**, **platinum (Pt)**, and **rhodium (Rh)**, as catalysts.

The elements contained within the converter are designed to convert the exhaust gases to harmless products such as water ($H_2O$), carbon dioxide ($CO_2$), nitrogen ($N_2$), and oxygen ($O_2$).

# THREE-WAY CATALYSTS

The **three-way catalyst (TWC)** is often called a **dual-bed converter** because it contains two individual substrates **(Figure 2)**. A TWC acts upon three harmful emission gases: oxides of nitrogen (NOx), hydrocarbons (HC), and carbon monoxide (CO). Each substrate within the "cat" behaves differently, depending on the emission gases treated.

The front section of the TWC is designed to act as a **reducing catalyst** for NOx. Reduction is a chemical process which causes oxygen to be removed from a compound. Using the catalyst elements platinum and rhodium, NOx is reduced into harmless $N_2$ and $O_2$.

The rear portion of the TWC substrate operates as an **oxidizing catalyst** for HC and CO. The chemical activity in the oxidizing catalyst provided by platinum and palladium causes HC and CO to revert to harmless $H_2O$ and $CO_2$. Oxidation causes the compounds to combine with oxygen to form new products—in this case, $H_2O$ and $CO_2$. The reaction that occurs within the converter is similar to a perfect combustion occurring inside the combustion chamber.

> **You Should Know** *Catalytic converters are located as close as practical to the exhaust manifold. This increases converter efficiency because the exhaust gases are not able to cool down before reaching the converter.*

Some TWCs use a type of air injection system, often called **secondary air**. This air supply is injected into the oxidizing catalyst, between the front and rear sections of the catalytic converter **(Figure 3)**. An electrically operated or belt-driven air pump supplies air, containing $O_2$. With an ample supply of $O_2$ available to the oxidation bed of the TWC, HC and CO are effectively oxidized into $H_2O$ and $CO_2$.

## TWC DESIGN IMPROVEMENTS

Newer designs combine the oxidation and reduction functions together in a single bed design by adding the chemical element **cerium (Ce)**. Cerium has a unique property in that it has the ability to store $O_2$. When the exhaust content is lean, $O_2$ is present. This causes the cerium within the converter to store $O_2$. When the exhaust content

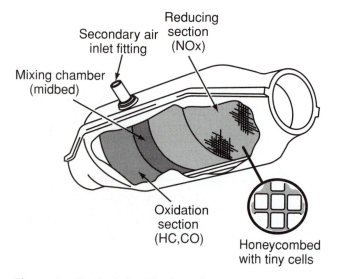

**Figure 2.** Typical dual-bed catalytic converter.

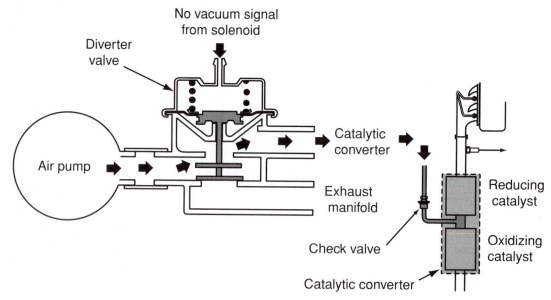

**Figure 3.** Secondary air system helps oxidize HC and CO in the rear section of converter.

becomes rich, the cerium within the converter releases O$_2$ into the converter, allowing oxidation of HC and CO.

An air pump is often used with cerium-based converter technology to enhance the performance of the catalyst.

## CATALYST OPERATING TEMPERATURES

Catalytic converters are ineffective in treating exhaust when cold. It is desirable to increase catalyst temperatures as rapidly as possible, to reduce harmful tailpipe emissions.

Before the catalyst reaches its "light-off" temperature, little chemical activity within the converter occurs. When the efficiency reaches 50 percent, the light-off temperature has been reached. This occurs between 400 and 500°F. Internal catalyst operating temperatures range from 800 to 1600°F during normal operation. Ignition misfire can cause catalyst temperatures to increase even higher. For this reason, use extreme caution when working near catalytic converters.

## LIGHT-OFF CATALYSTS

To rapidly increase the main catalyst temperature, a light-off catalytic converter is often used **(Figure 4)**. Light-off converters are sometimes called **warm up–three-way catalysts (WU-TWC)**. Other terms to describe the light-off catalyst are "pup" catalysts, "warm-up cats," or "mini-cats." Their function is the same: to promote early reaction before the TWC has reached operating temperature. The presence of a light-off catalyst reduces NOx, HC, and CO during warmup. Warmup cats also increase exhaust temperature

into the main converter, helping to increase its temperature rapidly after engine startup.

> **You Should Know** *Engine misfire can seriously damage a catalytic converter. In fact, if left to occur, misfire can destroy the converter in a short time. High concentrations of unburned fuel will cause the temperature of the catalyst to increase past its maximum safe operating temperature. This can cause the substrate to melt, rendering the converter useless.*

## CATALYST HEAT SHIELDS

As a result of the high operating temperatures, most catalytic converters are equipped with a heat shield. The heat shield protects the passenger compartment floor area from excessive temperatures. It also helps prevent leaves and other debris from igniting if the vehicle is parked over such terrain.

## CATALYST POISONING

Excessive antifreeze consumption resulting from a leaking cylinder head gasket can contaminate the honeycomb substrate within the converter. When this occurs, the catalyst is said to be "poisoned." The converter often requires replacement after repairing a leaking cylinder head.

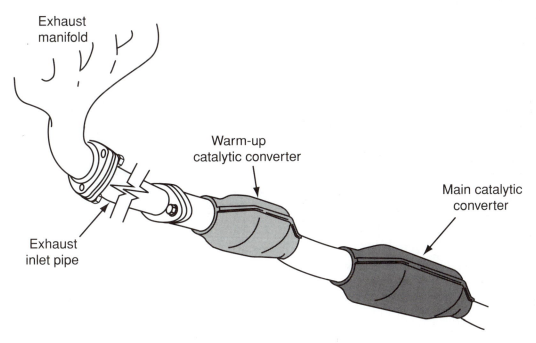

**Figure 4.**   Light-off converter helps promote early reaction of NOx, CO, and HC when main converter is cold.

Other causes of converter poisoning include the use of leaded gasoline and excessive oil consumption.

> ⚠️ **You Should Know** *Engines equipped with a catalyst must never use leaded fuel. Otherwise, the catalyst will be permanently damaged. Several tanks of leaded fuel are required to poison a catalytic converter. If a converter is exposed to leaded fuels, it is often possible to restore catalyst efficiency by switching immediately back to unleaded fuel.*

## REPAIRING CATALYTIC CONVERTERS

Catalytic converters are sealed units that are not serviceable. Once the converter is poisoned from leaded fuel, physically damaged, or corroded beyond use, it must be replaced. The honeycomb substrate within the converter is easily damaged from severe shock. Never drop or strike the converter.

Catalytic converter testing consists of temperature measurements and is covered in Chapter 36.

## AIR INJECTION REACTION SYSTEMS

Some vehicle applications use an air pump to help promote oxidation within the catalytic converter **(Figure 5)**. The air pump supplies a high-volume, low-pressure source of air through one or more check valves. The pump can be belt-driven off the engine accessory drive, or by a small electric motor. The powertrain control module (PCM) uses solenoids to control airflow generated by the pump **(Figure 6)**. Depending upon the specific strategy, air is directed to the converter, exhaust manifold, or atmosphere **(Figure 7)**.

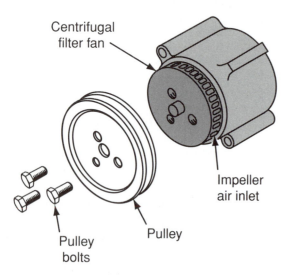

**Figure 5.** Typical belt-driven air pump.

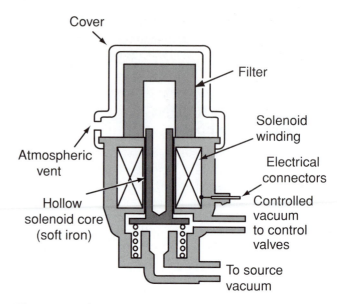

**Figure 6.** The vacuum solenoid control's engine vacuum to air by-pass and air diverter valves.

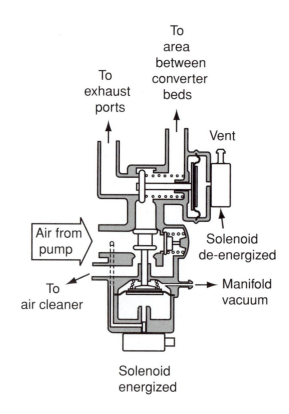

**COMBINATION BYPASS AND DIVERTER VALVE**

**Figure 7.** Using a combination air by-pass valve and diverter valve, secondary air can be directed to the converter, exhaust manifold, or atmosphere.

## PCM STRATEGIES

The PCM uses a variety of strategies to provide lowest emissions under varying vehicle operating conditions. In particular, cold engine operation presents the greatest challenge for the engineer because a rich air/fuel mixture is needed to keep the engine running smoothly. Rich air/fuel ratios cause increased CO and HC exhaust emissions, which must be minimized using special PCM strategies designed to afford quick engine and catalyst warmup.

### Cold Engine Operation

Under cold engine operation, the air pump supplies air upstream of the converter, into the exhaust manifold (**Figure 8**). At this point, the catalyst temperature is not high enough to reach light-off. The additional $O_2$ supplied to the exhaust stream serves to further oxidize HC and CO into $H_2O$ and $CO_2$ after the combustion process.

The added $O_2$ content in the exhaust stream increases the exhaust temperature. This helps to bring the temperature of the $O_2$ sensor and catalytic converter up to operating temperatures more quickly, reducing emissions.

### Hot Engine Operation

After the engine reaches operating temperature, the PCM switches airflow from upstream of the converter to the downstream mode (**Figure 9**). In the downstream mode, air from the pump is injected into the oxidizing portion of the converter. The air supplied by the air pump contains $O_2$, which helps promote oxidation of HC and CO into $H_2O$ and $CO_2$ within the converter.

### Deceleration

To prevent exhaust backfire, the PCM directs the airflow from the air pump to the atmosphere under conditions of rapid vehicle deceleration. Otherwise, the added oxygen content from the pump would cause an explosion inside the converter.

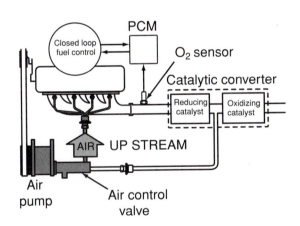

**Figure 8.** Under cold engine operation, the air pump supplies air upstream of the converter. This helps oxidize HC and CO into $H_2O$ and $CO_2$.

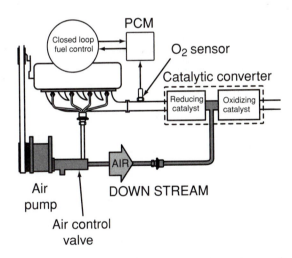

**Figure 9.** Once operating temperature is reached, air from the pump is directed downstream into the converter.

## *Summary*

- Exhaust aftertreatment consists of using one or more catalytic converters to reduce tailpipe emissions.
- The catalytic converter uses noble metals, such as palladium, platinum, and rhodium, as catalysts.
- The TWC is often called a dual-bed converter because it contains two individual substrates.
- A TWC acts upon three harmful emissions gases: NOx, HC, and CO.
- Catalytic converters are ineffective in treating exhaust when cold. It is desirable to increase catalyst temperatures as rapidly as possible to reduce harmful tailpipe emissions.

# Review Questions

1. Technician A says that a catalytic converter is needed only to eliminate NOx. Technician B says that a catalytic converter improves fuel economy. Who is correct?
   A. Technician A
   B. Technician B
   C. Both A and B
   D. Neither A nor B

2. Technician A says that HC and CO are reduced into $H_2O$ and $CO_2$. Technician B says that HC and CO are oxidized into $H_2O$ and $CO_2$. Who is correct?
   A. Technician A
   B. Technician B
   C. Both A and B
   D. Neither A nor B

3. Technician A says that when NOx is reduced, only $O_2$ and $N_2$ remain. Technician B says that when HC and CO are oxidized, $H_2O$ and $CO_2$ are produced. Who is correct?
   A. Technician A
   B. Technician B
   C. Both A and B
   D. Neither A nor B

4. Explain how a dual-bed converter works.

5. Explain how a reducing catalyst functions.

6. Explain why maintaining a minimum catalyst temperature is necessary.

7. Technician A says that leaded fuel can contaminate a catalyst. Technician B says that excessive oil consumption can contaminate a catalyst. Who is correct?
   A. Technician A
   B. Technician B
   C. Both A and B
   D. Neither A nor B

8. Technician A says that a light-off catalyst helps to increase the main converter's temperature. Technician B says that a light-off catalyst is used to improve fuel vaporization. Who is correct?
   A. Technician A
   B. Technician B
   C. Both A and B
   D. Neither A nor B

9. Technician A says that an air pump acts like a turbocharger and increases engine power. Technician B says that an air pump improves oxidation of harmful emission gases in the catalytic converter. Who is correct?
   A. Technician A
   B. Technician B
   C. Both A and B
   D. Neither A nor B

10. Technician A says that no air is sent into the exhaust stream during deceleration. Technician B says that air is always injected upstream of the converter once the engine reaches operating temperature. Who is correct?
    A. Technician A
    B. Technician B
    C. Both A and B
    D. Neither A nor B

# Chapter 36

# Basic Emissions System Diagnosis

## Introduction

Diagnosing emissions system concerns is important in terms of maintaining air quality standards. The service technician must follow manufacturer-recommended service procedures to ensure repairs made will maintain the appropriate levels of exhaust and evaporative emissions.

Because of the interaction among the fuel, ignition, and emissions systems, the technician must understand how components of one system interact with the others. Effective diagnosis begins with a thorough understanding of basic troubleshooting techniques. The shop manual is an important tool in terms of isolating emission concerns, as well as performing the appropriate repair.

The techniques included in this chapter are designed to make use of basic test equipment found in almost all repair facilities. In Chapter 37, electronic diagnosis of emissions equipment is presented for a variety of emissions systems. This includes using both scan tools and conventional electronic test equipment.

On-Board Diagnostics has improved repair effectiveness immensely since its introduction in 1993. Although OBDII was not mandated until 1996-model vehicles, manufacturers began installing the system in limited applications on 1994 models.

The enhanced emissions diagnosis offered by this powerful protocol has increased technician awareness in regard to both vehicle emissions and driveability. Although it will probably never replace the basic troubleshooting techniques presented in this chapter, OBDII will enhance your abilities to effectively analyze and repair emissions-related concerns on late-model vehicles. Soon you will ask yourself how you ever survived without it.

## EXHAUST GAS RECIRCULATION SYSTEM DIAGNOSIS AND REPAIR

Problems with the EGR system include little or no flow and a sticky EGR valve. In some cases, the valve is completely stuck open, resulting in an engine that will not idle. OBDII monitoring of the EGR system can also trigger a "check engine" warning lamp, if a fault occurs.

### Little or No EGR System Flow

An EGR valve that fails to open or a restriction in the EGR system can cause the following symptoms:
- Elevated NOx emissions, as measured with a gas analyzer.
- Spark knock, when accelerating, climbing hills, or pulling heavy loads.

### Testing a Vacuum-Operated EGR System With Little or No Flow

Use the following **Test Procedure (TP):**
- Perform a thorough visual inspection.
- Replace cracked or deteriorated vacuum lines.
- Using a hand-operated vacuum pump, verify the EGR valve diaphragm can hold vacuum **(Figure 1)**. If it fails this test, replace the valve.
- Start the engine and activate the valve with the vacuum pump. If the engine idle speed is relatively unchanged, check for restrictions in the exhaust passages. Most engines will stall when the EGR valve is fully opened at idle. Repair as necessary.
- Replace any faulty components.
- Retest the system and check for an idle speed change.
- Verify proper operation of the system before returning the vehicle to the customer.

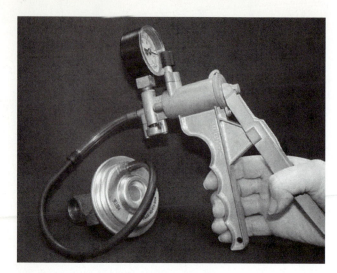

**Figure 1.** Using a hand-operated vacuum pump to check EGR valve operation.

## EGR Valve Stuck Wide Open or Partially Open

Symptoms of a stuck EGR valve include:
- Poor idle quality or the inability to idle.
- Poor performance, especially when accelerating.

## Testing a Vacuum-Operated EGR System for a Sticky Valve

Use the following TP:
- Perform a thorough visual inspection.
- Using a hand-operated vacuum pump, verify the EGR valve diaphragm operates when activating the pump. Note if the idle quality changes when repeatedly activating the valve. If so, replace the valve.
- Remove the EGR valve and inspect for carbon deposits or erratic plunger operation. Replace the valve if either condition is found.
- Replace any faulty components.
- Verify proper operation of the system before returning the vehicle to the customer.

## PCV SYSTEM DIAGNOSIS AND REPAIR

A properly operating PCV system applies a slight vacuum to the crankcase. Problems with the PCV system include little or no flow. Rarely does excessive flow result, unless the wrong PCV valve is installed on the engine.

## Little or No PCV System Flow

Little or no PCV flow results in the following:
- Excessive oil consumption
- Blown-out oil seals and gaskets
- Rich fuel mixture

## Testing a PCV System With Little or No Flow

Use the following TP:
- Perform a thorough visual inspection.
- Use an index card to check for crankcase suction at the oil filler cap. If the card fails to stay in place, it might mean that low vacuum is present at the valve. Proceed with more testing.
- Remove the PCV valve and shake it. It should rattle freely. If it does not, *do not* attempt to clean the valve. Replace it.
- If the valve rattles, it might still be defective. Replace it if the PCV valve has been installed longer than the manufacturer-recommended interval, usually 50,000 mi. A high-flow replacement valve is recommended in some high-mileage applications. Blowby gases tend to increase as the piston rings wear.
- Start the engine and remove the PCV valve from the engine, leaving it connected to the intake manifold. A strong vacuum should be felt at the inlet orifice **(Figure 2)**.
- You should be able to hear the plunger move within the valve as you rapidly place and remove your finger from the orifice. If not, check for low vacuum at the valve or a defective valve.
- Carefully check for a cracked vacuum line or hose leading to the intake manifold. Replace any defective or damaged vacuum lines to the PCV valve.
- Verify that vacuum is present at the intake manifold port of the PCV line. If not present, determine the cause of low or no engine vacuum at this port. Carbon and sludge are likely causes. The passages will need to be cleaned. A vacuum gauge is helpful in locating vacuum leaks and plugged lines.
- Replace any faulty components.
- Verify proper operation of the system before returning the vehicle to the customer.

**Figure 2.** Testing a PCV valve for strong vacuum.

# EVAP EMISSIONS SYSTEM DIAGNOSIS AND REPAIR

The evaporative emissions system is often neglected because problems often go unnoticed unless the vehicle is subject to an enhanced **IM240 emissions test**. OBDII monitoring of the EVAP system can also trigger a "check engine" warning lamp, if a fault occurs.

# EVAP EMISSIONS SYSTEM FAILURES

Failures in the EVAP emissions system can cause:
- Poor fuel economy
- Poor engine performance

# TESTING THE EVAP EMISSIONS SYSTEM FOR PROPER OPERATION

Use the following TP:
- Perform a thorough visual inspection. Pay special attention to physical damage of the carbon canister, fuel cap, and EVAP hoses.
- Check the EVAP system as outlined in the shop manual.
- Leaks in CANP solenoids can cause a constant flow of HCs into the engine's intake system. This results in an excessively rich air/fuel mixture. A subtle drop in fuel economy is often the result of leaky solenoids. Defective solenoids are not serviceable and must be replaced. Check solenoids using a vacuum pump and vacuum gauge.
- Vacuum leaks can cause unmetered air to enter the combustion chambers, causing an excessively lean air/fuel mixture. Replace cracked or broken vacuum lines. Such lines often deteriorate because of high underhood temperatures.
- Use a purge flow tester to check for adequate purge flow. Most vehicles must be driven for purging of HC vapors to occur. To meet state emissions test requirements, at least 1L of flow during the IM-240 test must occur. Otherwise, the vehicle will fail the evaporative emissions test. Consult the shop manual for specific details about how to perform a purge flow test and typical flow rates, which can exceed 10L or more.

# CATALYTIC CONVERTER DIAGNOSIS AND REPAIR

The catalytic converter is a nonserviceable component. If it is determined the converter is defective, it must be replaced. Problems with the catalytic converter include converter poisoning, overheating, corrosion, and physical damage.

Failure to pass an emissions test is a strong indicator of a faulty converter, especially if the vehicle has been driven over 50,000 mi. OBD-II system monitoring of the catalyst monitoring system can also trigger a "check engine" warning lamp, if a fault occurs.

## Improper Catalytic Converter Operation

An inoperative or faulty catalytic converter can cause:
- Elevated levels of HC, CO, and NOx emissions, as measured with a gas analyzer
- Poor engine performance

## Testing the Catalytic Converter for Proper Operation

Use the following TP:
- Perform a thorough visual inspection. Pay special attention to physical damage.
- Check the converter as outlined in the shop manual.
- Using a gas analyzer, observe HC, CO, and NOx tailpipe emissions. The readings should decrease as the converter reaches operating temperature.
- Use an infrared thermometer to verify that the outlet temperature is at least 10 percent greater than the inlet temperature **(Figure 3)**.
- Verify proper operation of the system before returning the vehicle to the customer.

# AIR INJECTION REACTION DIAGNOSIS AND REPAIR

Problems with the AIR system include low or no airflow into the exhaust stream or catalytic converter. Often, hoses or solenoids are to blame. Other problems include improper switching of airflow to the exhaust stream or catalytic converter.

## Improper AIR Operation

An inoperative or faulty AIR system can cause:
- Elevated levels of HC and CO emissions, as measured with a gas analyzer
- Exhaust backfire

## Testing an AIR System for Proper Operation

Use the following TP:
- Perform a thorough visual inspection. Pay special attention to hoses, fittings, solenoids, and drive belts **(Figure 4)**.
- Check the air pump as outlined in the shop manual.
- Using a gas analyzer, observe HC and CO tailpipe emissions. The reading should increase when the air pump is disabled or when hoses are restricted using pinch-off pliers.
- Replace any faulty components.
- Verify proper operation of the system before returning the vehicle to the customer.

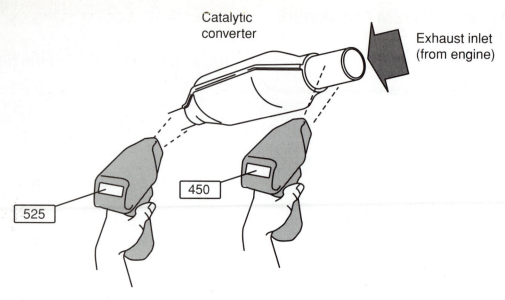

**Figure 3.**   Testing a catalytic converter using an infrared pyrometer.

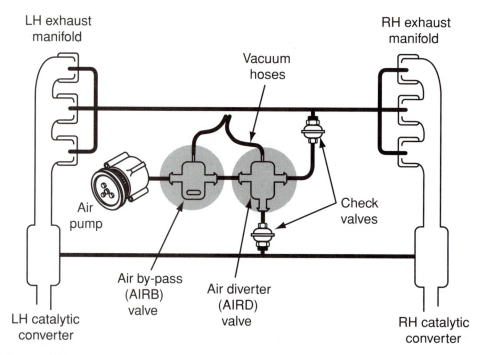

**Figure 4.**   Checking an AIR system.

# *Summary*

- Diagnosing emissions control system concerns is important in terms of maintaining air quality standards.
- Problems with the EGR system include little or no flow and a sticky EGR valve.
- Symptoms of an EGR valve that is stuck open include poor idle quality and poor engine performance, especially when accelerating.
- If an EGR valve is stuck closed, it can cause spark knock and elevated NOx emissions.
- A vacuum-operated EGR valve can be tested using a hand-held vacuum pump.
- Little or no PCV flow causes excessive oil consumption, blown-out gaskets, and a rich air/fuel mixture.

- The PCV systems are often checked using an index card held over the oil filler cap opening.
- Never attempt to clean a PCV valve.
- Failures in the EVAP system can cause poor fuel economy and an increase in the amount of evaporated HCs into the atmosphere.
- A purge-flow tester is used to diagnose EVAP system concerns.
- The catalytic converter is a nonserviceable component.

- A defective catalytic converter can cause increased HC, CO, and NOx emissions.
- A plugged catalytic converter can cause poor engine performance.
- An infrared thermometer is used to diagnose a catalytic converter.
- Faulty AIR systems can cause elevated HC and CO emissions.
- A defective AIR valve can cause exhaust backfire.

# Review Questions

1. Technician A says that a faulty emissions control system can be disregarded as long as the engine runs smoothly. Technician B says that all emissions control systems must function properly if air quality standards are to be met. Who is correct?
   A. Technician A
   B. Technician B
   C. Both A and B
   D. Neither A nor B

2. Technician A says that an EGR valve that sticks open can cause poor idle quality. Technician B says that an EGR valve that sticks open can cause spark knock. Who is correct?
   A. Technician A
   B. Technician B
   C. Both A and B
   D. Neither A nor B

3. Technician A says that a sticky EGR valve can be cleaned to restore proper operation. Technician B says to replace the EGR valve gasket whenever the valve is installed. Who is correct?
   A. Technician A
   B. Technician B
   C. Both A and B
   D. Neither A nor B

4. Technician A says to check for broken or deteriorated vacuum lines when diagnosing an inoperative EGR system. Technician B says to replace all vacuum lines whenever the EGR system is serviced. Who is correct?
   A. Technician A
   B. Technician B
   C. Both A and B
   D. Neither A nor B

5. Technician A says that vacuum-operated EGR valves can be tested using a hand-held vacuum pump. Technician B says that some EGR valves have a sensor mounted on the valve. Who is correct?
   A. Technician A
   B. Technician B
   C. Both A and B
   D. Neither A nor B

6. Technician A says that an inoperative PCV system can cause excessive oil consumption. Technician B says that the PCV valve should never be cleaned. Who is correct?
   A. Technician A
   B. Technician B
   C. Both A and B
   D. Neither A nor B

7. Technician A says that a lean air/fuel mixture can be caused by an inoperative PCV system. Technician B says that an index card can be used to verify operation of the PCV system. Who is correct?
   A. Technician A
   B. Technician B
   C. Both A and B
   D. Neither A nor B

8. Technician A says that poor engine performance can be the result of a faulty EVAP system. Technician B says that a cracked carbon canister can cause the EVAP system to operate improperly. Who is correct?
   A. Technician A
   B. Technician B
   C. Both A and B
   D. Neither A nor B

9. Technician A says that a faulty catalytic converter can cause excessive HC tailpipe emissions. Technician B says that a faulty catalytic converter can cause an increase in NOx tailpipe emissions. Who is correct?
   A. Technician A
   B. Technician B
   C. Both A and B
   D. Neither A nor B

10. Technician A says that a faulty AIR pump can cause decreased fuel economy. Technician B says that a faulty AIR valve can cause exhaust backfire to occur. Who is correct?
    A. Technician A
    B. Technician B
    C. Both A and B
    D. Neither A nor B

# Chapter 37

# Specialized Emissions Diagnostics Equipment

## Introduction

To accurately analyze emissions-related failures, the skilled technician uses a variety of instruments. In addition to basic tools and test equipment, effective emissions diagnosis includes the use of an **exhaust gas analyzer**, scan tool, and other specialized instruments.

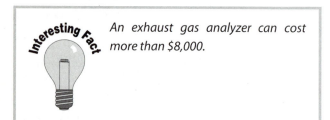

**Interesting Fact**

*An exhaust gas analyzer can cost more than $8,000.*

Emissions diagnosis equipment is often very expensive, requiring a considerable investment on the part of the company or shop owner. In fact, many repair shops choose not to engage in emissions-related services because of the high costs involved, and the need for skilled technician training.

## SERVICE PUBLICATIONS

Perhaps the most overlooked resource when it comes to emissions diagnosis is the shop manual. Depending upon the manufacturer, a special emissions diagnosis manual might be available. Repair shops performing emissions diagnosis will find such publications indispensable when diagnosing excessive tailpipe emissions concerns.

It is generally regarded that detailed information is usually available only from the vehicle manufacturer. However, with the increasing interest in diagnosing emissions failures, aftermarket information systems have enhanced the amount of information available to the service technician. This is especially beneficial to independent repair shops that have little access to factory publications.

## THE EXHAUST GAS ANALYZER

Perhaps the most useful instrument available to diagnose emissions-related concerns is the exhaust gas analyzer **(Figure 1)**. The exhaust gas analyzer, or gas analyzer, allows the technician to measure exhaust emissions directly.

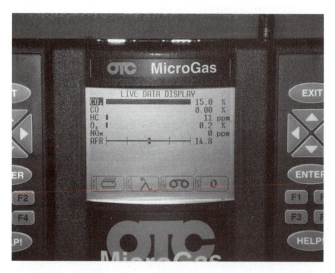

**Figure 1.** Exhaust gas analyzer.

**Figure 2.** Exhaust gas sampling probe.

Gas analyzers can be of the four-gas or five-gas variety. The four-gas analyzer measures CO, HCs, $O_2$, and $CO_2$. A five-gas analyzer not only measures CO, HCs, $O_2$, and $CO_2$, but also NOx as well. Because of the concern over NOx, the five-gas analyzer is more popular.

Exhaust gas analyzers operate by sampling a portion of the vehicle exhaust gases using a sampling probe **(Figure 2)**. Most portable analyzers operate by using an infrared measurement technique that isolates the various exhaust gases based upon temperature. The specific concentrations of each gas are then determined and displayed for the technician to interpret.

In the case of the infrared gas analyzer, the various gases are displayed as follows:

- CO is displayed as a percent.
- HC is displayed in parts per million (ppm).
- $O_2$ is displayed as a percent.
- $CO_2$ is displayed as a percent.
- NOx is displayed in ppm.

Obviously, CO, HC, and NOx should be as low as possible, whereas $CO_2$ approaches 15 percent. Although $O_2$ is not a harmful tailpipe emissions gas, its presence indicates the relative combustion efficiency and is helpful in isolating combustion failures.

Some gas analyzers use a technique called **constant volume sampling (CVS)**. With CVS, the analyzer uses a special technique to break down the various harmful emissions gases by weight. A chassis dynamometer is required to determine the amounts of pollutants emitted for every mile the vehicle is driven. In this case, HC, CO, and NOx are measured in grams per mile. This provides a more accurate account of how much the vehicle is emitting. Many state emissions testing programs use this method

to determine whether a vehicle passes or fails an emissions test.

More information about interpreting exhaust emissions is presented in Chapter 38.

## SCAN TOOL

The scan tool can be useful in diagnosing emissions failures, depending upon the model year of the vehicle. Pre-OBDII vehicles are limited in terms of emissions diagnosis using a scan tool. Although DTCs are available on many pre-OBDII systems, no provisions exist for diagnosing catalytic converter efficiency or EVAP systems, for example.

Because electronic engine control is closely aligned with vehicle emissions, the parameters available to diagnose engine control problems can be successfully used to carry out emissions diagnostics. The technician should not entirely rely on DTCs alone because by themselves they are of limited value. Instead, diagnosis should proceed using the PIDs available from the engine control computer.

As computer technology progressed through the 1980s, vehicle manufacturers began making increasingly more information available to the service technician. This information became available not only in terms of more DTCs, but also in the form of data streams. Looking at PID data for short- and long-term fuel trim, for example, is an effective means of diagnosing excessive fuel pressure concerns that can cause excessive HC and CO tailpipe emissions. Unfortunately, many of the monitors used to detect faults that can cause increased levels of tailpipe emissions were not available until OBDII was implemented in 1996.

## OBDII SYSTEM MONITORS

The OBDII system uses a number of monitors to keep track of various engine operating conditions. These monitors are a part of the PCM diagnostic strategies that help the technician in locating faults pertaining to the emissions control system.

### Comprehensive Component Monitor

The comprehensive component monitor continuously checks circuitry related to emissions control components. Electrical and component faults that can cause an emissions control component to operate incorrectly—or not at all—are checked by the PCM. Such faults can cause increased HC, CO, and NOx tailpipe emissions to occur.

### Misfire Monitor

The misfire monitor available on OBDII systems can be used to diagnose elevated HC emissions caused by a cylinder misfiring. Misfire fault codes are located in the ignition system category of P03xx (generic OBDII) or P13xx (manufacturer-specific) DTCs.

Because of the way the misfire monitor is designed, high-speed misfire might not be detected. In this case, the technician can easily locate the problem using an oscilloscope equipped with the appropriate secondary ignition adapter.

## Catalyst Performance Monitor

The catalyst performance monitor uses one or two additional oxygen sensors to verify that the catalytic converter is performing satisfactorily. The addition of a downstream sensor allows the PCM to make comparisons between the upstream and downstream sensors. If the downstream sensor reports activity greater than that considered normal, a DTC is set and the MIL is illuminated to alert the driver.

Because catalyst failures can cause elevated tailpipe emissions of HC, CO, and NOx, using the data available through the scan tool is an effective method for diagnosing catalyst concerns. Catalyst performance DTCs are located in the emissions control system category of P04xx (generic OBDII) or P14xx (manufacturer-specific) DTCs. P04xx and P14xx DTCs always relate to a fault in the emissions control system.

## EVAP System Monitor

The EVAP system monitor is designed to verify proper operation of the EVAP system, both in terms of system integrity (no leaks) and purge operation. By using DTCs and data stream information, the technician can begin a diagnostics routine that effectively points to the exact nature of the fault.

Faults in the EVAP system can cause the release of unburned HC vapors from the fuel tank or fuel system. The EVAP system DTCs are located in the emissions control system category of P04xx (generic OBDII) or P14xx (manufacturer-specific) DTCs.

## Secondary AIR System Monitor

OBDII regulations require monitoring of the secondary AIR system, if the vehicle is so equipped. The system must be monitored for activity of both the AIR valves and the presence of airflow in the exhaust stream. The latter is easily accomplished using the upstream $O_2S$.

Faults in the secondary AIR system can cause increased levels of HC and CO in the exhaust. Secondary AIR system DTCs are located in the emissions control system category of P04xx (generic OBDII) or P14xx (manufacturer-specific) DTCs.

## Oxygen Sensor Monitor

The $O_2Ss$ are monitored for proper operation to meet OBDII requirements. Response time, sensor activity, and inadequate switching are some of the parameters monitored by the PCM. If a fault is detected, the PCM can set a DTC, making it easier for the technician to isolate the problem.

Oxygen sensor faults can lead to incorrect air/fuel mixtures, causing an increase in HC and CO tailpipe emissions. The $O_2S$ monitor DTCs are located in the air metering and fuel system category of P01xx (generic OBDII) or P11xx (manufacturer-specific) DTCs.

## Oxygen Sensor Heater Monitor

To verify that the heater elements of all $O_2Ss$ are operating correctly, OBDII regulations require the PCM to monitor the heater elements for activity. One method of accomplishing this is to program the PCM to check how long it takes until the sensor begins to switch. If the time exceeds a certain threshold, a DTC is set, and the MIL is illuminated.

If the $O_2S$ heater is inoperative, it can cause increased HC and CO tailpipe emissions. The $O_2S$ heater monitor DTCs are located in the air metering and fuel system category of P01xx (generic OBDII) or P11xx (manufacturer-specific) DTCs.

## EGR System Monitor

The EGR system is monitored for abnormal flow rates. One method of accomplishing this is to monitor the pressure drop across a calibrated orifice located in the exhaust stream. As the EGR valve opens, the pressure drop across the orifice increases, providing the PCM with flow data.

If the EGR flow is not within preprogrammed parameters, the PCM sets a DTC and illuminates the MIL. Abnormal EGR flow can cause NOx emissions to increase. EGR system DTCs are located in the emissions control category of P04xx (generic OBDII) or P14xx (manufacturer-specific) DTCs.

## Fuel System Monitor

Long-term fuel trim is monitored by the fuel system monitor for acceptable activity. If long-term fuel trim shifts beyond preprogrammed limits, it often indicates a problem with fuel or air delivery. The PCM will set a DTC and illuminate the MIL.

Because long-term fuel trim is an indication of how well the PCM is able to control the air/fuel ratio, it is an important parameter in maintaining low HC and CO tailpipe emissions. Fuel system DTCs are located in the air metering and fuel system category of P01xx (generic OBDII) or P11xx (manufacturer-specific) DTCs.

Other fuel system DTCs are located in the fuel system category of P02xx (generic OBDII) or P12xx (manufacturer-specific) DTCs. The DTCs in this category are often related to fuel injector faults.

## PYROMETER

The **pyrometer** is an infrared thermometer that is useful in measuring the surface temperature of an object without actually coming in contact with it. Pyrometers are particularly useful in isolating dead cylinders and defective catalytic converters. A digital pyrometer is shown in **Figure 3**.

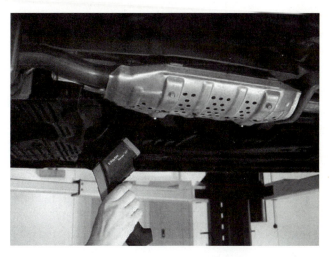

**Figure 3.** Infrared hand-held pyrometer measures surface temperature of an object.

# *Summary*

- In addition to basic tools and test equipment, effective emissions diagnosis includes the use of an exhaust gas analyzer, scan tool, and other specialized instruments.
- The use of service publications is important in diagnosing emissions-related failures.
- Gas analyzers can be of the four-gas or five-gas variety.
- The four-gas analyzer measures CO, HC, $O_2$, and $CO_2$.
- The five-gas analyzer measures CO, HC, $O_2$, $CO_2$, and NOx.

- The scan tool is useful in diagnosing emissions failures.
- Vehicles equipped with the OBDII system have enhanced monitoring that provides effective emissions diagnosis.
- Various monitors within the OBDII system can guide the technician in locating faults within the fuel, ignition, and emissions control systems.
- A pyrometer is useful in isolating a dead cylinder or testing a catalytic converter.

# *Review Questions*

1. Technician A says that a four-gas analyzer measures tailpipe emissions. Technician B says that a five-gas analyzer measures NOx. Who is correct?
   A. Technician A
   B. Technician B
   C. Both A and B
   D. Neither A nor B
2. Technician A says that service publications are important when diagnosing emissions-related failures. Technician B says that factory service publications contain detailed information regarding emissions troubleshooting. Who is correct?
   A. Technician A
   B. Technician B
   C. Both A and B
   D. Neither A nor B

3. Technician A says that all that is needed to diagnose an emissions-related failure is a gas analyzer. Technician B says that a gas analyzer can be used to diagnose excessive HC tailpipe emissions. Who is correct?
   A. Technician A
   B. Technician B
   C. Both A and B
   D. Neither A nor B
4. Technician A says that the comprehensive component monitor continuously monitors the $O_2$Ss for proper activity. Technician B says that catalyst efficiency is monitored by the upstream $O_2$S. Who is correct?
   A. Technician A
   B. Technician B
   C. Both A and B
   D. Neither A nor B

5. Technician A says that a P0443 DTC indicates a fault in the emissions control system. Technician B says that a P1433 DTC indicates a fault in the ignition system. Who is correct?
   A. Technician A
   B. Technician B
   C. Both A and B
   D. Neither A nor B

6. Technician A says that a fault in the EVAP system can cause CO emissions to increase. Technician B says that the EVAP system monitor checks for leaks in the fuel tank. Who is correct?
   A. Technician A
   B. Technician B
   C. Both A and B
   D. Neither A nor B

7. Technician A says that the OBDII EGR system monitor checks for proper EGR flow rates. Technician B says that OBDII EGR system failures do not have to set a DTC. Who is correct?
   A. Technician A
   B. Technician B
   C. Both A and B
   D. Neither A nor B

8. Technician A says that a faulty $O_2S$ heater can cause increased levels of NOx. Technician B says that a faulty $O_2S$ can cause higher-than-normal $CO_2$ levels. Who is correct?
   A. Technician A
   B. Technician B
   C. Both A and B
   D. Neither A nor B

9. Technician A says that if long-term fuel trim decreases beyond preprogrammed values, the vehicle might emit excessive HC tailpipe emissions. Technician B says that if long-term fuel trim goes out of range, a DTC is set and the MIL is illuminated. Who is correct?
   A. Technician A
   B. Technician B
   C. Both A and B
   D. Neither A nor B

10. Technician A says that long-term fuel trim can be used to diagnose a vehicle that has excessive CO emissions. Technician B says that only short-term fuel trim should be used to diagnose high CO concerns. Who is correct?
    A. Technician A
    B. Technician B
    C. Both A and B
    D. Neither A nor B

# Vehicle Emissions Testing

## Introduction

Many areas in the United States have come under close scrutiny of the Environmental Protection Agency because of poor air quality. Not all harmful emissions can be blamed on **mobile sources** such as passenger cars and trucks. Industrial air pollution accounts for a substantial amount of air quality concerns. However, mobile sources do contribute a significant portion of air pollutants. Even motorcycles, lawnmowers, and watercraft are included in the list of mobile sources of harmful emissions.

In fact, because of the serious nature of transportation air quality, the EPA has established a special division for regulating emissions arising from mobile sources. Called the **Office of Transportation Air Quality** (OTAQ), it oversees all of the state emissions testing programs proposed or in place nationally. Emissions testing standards are based upon specific vehicle model years for both tailpipe and evaporative emissions.

Initially, many states used a simple **idle test** to establish if a particular vehicle met state requirements. However, with increasingly tighter federal requirements, many state inspection facilities resorted to **transient emissions testing** to determine if a vehicle passed exhaust emissions when operating under simulated driving conditions. Problems that caused tailpipe emissions to increase under actual driving conditions were sometimes not apparent when using an idle test alone.

Emissions testing designed to spot troublesome vehicles in need of routine maintenance causing elevated tailpipe or evaporative emissions is often referred to as **Inspection and Maintenance (I/M)**.

Vehicle emissions testing is usually performed annually or every other year. If the state has an annual safety equipment inspection program, the emissions test usually coincides with the safety inspection.

Some states have no emissions testing program in place at this time. Additionally, some localities in crowded areas might be required to conduct vehicle emissions testing, while other less densely populated areas may be exempt. Other states might perform only a visual inspection of emissions equipment. An automatic failure occurs if tampering is evident.

Some emissions testing programs are centralized. In this case, only a state-approved vendor may perform the emissions test. In other centralized programs, tests are conducted at state-operated facilities. Customers are responsible for having their vehicles repaired at repair facilities of their choice. Once a vehicle is repaired, it must be retested before a decal can be issued.

Decentralized vehicle emissions testing programs permit licensed repair shops to perform emissions tests, provided state requirements are met. The customer may elect to have the inspection facility make repairs to the vehicle but is not bound to do so.

Most states conducting vehicle emissions testing have processes in place to prevent renewing the vehicle registration in the event the vehicle fails an emissions test. In certain hardship cases, the vehicle may be exempt from failing the test. If the owner spends more than the state-prescribed limit to repair the vehicle, yet it still fails the test, the vehicle can still be licensed and driven on public highways. The exemption is usually a one-time exclusion.

## TYPES OF EMISSIONS TESTS

Depending upon state regulations, different emissions tests are used to determine whether a vehicle conforms to

## TYPICAL EXHAUST GAS PARAMETERS
### (Pre 1982 vehicles at idle)

| Model years | Number of cylinders | Emission equipment | HC ppm | CO% |
|---|---|---|---|---|
| 1966 to 1970 | 6 and 8 | AIR | 400 | 4.5 |
| | | No AIR | 500 | 6.5 |
| 1971 to 1974 | 6 and 8 | AIR | 300 | 3.5 |
| | | No AIR | 400 | 6.5 |
| 1975 to 1979 | All | No converter | 200 | 3.5 |
| | | 3-way converter | 100 | 1.5 |
| | | 2-way converter (with AIR) | 150 | 1.5 |
| | | 2-way converter (without AIR) | 250 | 4.5 |
| 1980 to 1981 | All | All without AIR | 150 | 2.5 |
| | | All with AIR | 150 | 1.2 |

| 1982 through current year vehicles | | | | | | |
|---|---|---|---|---|---|---|
| Gases | Idle | | 1500 RPM | | 2500 RPM | |
| | converter | non converter | converter | non converter | converter | non converter |
| HC ppm | 0 - 150 | 75 - 135 | 0 - 135 | 50 - 200 | 0 - 75 | 25 - 150 |
| CO% | 0.1 - 1.5 | 0.5 - 3.0 | 0 - 1.1 | 0.5 - 2.0 | 0 - 0.8 | 0.1 - 1.5 |
| CO2% | 10 - 12 | 10 - 12 | 10 - 13 | 10 - 13 | 11 - 13 | 11 - 13 |
| O2% | 0.5 - 2.0 | 0.5 - 2.0 | 0.5 - 2.0 | 0.5 - 2.0 | 0.5. - 1.25 | 0.5 - 2.0 |

**Figure 1.** Tailpipe emissions standards vary according to vehicle model year.

emissions standards. These standards are often called "cut points." Depending on the model year involved, different cut points may be placed in effect **(Figure 1)**. Emissions testing in the form of simple idle tests or more dynamic methods involving a chassis dynamometer are used depending upon state regulations.

## Idle Test

The idle test was one of the first tests to become standard in many state emissions testing programs. In Portland, Oregon, for example, the idle test has been a mainstay inspection since 1975. Idle emissions tests are perhaps the most common form of vehicle emissions testing.

The idle test is usually accompanied by a visual inspection of under-hood and under-body emissions equipment. The visual inspection is designed to prevent potential tampering of emissions control components.

The idle test is easily performed using either a four-gas or five-gas exhaust analyzer **(Figure 2)**. After the vehicle has warmed to operating temperature, the analyzer probe is inserted into the tailpipe. After the reading stabilizes,

driving situations. Compared with the Federal Test Procedure (FTP) of 505 seconds, the IM240 test length is significantly shorter, being limited to about 240 seconds of driving.

*The exact length of the IM240 test is 239 seconds. The phrase "IM240" is based on a "rounded off" approximation of the test length.*

Enhanced emissions testing is required in many areas classified as having air quality problems. In some cases, the EPA classifies an area as a nonattainment area, meaning that significant emissions control measures must be implemented to improve air quality.

The FTP is a precisely defined, dynamic test used by automotive manufacturers to certify a motor vehicle for emissions compliance. This is a necessary procedure before the vehicle can be sold for use on public highways.

All IM240 testing must be conducted on a chassis dynamometer, using a sophisticated computer-controlled load at the drive wheels. The load simulates driving situations. While controlling the dynamics of the "dyne," the system computer is simultaneously collecting emissions gas data, storing it in memory, and comparing it with known good "cut points." An IM240 chassis dynamometer is shown in **Figure 3**.

The IM240 drive cycle is a standardized pattern of acceleration, steady-state driving, deceleration, and vehicle stop conditions that are designed to simulate combined city and highway driving. The IM240 trace is shown in **Figure 4**.

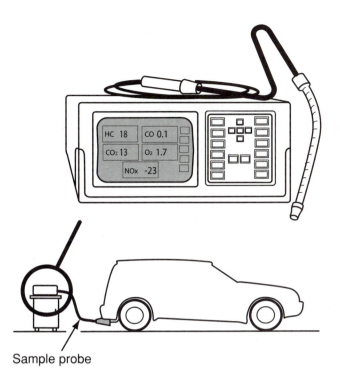

**Figure 2.** An idle test checks for HC and CO emissions output using a gas analyzer while the engine is at idle speed.

the technician determines if the vehicle HC and CO emissions are within specification for the particular model year. If so, the vehicle passes. If not, repairs must be made to bring the vehicle within compliance. NOx emissions readings are taken with this test.

## TWO-SPEED IDLE (BAR 90) TEST

This is an enhanced version of the basic idle test. Instead of taking emissions readings only at idle, the two-speed test also analyzes emissions output at an engine speed of 2500 rpm. NOx emissions readings are taken with this test.

The two-speed idle test is sometimes referred to as a BAR 90 emissions test, BAR being the acronym for California Bureau of Automotive Repair. Although two-speed idle testing has been around for years, the BAR 90 test defined specific parameters that were to be followed when conducting the test. The test was originally introduced in California in 1990.

## IM240 TRANSIENT EMISSIONS TESTING

IM240 refers to an enhanced Inspection and Maintenance test that has gained favorable recognition in terms of providing an accurate simulation of city and highway

**Figure 3.** Typical IM240 chassis dynamometer.

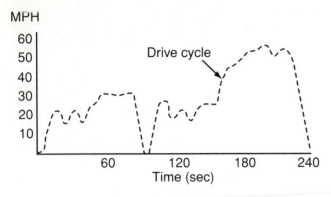

**Figure 4.** IM240 drive trace.

Although enhanced IM240 testing provides accurate information relating to HC, CO, $CO_2$, $O_2$, and NOx output, the system is expensive to install and maintain. It also places the testing facility at a greater liability risk because the customer's vehicle must be driven at speeds in excess of 50 mph.

If any malfunction occurs as a result of operator or equipment error, personal injury—and even death—can occur. In addition, the potential for damaging the customer's vehicle is significantly greater when the vehicle must be driven in a dynamic situation.

## ASM Testing

**Acceleration Simulation Mode (ASM)** testing uses a simple driving pattern to determine if the vehicle under test (VUT) meets emissions criteria. Also, ASM testing is much less rigorous than is IM240 testing. Although it provides less information than an IM240 test lane, ASM testing has been adopted by many states for their enhanced I/M programs.

An ASM test provides a reasonable compromise between a simple idle test and a full IM240 program. Two different types of ASM tests are used:
- ASM 50/15
- ASM 25/25

The first number represents the percent load that is applied to the vehicle, whereas the second number indicates the maximum speed during the test. A chassis dynamometer is still required when conducting ASM tests. Although NOx measurements are not always included in ASM tests, this is a function of the gas analyzer capability and not the chassis dyno itself.

## EMISSIONS FAILURE ANALYSIS

Vehicle emissions failure analysis is often a difficult task, owing to the complexity of the combustion process. Diagnosis should proceed according to the type of emissions failure.

## Failure Due to High HC Emissions

A vehicle that fails for high HC emissions is usually experiencing cylinder misfire. This can be caused by one or more of the following:
- Faulty secondary ignition system components, including spark plugs, ignition wires, cap, rotor, or coil
- Incorrect ignition timing
- Intermittent CKP or CMP sensors
- Intermittent ignition control module

Always follow the manufacturer's diagnostic procedures to isolate the cause. Replace damaged or worn components in accordance with shop manual procedures. Retest the vehicle after verifying proper engine operation.

## Failure Due to High CO Emissions

A vehicle that fails for high CO emissions usually results from an overly rich air/fuel mixture. This can be caused by:
- Excessive fuel pressure
- Faulty EVAP system operation
- Restricted combustion air intake

## Failure Due to High HC and High CO

A vehicle that fails for both HC and CO usually has fouled spark plugs caused by a rich air/fuel mixture. This elevates the CO levels. The excessive HC output occurs because of cylinder misfire. Unburned fuel is escaping into the exhaust stream. Follow the causes listed above for high CO and HC.

## Failure Due to High NOx

Excessive amounts of NOx are caused by high combustion chamber temperatures. This can be caused by:
- A lean air/fuel mixture
- A faulty cooling system
- Excessive carbon buildup in the combustion chamber
- Faulty EGR system operation

## RELATIONSHIP AMONG HC, CO, $CO_2$, $O_2$, AND NOx

**Figure 5** shows the complex relationship among concentrations of substances entering the combustion reaction and the corresponding emissions output. A significant drop in HC emissions occurs at lean air/fuel ratios, as expected. Although CO emissions continue to fall with increasingly leaner air/fuel ratios, HC emissions begin to increase substantially, as a result of increased lean misfire events.

NOx emissions output begins to increase at lean air/fuel ratios greater than 13:1. This occurs as a result of

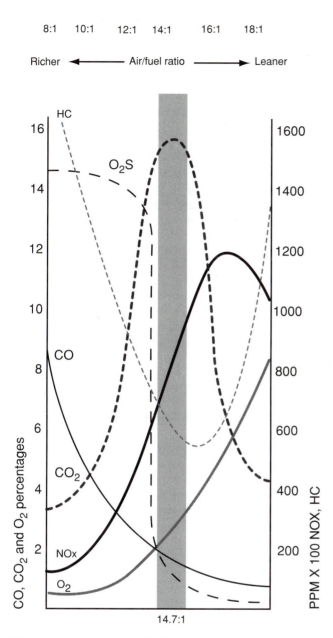

**Figure 5.** Complex relationship among substances entering the combustion reaction and the corresponding emissions output.

increased combustion chamber temperature associated with lean air/fuel ratios.

## ON-BOARD DIAGNOSTICS EMISSIONS TESTING

OBDII testing is rapidly replacing all other forms of vehicle emissions testing. It is becoming the accepted standard because of the sophisticated algorithms developed to isolate emissions-related failures. Most states have embraced enhanced I/M testing programs using OBDII protocols instead of idle and dynamometer testing. Almost all 1996 and newer model-year vehicles are OBDII compliant.

Relying on the information collected by the OBDII system, an accurate picture of engine operation is constantly being updated. If no DTCs are present at the time of the I/M test, the vehicle is allowed to pass.

Specific parameters must be checked, including verifying proper operation of the "check engine" lamp. Because several system monitors are used in the OBDII protocol, most state programs require that most or all have "run" before considering the vehicle as passing the I/M emissions test. The option to disregard one or more monitors is one currently under debate, as all were considered important under the original OBD program.

## EVAPORATIVE EMISSIONS TESTING

Most IM240 and ASM vehicle testing programs include a separate test for evaporative emissions. This includes a purge test and a pressure test. The fuel cap is often tested separately from other system components.

### Purge Flow Testing

Purge flow testing requires that either a vapor purge line be disconnected or a connection to the EVAP port be made, depending upon the vehicle. A flow meter is installed and the volume of vapor purge is measured during the IM240 test.

If the measured flow is less than 1.0 L, the vehicle fails the purge test. A variety of component failures can cause low purge volume, including:

- Faulty solenoid purge valves
- Disconnected, plugged, incorrectly routed, or damaged vapor lines
- A faulty carbon canister
- Electrical problems involving the control circuits between the PCM and solenoid valves
- A faulty PCM

### Pressure Testing

Pressure testing checks the integrity of the evaporative emissions system for leaks. Specialized equipment is required to pressure test the EVAP system. Always follow the manufacturer's service procedure to prevent injury and damage to system components.

A leak in the fuel tank, fuel lines, rollover valve, fuel cap, or vapor lines can cause pressure to escape from the EVAP emissions system. A gas analyzer is helpful in "sniffing" for released HC vapors. Follow repair or replacement procedures as outlined in the shop manual.

# Summary

- Many states have adopted I/M vehicle emissions testing programs to reduce air pollution from passenger cars and trucks.
- The simplest tailpipe emissions test is an idle test.
- The BAR 90 emissions test is a two-speed idle test designed to check tailpipe emissions at curb idle and 2500 rpm.
- An IM240 enhanced emissions test is a shortened version of the Federal Test Procedure (FTP) protocol used by manufacturers.
- The ASM test is an alternative to the IM240 test. Although it provides less data than an IM240 test, the ASM test is effective in isolating problematic vehicles.

- Both IM240 and ASM testing are of the transient variety because the tests are conducted under simulated driving conditions. Transient testing provides more information than a simple idle test.
- If a vehicle fails an emissions test for high HC output, it is often the result of ignition misfire.
- If a vehicle fails an emissions test for high CO output, it is usually because the air/fuel mixture is too rich.
- A complex relationship exists between air/fuel ratios and vehicle exhaust emissions.
- On-board diagnostics is rapidly replacing tailpipe emissions testing.
- Pressure and purge flow tests are used to verify proper operation of the EVAP system.

# Review Questions

1. Technician A says that an idle test is more effective than an IM240 test in identifying emissions-related faults. Technician B says that high CO readings are often the result of fuel quality problems. Who is correct?
   - A. Technician A
   - B. Technician B
   - C. Both A and B
   - D. Neither A nor B

2. Technician A says that a vacuum leak can cause high CO readings. Technician B says that a rich air/fuel mixture can cause increased HC readings. Who is correct?
   - A. Technician A
   - B. Technician B
   - C. Both A and B
   - D. Neither A nor B

3. Technician A says that an ASM test is performed to certify a motor vehicle for sale in the United States. Technician B says that an IM240 test must be performed before any new vehicle is offered for sale in the United States. Who is correct?
   - A. Technician A
   - B. Technician B
   - C. Both A and B
   - D. Neither A nor B

4. Explain how CO and HC emissions readings can both be high at the same time.

5. Technician A says that a vehicle that fails an emissions test for high HC will likely have low NOx readings. Technician B says that NOx readings increase at idle. Who is correct?
   - A. Technician A
   - B. Technician B
   - C. Both A and B
   - D. Neither A nor B

6. Technician A says that high combustion chamber temperatures are responsible for the formation of NOx. Technician B says that high fuel pressure can cause increased HC emissions. Who is correct?
   - A. Technician A
   - B. Technician B
   - C. Both A and B
   - D. Neither A nor B

7. Technician A says that a faulty EGR valve can cause high NOx readings. Technician B says that a faulty engine thermostat can cause high NOx readings. Who is correct?
   - A. Technician A
   - B. Technician B
   - C. Both A and B
   - D. Neither A nor B

8. Technician A says that a purge flow test can isolate EVAP system concerns. Technician B says that a faulty fuel filler cap can cause excessive EVAP emissions. Who is correct?
   - A. Technician A
   - B. Technician B
   - C. Both A and B
   - D. Neither A nor B

9. Technician A says that a cracked vapor vent hose can cause increased CO emissions at the tailpipe. Technician B says that a cracked vapor vent hose can cause increased HC emissions at idle. Who is correct?
   A. Technician A
   B. Technician B
   C. Both A and B
   D. Neither A nor B

10. Technician A says that a gas analyzer can be used to check for EVAP system leaks. Technician B says that a gas analyzer can be used to check for high CO emissions output. Who is correct?
    A. Technician A
    B. Technician B
    C. Both A and B
    D. Neither A nor B

# Appendix A

## ASE PRACTICE EXAM

1. Two technicians are discussing the diagnosis of a customer's vehicle concern. Technician A says that to properly diagnose the concern, you must first duplicate the concern. Technician B says that before you attempt to duplicate a vehicle concern you should gather as much data as possible from the customer. Who is correct?
   A. Technician A
   B. Technician B
   C. Both A and B
   D. Neither A nor B

2. Two technicians are discussing the availability of service information. Technician A says that technical service bulletins apply to specific vehicles and specific concerns. Technician B says that technical service bulletins can be found in the service manual. Who is correct?
   A. Technician A
   B. Technician B
   C. Both A and B
   D. Neither A nor B

3. All of the following information can be cross-referenced from the VIN except:
   A. The year model of the vehicle.
   B. The engine size.
   C. The factory-installed tire size.
   D. Which plant the vehicle was built in.

4. Two technicians are discussing the buildup of a milky substance found on the backside of the oil fill cap in a vehicle with a V6 engine. Technician A says that the substance indicates an internal coolant leak. Technician B says that a leaking intake gasket can cause the buildup of the substance on the oil fill cap. Who is correct?
   A. Technician A
   B. Technician B
   C. Both A and B
   D. Neither A nor B

5. Two technicians are discussing an engine vibration that increases with intensity as the engine warms up. Technician A says that a broken fan blade can cause a noticeable engine vibration. Technician B says that a broken fan blade can accelerate the wear of the bearings within the water pump. Who is correct?
   A. Technician A
   B. Technician B
   C. Both A and B
   D. Neither A nor B

6. Two technicians are discussing the emission of an abnormal amount of smoke from the exhaust system of a customer vehicle. Technician A says that a blown head gasket will result in the emission of black smoke from the exhaust system. Technician B says that a leaking fuel injector can cause the exhaust to emit black smoke. Who is correct?
   A. Technician A
   B. Technician B
   C. Both A and B
   D. Neither A nor B

7. Two technicians are discussing a vehicle that has low engine vacuum while under acceleration. Technician A says that a restricted exhaust can cause low engine manifold vacuum. Technician B says that a leaking intake manifold gasket will cause an internal vacuum leak on an in-line four-cylinder engine. Who is correct?
    A. Technician A
    B. Technician B
    C. Both A and B
    D. Neither A nor B

8. Two technicians are discussing the benefits of performing a cylinder balance test. Technician A says that a cylinder balance test can determine if the valves are sealing properly. Technician B says that a cylinder balance test is used to determine which cylinder is not producing adequate power. Who is correct?
    A. Technician A
    B. Technician B
    C. Both A and B
    D. Neither A nor B

9. Two technicians are discussing the results of a compression test that was performed on a customer vehicle. Technician A says that a severely worn camshaft lobe will cause a cylinder to have low compression. Technician B says that a leaking intake manifold gasket will cause a cylinder to have low compression. Who is correct?
    A. Technician A
    B. Technician B
    C. Both A and B
    D. Neither A nor B

10. Two technicians are discussing cylinder leakage testing. Technician A says that the cylinder being tested must be at TDC on the compression stroke. Technician B says that air leaking from the exhaust indicates a worn camshaft lobe. Who is correct?
    A. Technician A
    B. Technician B
    C. Both A and B
    D. Neither A nor B

11. All of the following tests can be performed with an oscilloscope except one:
    A. Test fuel pump current
    B. Monitor $HO_2S$ performance
    C. Monitor PCM injector signals
    D. Retrieve stored trouble codes

12. Two technicians are discussing exhaust gas emission testing. Technician A says that a restricted air filter will cause elevated HC readings. Technician B says that a thermostat stuck in the open position will cause elevated HC readings. Who is correct?
    A. Technician A
    B. Technician B
    C. Both A and B
    D. Neither A nor B

13. Two technicians are discussing how the cooling system affects engine performance. Technician A says that the lower the engine operating temperature, the more efficient the engine will operate. Technician B says that low engine coolant can cause inaccurate temperature gauge readings. Who is correct?
    A. Technician A
    B. Technician B
    C. Both A and B
    D. Neither A nor B

14. All of the statements about the cooling system service are true except:
    A. Pressure may be applied to the cooling system with a handheld pump.
    B. A leaking head gasket can cause an engine misfire.
    C. Some engine gaskets might leak only when they are cold.
    D. A radiator cap that does not hold pressure will not affect the boiling point of the engine coolant.

15. Two technicians are discussing the camshaft timing during a timing belt replacement. Technician A says that camshaft-timing marks place the camshaft operation in sequence with crankshaft events. Technician B says that when replacing a timing belt the engine should be turned by hand at least two revolutions to verify the alignment of the timing marks. Who is correct?
    A. Technician A
    B. Technician B
    C. Both A and B
    D. Neither A nor B

16. Two technicians are discussing the abilities of a typical automotive scan tool. Technician A says that a scan tool can be used to review stored DTCs. Technician B says that a scan tool can be used to clear stored DTCs. Who is correct?
    A. Technician A
    B. Technician B
    C. Both A and B
    D. Neither A nor B

17. Two technicians are discussing DTCs. Technician A says that a DTC is stored when the PCM detects a problem in a specific system. Technician B says that anytime a DTC is set, the MIL will illuminate. Who is correct?
    A. Technician A
    B. Technician B
    C. Both A and B
    D. Neither A nor B

18. Two technicians are discussing the information that is stored by the PCM when a DTC is set. Technician A says that freeze frame data is stored for all DTCs. Technician B says that freeze frame data is stored only for type C DTCs. Who is correct?
    A. Technician A
    B. Technician B
    C. Both A and B
    D. Neither A nor B

19. Two technicians are discussing a vehicle that was brought in with a no-code driveability concern. Technician A says that if a vehicle does not have a DTC set in memory, a problem does not exist with the vehicle. Technician B says that for a DTC to be set, the concern must meet all of the criteria for a specific DTC. Who is correct?
    A. Technician A
    B. Technician B
    C. Both A and B
    D. Neither A nor B

20. Two technicians are discussing the usefulness of a digital storage oscilloscope (DSO). Technician A says that a DSO can be used in any capacity in which you might use a voltmeter. Technician B says that a voltmeter will detect some problems that a DSO will not. Who is correct?
    A. Technician A
    B. Technician B
    C. Both A and B
    D. Neither A nor B

21. Two technicians are discussing the retrieval of engine data using a scan tool. Technician A says that to receive engine data, a trouble code must be set. Technician B says that to receive failure record information, a trouble code must have been set. Who is correct?
    A. Technician A
    B. Technician B
    C. Both A and B
    D. Neither A nor B

22. Trouble shooting information can be located in which of the following?
    A. Service manual
    B. Service bulletin
    C. Both A and B
    D. Neither A nor B

23. Two technicians are discussing a customer vehicle that was presented with an engine misfire. Technician A says that if the engine misfires with the A/C compressor engaged, but stops when the A/C is turned off, the compressor should be replaced. Technician B says that the problem might be an ignition system problem. Who is correct?
    A. Technician A
    B. Technician B
    C. Both A and B
    D. Neither A nor B

24. Two technicians are discussing a V6 engine with a distributorless ignition system; there is no spark on any cylinder. Technician A says that a faulty coil can be the cause. Technician B says that a faulty crankshaft sensor can be the cause. Who is correct?
    A. Technician A
    B. Technician B
    C. Both A and B
    D. Neither A nor B

25. Two technicians are discussing a no-spark condition in a vehicle with a V6 engine and distributor-type ignition system. Technician A says that a faulty coil might be the cause. Technician B says that a damaged ignition rotor might be the cause. Who is correct?
    A. Technician A
    B. Technician B
    C. Both A and B
    D. Neither A nor B

26. A pair of cylinders in a V6 engine with a distributorless ignition system has no spark at the coil tower as tested with a spark tester. Technician A says that a faulty coil might be the cause. Technician B says a faulty crankshaft sensor might be the cause. Who is correct?
    A. Technician A
    B. Technician B
    C. Both A and B
    D. Neither A nor B

27. Two technicians are discussing secondary ignition cables. All of the statements about secondary cables are true *except*:
    A. If a cable is routed too close to a low-voltage sensor circuit, corruption of the low-voltage signal can occur.
    B. Internal resistance of the cable should not exceed 10,000 ohms of resistance per foot.
    C. If a spark plug boot has signs of carbon tracking, the wire should be replaced.
    D. If the spark plug boot has signs of carbon tracking, the boot must be replaced.

28. Two technicians are discussing the diagnosis of an ignition coil for a distributorless ignition system. Both the primary and secondary resistance is within specifications. The resistance between the primary side of the coil and the secondary coil is 2 ohms. Technician A says that the ignition coil is good and should be reused. Technician B says that the ignition coil is shorted and should be discarded. Who is correct?
    A. Technician A
    B. Technician B
    C. Both A and B
    D. Neither A nor B

29. Two technicians are discussing electronic spark/timing advance and retard. Technician A says that a knocking noise from the A/C compressor can cause the ignition timing to be retarded. Technician B says that the amount of ignition timing is calculated based on engine speed and load. Who is correct?
    A. Technician A
    B. Technician B
    C. Both A and B
    D. Neither A nor B

30. Two technicians are discussing primary ignition sensors and triggering devices. Technician A says that sensors must be positioned so that they can detect crankshaft position. Technician B says that these sensors control when the ignition coil is grounded. Who is correct?
    A. Technician A
    B. Technician B
    C. Both A and B
    D. Neither A nor B

31. Two technicians are discussing a misfire condition on a fuel-injected engine. Technician A says that a clogged fuel injector can be the cause. Technician B says that a fuel injector that is stuck open can be the cause. Who is correct?
    A. Technician A
    B. Technician B
    C. Both A and B
    D. Neither A nor B

32. Two technicians are discussing an extended crank condition of a fuel-injected engine. Technician A says that a clogged fuel injector can be the cause. Technician B says that a fuel injector that is stuck open can be the cause. Who is correct?
    A. Technician A
    B. Technician B
    C. Both A and B
    D. Neither A nor B

33. Two technicians are discussing fuel quality. Technician A says that using a fuel with the highest octane possible will help the engine start faster. Technician B says that excessive amounts of alcohol can cause engine hesitation and detonation. Who is correct?
    A. Technician A
    B. Technician B
    C. Both A and B
    D. Neither A nor B

34. Two technicians have tested and found the fuel pressure for a port fuel engine to be low. Technician A says that a faulty fuel pump might be responsible for the low fuel pressure. Technician B says that a leaking vacuum diaphragm on the fuel pressure regulator might be the cause. Who is correct?
    A. Technician A
    B. Technician B
    C. Both A and B
    D. Neither A nor B

35. Two technicians are discussing in-line fuel filters on a fuel-injected engine. Technician A says that a clogged fuel filter will cause high fuel pressure between the filter and the fuel rail. Technician B says that a clogged fuel filter will cause low fuel pressure between the pump and the filter. Who is correct?
    A. Technician A
    B. Technician B
    C. Both A and B
    D. Neither A nor B

36. Two technicians are discussing external engine vacuum leaks. Technician A says that a vacuum leak at the intake manifold can cause a lean air/fuel ratio. Technician B says that external vacuum leaks can be found using a spray bottle with water. Who is correct?
    A. Technician A
    B. Technician B
    C. Both A and B
    D. Neither A nor B

37. Two technicians are discussing fuel injector diagnosis on an engine with port fuel injection. Technician A says that a fuel injector with a leaking nozzle will cause a lean air/fuel ratio. Technician B says that a fuel injector that is not opening can cause the PCM to increase the fuel delivered to the engine. Who is correct?
    A. Technician A
    B. Technician B
    C. Both A and B
    D. Neither A nor B

38. Two technicians are discussing the exhaust system. Technician A says that some amount of backpressure is desired. Technician B says that the installation of a free-flowing exhaust can adversely affect the operation of the EGR system. Who is correct?
    A. Technician A
    B. Technician B
    C. Both A and B
    D. Neither A nor B

39. Two technicians are discussing diagnosis of a suspected restriction in the exhaust system. Technician A says that when using a vacuum gauge to test for excessive backpressure, a decrease in manifold vacuum indicates a clogged exhaust. Technician B says that monitoring the $HO_2S$ can help identify a clogged exhaust. Who is correct?
    A. Technician A
    B. Technician B
    C. Both A and B
    D. Neither A nor B

40. Two technicians are discussing boost pressures of a turbo-charged engine. Technician A says that a clogged air filter can cause low boost pressures. Technician B says that a wastegate that is stuck in the closed position will cause low boost pressures. Who is correct?
    A. Technician A
    B. Technician B
    C. Both A and B
    D. Neither A nor B

41. Two technicians are discussing the operation of the PCV system. Technician A says that installation of an incorrect or low-quality PCV valve can result in various driveability symptoms. Technician B says that PCV valves are not application specific and the most important fact about PCV valve replacement is that the valve fits the location. Who is correct?
    A. Technician A
    B. Technician B
    C. Both A and B
    D. Neither A nor B

42. Two technicians are discussing the operation of the PCV system. Technician A says that a clogged PCV valve can result in various engine oil leaks. Technician B says that a broken PCV valve vacuum hose can result in a lean air/fuel ratio. Who is correct?
    A. Technician A
    B. Technician B
    C. Both A and B
    D. Neither A nor B

43. Which statement about the EGR system is correct?
    A. The EGR system increases the amount of fuel that can be burned in the engine.
    B. The introduction of burned exhaust gases increases combustion temperatures.
    C. The EGR system is installed only as an emissions control and can have no effect on engine performance.
    D. The introduction of burned exhaust gases reduces combustion temperatures.

44. All of the statements below about an electronic EGR valve system are correct except one:
    A. The PCM monitors operation of the EGR valve.
    B. An electronic valve can have an infinite number of positions.
    C. An electronic valve can have a specified number of positions.
    D. Excessive EGR flow will not result in a driveability concern.

45. Two technicians are discussing the operation of an electronic EGR system. Technician A says that the PCM can calculate EGR flow by measuring the amount of pressure within the EGR tube. Technician B says that the PCM can monitor EGR flow by observing the position of the valve pintle. Who is correct?
    A. Technician A
    B. Technician B
    C. Both A and B
    D. Neither A nor B

46. Two technicians are discussing the operation of the catalytic converter. Technician A says that excessive heat will damage the converter. Technician B says that water is a by-product of catalytic converter operation. Who is correct?
    A. Technician A
    B. Technician B
    C. Both A and B
    D. Neither A nor B

47. Two technicians are discussing the diagnosis of the catalytic converter. Technician A says that a pyrometer can help diagnose catalytic converter problems. Technician B says that catalytic converters do not require testing because they normally last for the life of the vehicle. Who is correct?
    A. Technician A
    B. Technician B
    C. Both A and B
    D. Neither A nor B

48. Two technicians are discussing the operation of the catalytic converter. Technician A says that a leaking fuel injector can damage the catalytic converter. Technician B says that a misfiring cylinder can cause the catalytic converter to overheat. Who is correct?
    A. Technician A
    B. Technician B
    C. Both A and B
    D. Neither A nor B

49. Two technicians are discussing the operation of the EVAP system. Technician A says that the charcoal canister stores excess liquid gasoline. Technician B says that only gasoline vapors are stored in the canister. Who is correct?
    A. Technician A
    B. Technician B
    C. Both A and B
    D. Neither A nor B

50. Two technicians are discussing the diagnosis of the EVAP system. Technician A says to use a high-pressure source of compressed air to locate any leaks in the EVAP system. Technician B says that a smoke machine is often used to locate leaks in the EVAP system. Who is correct?
    A. Technician A
    B. Technician B
    C. Both A and B
    D. Neither A nor B

# Appendix B

## INTERNET RESOURCES

| Manufacturers | Web Site |
|---|---|
| DaimlerChrysler | http://www.daimlerchrysler.com/dccom |
| Ford Motor Company | http://ford.com/en/default.htm |
| General Motors | http://gm.com/flash_homepage/ |
| Honda | http://honda.com/ |
| Hyundai | http://hyundai.com/ |
| Kia | http://kia.com/ |
| Mazda | http://www.mazda.com/ |
| Nissan | http://www.nissanusa.com |
| Toyota | http://www.toyota.com |
| Volkswagen | http://www.vw.com/ |

| Diagnostic Equipment | Web Site |
|---|---|
| AES Wave | http://aeswave.com/ |
| Ease Diagnostics | http://www.obd2.com/ |
| Fluke Corp. | http://www.fluke.com/ |
| Hickok Inc. | http://www.hickok-inc.com/ |
| OTC–SPX Corp. | http://www.otctools.com/ |
| Snap-On Inc. | http://www.snapon.com/ |
| Vetronix Corp. | http://www.vetronix.com/ |

| Automotive Parts and Tool Suppliers | Web Site |
|---|---|
| AC Delco | http://acdelco.com |
| Car Tools | http://www.cartools.com/ |
| CARQUEST | http://www.carquest.com/ |
| MAC Tools | http://www.mactools.com/ |
| MATCO | http://www.matcotools.com/ |
| Motorcraft | http://motorcraft.com/ |
| NAPA | http://www.napaonline.com/ |
| Pro Tool, Inc. | http://www.protoolinc.com/ |
| Sears–Craftsman | http://www.sears.com/sr/craftsman |
| Snap-On Inc. | http://www.snapon.com/ |

**Automotive Specialty Equipment Manufacturers**
Bosch
Edelbrock
Federal-Mogul (Champion Spark Plugs)
Holley Companies
NGK Spark Plugs (USA) Inc.
Prestolite Wire Corp.
Standard Motor Products, Inc.

**Web Site**
http://www.boschusa.com/
http://edelbrock.com/
http://www.federal-mogul.com/
http://holley.com/
http://www.ngksparkplugs.com/
http://www.prestolitewire.com/
http://www.smpcorp.com/

**Automotive Service Information Systems**
ALLDATA
Mitchell 1

**Web Site**
http://www.alldata.com/
http://www.mitchellrepair.com/

**Professional Development**
Automotive Service Excellence (ASE)
International Automotive Technicians Network
National Alternative Training Consortium
Society of Automotive Engineers

**Web Site**
http://www.ase.com
http://iatn.net/
http://naftp.nrcce.wvu.edu
http://www.sae.org/

**Training Resources**
AC Delco
Car Quest
Jendham Inc.
Motorcraft
NAPA
Snap-On Inc.
SYSPEC Inc.

**Web Site**
http://www.acdelcotechconnect.com/
http://www.ctitraining.org/
http://www.jendham.com/
http://www.motorcraftservice.com
http://www.napaautocare.com/education
http://www.snapontraining.com/
http://www.syspec.com/

**Motorsports Sanctioning Bodies**
CART
Formula One
IHRA
IMCA
Indy Racing League
NASCAR
NHRA
SCCA
USAC
World of Outlaws

**Web Site**
http://cart.com/
http://www.formula1.com/
http://ihra.com/
http://www.imca.com/
http://indyracingleague.com/
http://www.nascar.com/
http://nhra.com/
http://scca.com/
http://www.usacracing.com/
http://www.worldofoutlawsracing.com/

# Appendix C

## CONVERSION CHARTS

In order to calculate English measurement, divide by the number in the center column.
In order to calculate metric measurement, multiply by the number in the center column.

| English | Multiply/ Divide by | Metric | English | Multiply/ Divide by | Metric |
|---|---|---|---|---|---|
| | *Length* | | | *Acceleration* | |
| in | 25.4 | mm | ft/s$^2$ | 0.3048 | m/s$^2$ |
| ft | 0.3048 | m | in/s$^2$ | 0.0254 | m/s$^2$ |
| yd | 0.9144 | m | | *Torque* | |
| mi | 1.609 | km | lb in | 0.11298 | N•m |
| | *Area* | | lb ft | 1.3558 | N•m |
| sq in | 645.2 | sq mm | | *Power* | |
| sq in | 6.45 | sq cm | hp | 0.745 | kW |
| sq ft | 0.0929 | sq m | | *Pressure (Stress)* | |
| sq yd | 0.8361 | sq m | inches of H$_2$O | 0.2488 | kPa |
| | *Volume* | | lb/sq in | 6.895 | kPa |
| cu in | 16,387.0 | cu mm | | *Energy (Work)* | |
| cu in | 16.387 | cu cm | Btu | 1055.0 | J (J = one Ws) |
| cu in | 0.0164 | L | lb ft | 1.3558 | J (J = one Ws) |
| qt | 0.9464 | L | kW hour | 3,600,000.0 | J (J = one Ws) |
| gal | 3.7854 | L | | *Light* | |
| cu yd | 0.764 | cu m | foot candle | 10.764 | lm/m$^2$ |
| | *Mass* | | | *Velocity* | |
| lb | 0.4536 | kg | mph | 1.6093 | km/h |
| ton | 907.18 | kg | | *Temperature* | |
| ton | 0.907 | tonne (t) | (°F −32) 5/9 | = | °C |
| | *Force* | | °F | = | (9/5 °C +32) |
| kg F | 9.807 | newtons (N) | | *Fuel Performance* | |
| oz F | 0.2780 | newtons (N) | 235.215/mpg | = | 100 km/L |
| lb F | 4.448 | newtons (N) | | | |

# Bilingual Glossary

**Absorption**    To take in by capillary action, as a sponge.
*Absorción*    *Incorporar mediante acción capilar, como una esponja.*

**Acceleration Simulation Mode (ASM)**    A vehicle emissions test using a chassis dynamometer that follows a specific pattern of acceleration and deceleration.
*Modo de simulación de aceleración (ASM)*    *Prueba de emisiones de un vehículo mediante un dinamómetro de chasis que sigue un patrón específico de aceleración y desaceleración.*

**Accelerator pedal position (APP) Sensor**    A sensor that sends signals to the throttle actuator control (TAC) indicating when the throttle is being opened or closed and how much and at what rate the position is being changed.
*Sensor de posición del pedal del acelerador (APP)*    *Sensor que envía señales al control del actuador del acelerador (TAC) que indica cuando el acelerador se abre o se cierra, así como la cantidad y proporción en que cambia la posición.*

**Actuator**    An electromechanical device that changes electrical signals into mechanical actions.
*Actuador*    *Dispositivo electromecánico que convierte las señales eléctricas en acciones mecánicas.*

**Adaptive strategy**    A strategy that allows the computer to learn and remember certain aspects of an operational condition.
*Estrategia adaptable*    *Estrategia que permite que la computadora aprenda y recuerde ciertos aspectos de una condición de funcionamiento.*

**Adsorption**    The use of solids to remove substances from either gaseous or liquid solutions.
*Adsorción*    *Uso de sólidos para retirar sustancias de soluciones gaseosas o líquidas.*

**Air/fuel ratio (A/F ratio)**    Numerical comparison of the amount of air to the amount of fuel, both measured by weight.
*Proporción de aire y combustible (proporción A/C)*    *Comparación numérica de la cantidad de aire en relación con la cantidad de combustible, medida por peso.*

**Air injection reaction (AIR) system**    The system used to inject ambient air into the exhaust stream or catalytic converter.
*Sistema de reacción de inyección de aire (AIR)*    *Sistema que se usa para inyectar aire del medio ambiente en el caudal de escape o convertidor catalítico.*

**Air pollution**    The introduction of impurities and contaminants into the atmosphere, many of which are caused by industrial activities.
*Contaminación del aire*    *Introducción de impurezas y contaminantes en la atmósfera, muchos de los cuales se deben a actividades industriales.*

**Alternating current (AC)**    A current in which electrons can flow in either a positive or negative direction.
*Corriente alterna (AC)*    *Corriente en la cual los electrones pueden fluir en dirección positiva o negativa.*

**Amperage**    A unit of measure expressing the cumulative flow of electrons within a circuit.
*Amperaje*    *Unidad de medida que expresa el flujo acumulativo de electrones dentro de un circuito.*

**Analog**    A signal in which the voltage can be any value within a range.
*Análoga*    *Señal en la cual el voltaje puede ser cualquier valor dentro de un rango.*

**Antilock brake system (ABS)**    A system installed on the vehicle to prevent the wheel brakes from locking up in a panic or inclement weather condition.
*Sistema de frenos antibloqueo (ABS)*    *Sistema instalado en el vehículo para evitar que los frenos de las ruedas se bloqueen en un momento de pánico o bajo condiciones de clima inclemente.*

**Atmosphere**    The air enveloping the earth.
*Atmósfera*    *Aire que rodea a la tierra.*

**Atmospheric pressure**    The amount of pressure that the atmosphere exerts on the surface of the earth.
*Presión atmosférica*    *Cantidad de presión que la atmósfera ejerce sobre la superficie de la tierra.*

**Autotransformer**   A transformer using a combined primary/secondary winding.
*Autotransformador*   *Transformador que usa un embobinado combinado principal/secundario.*

**Back probing**   A method of testing a circuit by touching the connector's terminals on the backside without damage to the terminal.
*Sondeo trasero*   *Método de probar un circuito al tocar las terminales de los conectores en la parte trasera, sin dañar la terminal.*

**Balance shaft**   A shaft used to counter the vibrations that are naturally created within an engine.
*Flecha de balance*   *Flecha que se usa para contrarrestar las vibraciones que se crean naturalmente dentro de un motor.*

**Bank**   A row of cylinders.
*Banco*   *Fila de cilindros.*

**Barometric pressure sensor (BARO)**   A sensor that is similar in operation to a MAP sensor that is dedicated to measuring barometric pressure.
*Sensor de presión barométrica (BARO)*   *Sensor con funcionamiento similar al sensor MAP que se usa para medir la presión barométrica.*

**Base timing**   The initial ignition timing setting without any mechanical or electronic advance added.
*Sincronización base*   *Ajuste de la sincronización del encendido inicial sin ningún avance mecánico o electrónico agregado.*

**Bearing saddles**   The cradle in which the crankshaft sits centered lengthwise in the block.
*Monturas de balero*   *Soporte en el que se asienta el cigüeñal de manera transversal y centrada.*

**Bidirectional control**   The programming that allows manual control of specific actuators.
*Control bidireccional*   *Programación que permite el control manual de actuadores específicos.*

**Body control module (BCM)**   A control module used in some vehicles to control functions of various body components.
*Módulo de control de la carrocería (BCM)*   *Módulo de control que se usa en algunos vehículos para controlar las funciones de varios componentes de la carrocería.*

**Boost pressure**   The amount of pressure provided to an engine by a forced induction component such as a supercharger or turbocharger.
*Presión de supercargador*   *Cantidad de presión que se proporciona a un motor mediante un componente de inducción forzada como un supercargador o turbocargador.*

**Boot**   A heavily insulated connection that covers the top of the spark plug to prevent arcing and the intrusion of water and other contaminants.
*Envoltura*   *Conexión altamente aislada que cubre la parte superior de la bujía para evitar la formación de arcos y la intrusión de agua y otros contaminantes.*

**Bottom dead center (BDC)**   The point of a piston's stroke in which it has traveled to its lowest point within the cylinder and can only move upward.
*Punto muerto inferior (BDC)*   *Punto de la carrera de un pistón en el cual ha viajado a su punto más bajo dentro del cilindro y sólo puede moverse hacia arriba.*

**Brake pedal position (BPP) switch**   The brake pedal position switch is located near the brake pedal to provide the PCM with information concerning the application of the vehicle brakes.
*Interruptor de posición del pedal del freno (BPP)*   *Ubicado cerca del pedal del freno para proporcionar información al PCM relacionada con la aplicación de los frenos del vehículo.*

**Branches**   Individual paths in which current flows within a larger circuit.
*Ramales*   *Rutas individuales en las cuales la corriente fluye dentro de un circuito más grande.*

**Breaker points**   An electrical switch that opens to interrupt current flow in the primary circuit.
*Puntos de corte*   *Interruptor eléctrico que se abre para interrumpir el flujo de corriente en el circuito principal.*

**Bulb check**   A sequence in which the bulbs located on the instrument panel are illuminated to verify their operation.
*Verificación de bombilla*   *Secuencia en la que las bombillas situadas en el panel de instrumentos se iluminan para verificar su funcionamiento.*

**Burn time**   Time required for the air/fuel mixture in the combustion chamber to burn.
*Tiempo de quemado*   *Tiempo requerido para que se queme la mezcla aire y combustible en la cámara de combustión.*

**Camshaft position (CMP) sensor**   A Hall effect or variable reluctance sensor that detects camshaft position.
*Sensor de posición del árbol de levas (CMP)*   *Sensor de reluctancia o efecto Hall que detecta la posición del árbol de levas.*

**Cam timing**   See Valve timing.
*Sincronización del árbol de levas*   *Véase Sincronización de las válvulas.*

**Canister purge (CANP) solenoid**   An electrical solenoid valve that admits fuel vapors to the canister for processing.
*Solenoide de purga del depósito (CANP)*   *Válvula solenoide eléctrica que admite vapores de combustible en el depósito para su procesamiento.*

**Capacitor**   An electronic component that blocks DC, yet allows AC to pass.
*Capacitor*   *Componente eléctrico que bloquea la corriente directa pero permite pasar la corriente alterna.*

**Carbon monoxide**   A colorless, odorless but highly poisonous gas found in engine exhaust gases consisting of partially burned fuel.
*Monóxido de carbono*   *Gas incoloro e inodoro pero altamente venenoso que se encuentra en los gases de escape y que está parcialmente compuesto de combustible quemado.*

**Catalyst efficiency monitor**　A passive test in which the PCM uses two HO$_2$Ss to monitor catalyst efficiency.
*Monitor de eficiencia del catalizador*　*Prueba pasiva en la cual el PCM usa dos HO$_2$S para monitorear la eficiencia del catalizador.*

**Catalytic converter**　Converts harmful exhaust pollutants created by the combustion process into harmless products.
*Convertidor catalítico*　*Convierte los contaminantes dañinos del escape creados por el proceso de combustión en productos inofensivos.*

**Center electrode**　The insulated part of a spark plug that conducts electricity toward the electrode gap to ground.
*Electrodo central*　*Parte aislada de una bujía que conduce electricidad hacia la distancia de electrodo a tierra.*

**Central port injection**　CPI injection is a version of PFI that uses a single injector to provide fuel to all of the cylinders.
*Inyector de puerto central*　*Versión de PFI que usa un solo inyector para proporcionar combustible a todos los cilindros.*

**Central sequential fuel injection (CSFI)**　A version of the CPI unit that uses individual, centrally located fuel injectors rather than one large injector.
*Inyección de combustible secuencial central (CSFI)*　*Versión de la unidad CPI que usa inyectores de combustibles individuales, ubicados en la posición central, en lugar de un inyector grande.*

**Cerium**　A rare-earth metal used in the fabrication of catalytic converters to assist in oxygen storage.
*Cerio*　*Metal de tierras raras que se usa en la fabricación de convertidores catalíticos para asistir en el almacenamiento del oxígeno.*

**Chemical energy**　The energy produced when chemicals such as gasoline, diesel, liquid propane, or natural gas are ignited.
*Energía química*　*Energía que se produce cuando se queman químicos como gasolina, diesel, propano líquido o gas natural.*

**Clear flood mode**　Decreases injector pulse width to create a very lean A/F ratio when the PCM senses a high TPS signal while the engine is cranking.
*Modo de inundado despejado*　*Reduce el ancho de pulso del inyector para crear una relación aire y combustible muy reducida cuando el PCM detecta una señal TPS alta mientras se hace girar el motor.*

**Closed loop**　An operational mode in which the PCM has complete authority of fuel control and can make adjustments that are based on sensor readings and fuel map specifications.
*Ciclo cerrado*　*Modo operativo en el cual el PCM tiene autoridad completa del control de combustible y puede hacer ajustes basándose en las lecturas del sensor y las especificaciones de mapa de combustible.*

**Coil driver**　An electric circuit within the PCM or ignition control module that switches coil primary current on and off as required.
*Impulsor de bobina*　*Circuito eléctrico dentro del PCM o módulo de control de encendido que enciende o apaga la corriente primaria de la bobina según se requiera.*

**Coil-near-plug (CNP)**　An ignition system that uses one ignition coil per cylinder. Each coil is located very near to its respective spark plug and is connected using a very short secondary cable.
*Bobina cerca de la bujía (CNP)*　*Sistema de encendido que usa una bobina de encendido para cada cilindro. Cada bobina está ubicada muy cerca de su bujía respectiva y está conectada mediante un cable secundario muy corto.*

**Coil-on-plug (COP)**　An ignition system that uses one ignition coil for every cylinder with the coil being located directly above the spark plug.
*Bobina sobre la bujía (COP)*　*Sistema de encendido independiente que usa una bobina de encendido para cada cilindro con la bobina ubicada directamente encima de la bujía.*

**Cold start stall**　A condition in which the engine is started from a cold start but dies only a few seconds after it initially started.
*Parada por arranque en frío*　*Condición en la cual el motor arranca a partir de un arranque en frío pero se apaga unos segundos después del arranque inicial.*

**Combustion**　The burning of fuel in an engine.
*Combustión*　*Quema de combustible en un motor.*

**Combustion chamber**　Pockets provided in the cylinder head for combustion to take place.
*Cámaras de combustión*　*Compartimientos proporcionados en la cabeza del cilindro para que tenga lugar la combustión.*

**Comebacks**　A situation in which a customer makes a return trip to the service facility because the concern was not corrected.
*Regresos*　*Situación en la cual el cliente regresa a la estación de servicio porque el problema no se corrigió.*

**Comprehensive component monitoring (CCM)**　A monitoring process in which the PCM monitors the sensors and actuators to verify that they are still on-line in the system.
*Monitoreo comprensivo de componentes (CCM)*　*Proceso de monitoreo en el cual el PCM monitorea los sensores y actuadores para verificar que sigan "en línea" en el sistema.*

**Compression ignition**　Initiating the combustion process using the heat of compression.
*Encendido de compresión*　*Inicio del proceso de combustión mediante el calor de compresión.*

**Computer**　An electronic device that gathers electronic data, analyzes the information, and sends signals to make changes to the actuators of a system.
*Computadora*　*Dispositivo electrónico que reúne datos electrónicos, analiza la información y envía señales para hacer cambios en los actuadores del sistema.*

**Conductors**   The wires and connections in which current flows.
**Conductores**   *Cables y conexiones a través de los cuales fluye la corriente.*

**Connecting rods**   Metal beams that connect the piston to the crankshaft.
**Bielas**   *Vigas metálicas que conectan el pistón con el cigüeñal.*

**Constant volume sampling (CVS)**   A specialized sampling method used to detect the quantities of various emissions gases during a vehicle emissions test.
**Muestreo de volumen constante (CVS)**   *Método de muestreo especializado que se usa para detectar las cantidades de varios gases de emisiones durante una prueba de emisiones del vehículo.*

**Control maps**   Computer programs stored within the memory of the PCM that are used to control the operation of various actuators in various situations.
**Mapas de control**   *Mapas de computadora almacenados dentro de la memoria del PCM que se usan para controlar el funcionamiento de actuadores diversos en situaciones diversas.*

**Crankshaft position (CKP) sensor**   The CKP is used by the PCM to determine engine RPM and firing order and to calculate ignition timing positions.
**Sensor de posición del cigüeñal (CKP)**   *Sensor que usa el PCM para determinar las RPM del motor y el orden de encendido y para calcular las posiciones de sincronización del encendido.*

**Cross counts**   The number of times that the $O_2S$ voltage cycles above and below 0.450V in a specific period.
**Cuenta cruzada**   *Número de veces que el voltaje de $O_2S$ alterna por encima o por debajo de 0.450 V en un periodo específico.*

**Crude oil**   A complex mixture of many different hydrocarbon compounds.
**Petróleo crudo**   *Mezcla compleja de muchos compuestos de hidrocarburo diferentes.*

**Current**   The cumulative flow of electrons from one atom to another.
**Corriente**   *Flujo acumulativo de electrones de un átomo a otro.*

**Cycle**   See Stroke.
**Ciclo**   *Véase Carrera.*

**Cylinder block**   The main structure of the engine in which all other components are fastened.
**Bloque de cilindros**   *Estructura principal del motor a la cual están sujetos todos los demás componentes.*

**Cylinder bore**   The space in which the piston is located.
**Diámetro interior del cilindro**   *Espacio en el cual está ubicado el pistón.*

**Decelerating**   A state in which the accelerator is released and the engine and vehicle are slowing.
**Desaceleración**   *Estado en el cual el acelerador se libera y el motor y el vehículo reducen la velocidad.*

**Deck**   The flat surface located at the top of the block above the cylinders that provides a perfectly flat surface onto which the cylinder heads attach.
**Plataforma**   *Superficie plana localizada en la parte superior del bloque encima de los cilindros que proporciona una superficie perfectamente plana a la cual se fijan las cabezas de cilindro.*

**Detonation**   A pinging or knocking in the engine caused by the collision of opposing flame fronts.
**Detonación**   *Sonido metálico o cascabeleo en el motor causado por la colisión de frentes de flama opuestos.*

**Diagnostic link connector (DLC)**   The interface connector in which a scan tool can communicate with the PCM and other electronic modules.
**Conector de enlace de diagnóstico (DLC)**   *Conector de interfaz en el cual una herramienta examinadora de diagnóstico se puede comunicar con el PCM y otros módulos electrónicos.*

**Diagnostic trouble code (DTC)**   An alphanumerical code that is stored in the PCM's memory; each monitored circuit will have one or more numeric codes specifying a condition.
**Código de problema de diagnóstico (DTC)**   *Código alfanumérico que se almacena en la memoria del PCM donde cada circuito monitoreado tendrá uno o más códigos numéricos que especifiquen una condición.*

**Differential pressure feedback EGR (DPFE) sensor**   A solid-state differential pressure sensor used to monitor EGR operation.
**Sensor de retroalimentación de presión diferencial EGR (DPFE)**   *Sensor de presión diferencial de estado sólido que se usa para monitorear la operación del EGR.*

**Digital multimeter (DMM)**   A piece of test equipment that combines functions of a voltmeter, ohmmeter, and ammeter.
**Multímetro digital (DMM)**   *Pieza de equipo de prueba que combina las funciones de un voltímetro, ohmiómetro y amperímetro.*

**Digital volt-ohm meter (DVOM)**   See Digital multimeter (DMM).
**Ohmiómetro-voltímetro digital (DVOM)**   *Véase Multímetro digital (DMM).*

**Direct current (DC)**   A current in which electrons can flow only in a positive direction.
**Corriente directa (DC)**   *Corriente en la cual los electrones pueden fluir exclusivamente en dirección positiva.*

**Displacement**   The volume that an individual cylinder can hold between bottom dead center and top dead center of piston travel.
**Desplazamiento**   *Volumen que puede contener un cilindro individual entre el punto muerto inferior y el punto muerto superior de la carrera del pistón.*

**Distributor**   See Ignition distributor.
**Distribuidor**   *Véase Distribuidor de encendido.*

**Distributor cap**   A cover for the conventional ignition system distributor, having a central terminal that receives secondary voltage from the coil and four, six, or eight

peripheral terminals to transmit the high-voltage signal to the spark plugs.
**Tapa del distribuidor**  *Cubierta para el distribuidor de sistema de encendido convencional, con una terminal central que recibe voltaje secundario de la bobina y cuatro, seis u ocho terminales secundarias para transmitir la señal de alto voltaje a las bujías.*

**Distributor ignition (DI)**  An ignition system that relies on a conventional distributor for proper operation.
**Encendido del distribuidor (DI)**  *Sistema de encendido que usa un distribuidor convencional para funcionar apropiadamente.*

**Distributorless ignition**  An ignition system that does not use a distributor; coils are fastened directly to the module and can be located anywhere on the engine block or on an accessory bracket.
**Encendido sin distribuidor**  *Sistema de encendido que no usa distribuidores; las bobinas se fijan directamente al módulo y se pueden ubicar en cualquier parte en el bloque del motor o en un soporte adicional.*

**Drive by wire**  Throttle activation system that uses no mechanical linkage.
**Impulsado por cable**  *Sistema de activación del acelerador que no usa articulación mecánica.*

**Drive cycle**  Occurs when a vehicle is driven in such conditions that all of the diagnostic monitors have been run.
**Ciclo impulsor**  *Ocurre cuando un vehículo se maneja bajo condiciones en las que se han ejecutado todos los monitores de diagnóstico.*

**Dual-bed converter**  See Three-way catalyst.
**Convertidor de cama doble**  *Véase Catalizador de tres vías.*

**Dual overhead cam (DOHC)**  An overhead valve design in which two camshafts per cylinder head are supported above the valves. The valves are actuated through the use of followers, tappets, or rocker arms.
**Doble árbol de levas (DOHC)**  *Diseño de válvula encima de la culata en el cual dos árboles de levas por cabeza de cilindro se apoyan sobre las válvulas. Las válvulas se accionan mediante el uso de rodillos de leva, varillas de empuje o balancines.*

**Duration**  The amount of time expressed in degrees of crankshaft rotation that the valves are held open.
**Duración**  *Cantidad de tiempo expresada en grados de giro del cigüeñal en los cuales permanecen abiertas las válvulas.*

**Duty cycle**  A method of controlling an actuator by controlling the amount of time that a component is turned on within a cycle.
**Ciclo de trabajo**  *Método de controlar un actuador al controlar la cantidad de tiempo en el cual se enciende un componente dentro de un ciclo.*

**Dwell**  The degree of distributor shaft rotation while the ignition breaker points are closed.
**Reposo**  *Grado de rotación de la flecha de distribuidor mientras los puntos del interruptor de encendido están cerrados.*

**Dwell angle**  A term used for dwell.
**Ángulo de reposo**  *Término que se usa para reposo.*

**Electrical energy**  The energy used to push electrons through an electrical circuit.
**Energía eléctrica**  *Energía que se usa para empujar electrones a través de un circuito eléctrico.*

**Electrode**  In an electric circuit, one element of an air gap that allows an arc to form between it and a conductor of opposite polarity. In the case of a spark plug, one electrode is welded to the shell of the plug; the center rod forms the mating electrode.
**Electrodo**  *En un circuito eléctrico, elemento de un espacio vacío que permite que se forme un arco entre el elemento y el conductor de polaridad opuesta. En el caso de una bujía, el electrodo está soldado a la cubierta de la bujía; la varilla central forma el electrodo de correspondiente.*

**Electromagnetic interference (EMI)**  Extent to which the vehicle electrical system is affected by external electromagnetic fields.
**Interferencia electromagnética (EMI)**  *Punto hasta el cual se ve afectado el sistema eléctrico de un vehículo debido a campos electromagnéticos.*

**Electronic fuel injection (EFI)**  Electronic fuel injection uses an electrically actuated injector to precisely control the amount of fuel that is allowed into the engine.
**Inyección electrónica de combustible (EFI)**  *Usa un inyector actuado electrónicamente para controlar con precisión la cantidad de combustible que puede entrar al motor.*

**Electronic ignition (EI)**  See Distributorless ignition.
**Encendido electrónico (EI)**  *Véase Encendido sin distribuidor.*

**Electronic spark timing (EST)**  The timing advance signal developed by the PCM. Also see SPOUT.
**Sincronización electrónica de la chispa (EST)**  *Señal de avance de sincronización desarrollada por el PCM. Véase también SPOUT.*

**Enable criteria**  The conditions that must be met in order for a specific diagnostic monitor to run.
**Criterio de activación**  *Condiciones que se deben cumplir para que funcione un monitor de diagnóstico específico.*

**Energy conversion**  Occurs when one type of energy is changed to another type.
**Conversión de energía**  *Ocurre cuando un tipo de energía se cambia por otro.*

**Engine coolant temperature (ECT)**  The temperature of the coolant in the engine.
**Temperatura del refrigerante del motor (ECT)**  *Temperatura del refrigerante en el motor.*

**Engine coolant temperature (ECT) sensor**  The coolant temperature sensor is a thermistor that is typically screwed into a coolant passage in the intake manifold or thermostat housing and is used to provide the PCM with information about engine temperature.
**Sensor de temperatura del enfriador del motor (ECT)**  *Termistor que habitualmente se atornillan en un pasaje del enfriador en el múltiple de admisión o cubierta del termostato y se usa para proporcionar información al PCM acerca de la temperatura del motor.*

**Engine ping**    A mild rattling noise from the engine.
*Sonido metálico del motor*    *Sonido ligero de cascabeleo de un motor.*

**Evaporative emissions control system (EVAP system)**
A system that prevents fuel vapors from escaping to the atmosphere.
*Sistema de control de emisiones evaporativas (EVAP)*
*Sistema que impide que los vapores de combustible escapen a la atmósfera.*

**Exhaust gas analyzer**    An instrument used to measure and display the composition and concentration of engine exhaust gases.
*Analizador de gases de escape*    *Instrumento que se usa para medir y mostrar la composición y concentración de gases de escape del motor.*

**Exhaust gas recirculation (EGR)**    An emissions control system that reduces an engine's production of oxides of nitrogen by diluting the air/fuel mixture with exhaust gas so that peak combustion temperature in the cylinders is lowered.
*Recirculación de los gases de escape (EGR)*    *Sistema de control de emisiones que reduce la producción de óxidos de nitrógeno del motor al diluir la mezcla aire y combustible con gas de escape de manera que se reduzca la temperatura máxima de combustión en los cilindros.*

**Exhaust gas recirculation temperature (EGRT) sensor**
The EGRT provides the PCM with information about the temperature of the gases within the EGR passages.
*Sensor de temperatura de recirculación de gas del escape (EGRT)*    *Sensor que proporciona información al PCM acerca de la temperatura de los gases dentro de los pasajes del EGR.*

**Exhaust gas recirculation (EGR) valve**    A valve installed on an engine to recirculate spent exhaust gases back to the intake manifold, reducing combustion chamber temperatures.
*Válvula de recirculación de los gases de escape (EGR)*
*Se instala en un motor para recircular los gases del escape gastados, de regreso al múltiple de admisión para reducir la temperatura de la cámara de combustión.*

**Exhaust gas recirculation valve position (EVP) sensor**
A linear potentiometer that measures the position of the EGR valve pintle position.
*Sensor de posición de la válvula de recirculación de los gases de escape (EVP)*    *Potenciómetro lineal que mide la posición del pivote central de la válvula EGR.*

**Extended crank**    A condition in which the engine spins over longer than normal before starting.
*Arranque extendido*    *Condición en la cual el motor gira más de lo normal antes de arrancar.*

**Fan shroud**    An enclosure around the fan that increases the fan's efficiency by sealing the area around the fan.
*Envoltura del ventilador*    *Cubierta de un ventilador que aumenta su eficacia al sellar el área que lo rodea.*

**Firing order**    The number sequence in which the cylinders of a multicylinder engine fire.
*Orden de disparo*    *Secuencia de números en la que se disparan los cilindros de un motor con cilindros múltiples.*

**Flame front**    The leading edge of the burning air/fuel mixture.
*Frente de flama*    *Borde de entrada de la mezcla de aire y combustible.*

**Flexible fuel vehicles (FFV)**    Vehicles that have the capacity to use multiple fuel types.
*Vehículos de combustible flexible (FFV)*    *Vehículos que tienen la capacidad de usar múltiples tipos de combustible.*

**Flywheel**    Large steel plate that is bolted to the rear of the crankshaft.
*Volante*    *Placa de tamaño grande que está atornillada a la parte trasera del cigüeñal.*

**Force**    The act of applying power to an object.
*Fuerza*    *Acto de aplicar potencia a un objeto.*

**Forced induction**    Filling the engine's cylinders using an external pump to force air through the induction system.
*Inducción forzada*    *Llenado de los cilindros del motor mediante una bomba externa para forzar aire a través del sistema de inducción.*

**Fossil fuel–burning**    The combustion of fossil fuels, such as coal or oil.
*Quemado de combustible fósil*    *Combustión de combustibles fósiles como carbón o petróleo.*

**Frequency**    The controlled rate at which a change of direction takes place in an AC circuit; also the rate at which a DC circuit is cycled from off to on and back to off in 1 second.
*Frecuencia*    *Tasa controlada a la cual se lleva a cabo un cambio de dirección en un circuito de CA; también, la tasa a la cual un circuito de CD cambia de apagado a encendido y de nuevo a apagado en un segundo.*

**Friction**    Resistance to motion that occurs when two surfaces touch one another and move in different directions or at different speeds; can be found in liquids, gases, and solids.
*Fricción*    *Resistencia al movimiento que ocurre cuando dos superficies se tocan entre sí y se mueven en direcciones diferentes o a velocidades diferentes; se le puede encontrar en líquidos, gases y sólidos.*

**Fuel rail**    A component that serves the purpose of delivering fuel to the injectors.
*Carril de combustible*    *Componente que sirve para hacer llegar combustible a los inyectores.*

**Fuel system monitor**    An adaptive strategy in which the PCM monitors the active deviation of the air/fuel ratio from that of the PCM program.
*Monitor de sistema de combustible*    *Estrategia adaptable en la cual el PCM monitorea la desviación activa de la proporción de aire y combustible del programa PCM.*

**Fuel tank pressure (FTP) sensor**    The FTP sensor is used to monitor the pressure within the fuel storage system.
*Sensor de presión del tanque de combustible (sensor FTP)*    *Se usa para monitorear la presión dentro del sistema de almacenamiento de combustible.*

**Gaskets** Materials used to fill a gap between two connected components that hold a fluid or pressure.
*Juntas* *Materiales que se usan para llenar el espacio entre dos componentes conectados que contienen un fluido o presión.*

**Gasoline direct injection (GDI)** A method of fuel injection in which fuel is injected directly into the cylinder.
*Inyección directa de gasolina (GDI)* *Método de inyección de combustible en el cual el combustible se inyecta directamente en el cilindro.*

**Greenhouse gas** Infrared absorbing gases, including carbon dioxide and chlorofluorocarbons, that contribute to the greenhouse effect.
*Gas de efecto invernadero* *Gases de absorción de infrarrojo que contribuyen al efecto invernadero, incluidos dióxido de carbono y clorofluorocarbonos.*

**Ground (side) electrode** The electrical part of the spark plug that is connected to the cylinder head.
*Electrodo a tierra (lateral)* *Parte eléctrica de la bujía que está conectada a la cabeza del cilindro.*

**Group fired** A method of pulsing the injectors in pairs or groups.
*Disparo en grupo* *Método de pulsación de inyectores en pares o grupos.*

**Hall effect** A semiconductor magnetic sensor that is used to detect the presence of a magnetic field.
*Efecto Hall* *Sensor magnético del semiconductor que se usa para detectar la presencia de un campo magnético.*

**Heat range** The measure of a spark plug's ability to transfer heat from the tip of the insulator to the cylinder head.
*Rango de calor* *Medida de la habilidad de la bujía de transferir calor desde la punta del aislante hasta la cabeza del cilindro.*

**Heated oxygen sensors ($HO_2S$)** See Oxygen sensor.
*Sensores de oxígeno calentado ($HO_2S$)* *Véase Sensor de oxígeno.*

**High-voltage terminal** A secondary ignition connection.
*Terminal de alto voltaje* *Conexión de encendido secundaria.*

**Hydrocarbon (HC)** A compound, such as gasoline, containing hydrogen and carbon.
*Hidrocarburo* *Compuesto, como la gasolina, que contiene hidrógeno y carbono.*

**Idle air control (IAC) valve** An electronic valve that allows the PCM to control the idle speed of the engine.
*Válvula de control de aire en marcha mínima (IAC)* *Válvula electrónica que permite que el PCM controle la velocidad en marcha mínima del motor.*

**Idle test** A simple tailpipe emissions test conducted at curb idle using an exhaust gas analyzer.
*Prueba en marcha mínima* *Prueba sencilla de emisiones del tubo de escape que se lleva a cabo en marcha mínima mediante un analizador de gas de escape.*

**Ignition coil** A transformer containing a primary and secondary winding that acts to increase the battery voltage of 12V to as much as 30,000V to fire the spark plugs.
*Bobina de encendido* *Transformador que contiene embobinado primario y secundario que actúa para aumentar el voltaje de la batería de 12 V a hasta 30,000 V para disparar las bujías.*

**Ignition control module (ICM)** See Ignition module.
*Módulo de control de encendido (ICM)* *Véase Módulo de encendido.*

**Ignition diagnostic monitor (IDM)** A signal developed by the PCM to indicate a cylinder misfire occurred. The IDM signal is displayed as a PID on the scan tool.
*Monitor de diagnóstico de encendido (IDM)* *Señal desarrollada por el PCM que indica que ha ocurrido una falla de chispa del cilindro. La señal IDM se presenta como un PID en la herramienta de diagnóstico.*

**Ignition distributor** A device used to send spark energy to the appropriate cylinder at the proper time.
*Distribuidor de encendido* *Dispositivo que se usa para enviar energía de encendido al cilindro apropiado en el momento apropiado.*

**Ignition module** An electronic module that controls coil charging and coil firing.
*Módulo de encendido* *Módulo electrónico especial que controla la carga y el disparo de la bobina.*

**IM240 emissions test** A 240-second dynamic emissions test following a specific drive cycle that simulates driving conditions.
*Prueba de emisiones IM240* *Prueba de emisiones dinámicas de 240 segundos de duración que sigue un ciclo de manejo específico que simula las condiciones de manejo.*

**Induction** A process whereby an electromagnet or electrically charged object transmits a magnetic force or electrical current to a nearby object without making physical contact.
*Inducción* *Proceso mediante el cual un objeto con carga electromagnética o eléctrica transmite una fuerza magnética o corriente eléctrica a un objeto cercano sin hacer contacto físico.*

**Initial timing** See Base timing.
*Sincronización inicial* *Véase Sincronización base.*

**Inspection and maintenance (I/M)** The periodic and systematic inspection and maintenance of a vehicle's ignition, fuel, and emissions control system.
*Inspección y mantenimiento (I/M)* *Inspección y mantenimiento periódicos y sistemáticos de los sistemas de encendido, combustible y control de emisiones de un vehículo.*

**Insulator** A material that prevents the flow of electrons.
*Aislante* *Material que evita el flujo de electrones.*

**Intake air temperature (IAT) sensor** The Intake Air Temperature sensor provides the PCM with information about the temperature of the air that is entering the engine.
*Sensor de temperatura del aire de admisión (IAT)* *Sensor que proporciona información al PCM acerca de la temperatura del aire que entra al motor.*

**Intake manifold runner control (IMRC)** A PCM controlled system in which the effective length of intake manifold runners can be changed.
*Control de pasaje del múltiple de admisión (IMRC)* Sistema controlado por el PCM en el cual la longitud efectiva de los conductos del múltiple de admisión pueden cambiarse.

**Interference suppression** The reduction of electrical noise caused by arcing, fast-rising signals, and strong radio signals.
*Supresión de interferencia* Reducción de ruido eléctrico causado por formación de arcos eléctricos, señales de elevación rápida y señales fuertes de radio.

**Intermittent problem** A problem that happens for a short period of time and then stops.
*Problema intermitente* Problema que ocurre durante un periodo corto y luego se detiene.

**Interrupter ring** A thin metal ring used with a Hall effect switch that has windows or gaps cut out of it either blocking or allowing the passage of a magnetic field.
*Anillo de interruptor* Anillo delgado de metal que se usa con un interruptor de efecto Hall con ventanas o espacios cortados ya sea para bloquear o permitir el paso de un campo magnético.

**Ionizes** When the air between an air gap breaks down, allowing electrical current to flow.
*Ionizado* Cuando el aire en un espacio vacío se descompone y permite que fluya la corriente eléctrica.

**Key-on–engine-off** The ignition switch position where electrical circuits are energized while the engine is not operating.
*Llave en encendido–motor apagado* Posición del interruptor de encendido donde se energizan los circuitos mientras el motor no está funcionando.

**Knock sensor (KS)** A sensor that detects the onset of spark knock during combustion.
*Sensor de cascabeleo* Sensor que detecta el efecto inicial del golpeo por encendido durante la combustión.

**Lean** An air/fuel ratio in which there is less than 1 pound of fuel provided for every 14.7 pounds of air.
*Pobre* Relación de aire y combustible en la cual no hay más de una libra de combustible por cada 14.7 libras de aire.

**Lift** The distance that the camshaft is able to move the valves off their respective seats.
*Elevación* Distancia a la que el árbol de levas puede mover las válvulas fuera de sus asientos respectivos.

**Light-emitting diode (LED)** A semiconductor component that allows current to flow in only one direction and emits a glow whenever current flow is present.
*Diodo emisor de luz (LED)* Componente de semiconductor que permite que la corriente fluya solo en una dirección y emite un brillo incandescente cada vez que está presente el flujo de corriente.

**Magnetic flux** Invisible magnetic lines of force.
*Flujo magnético* Líneas invisibles de fuerza magnética.

**Malfunction indicator lamp (MIL)** The light on the dash that warns the driver that there has been a failure in one of the monitored powertrain circuits.
*Lámpara indicadora de mal funcionamiento (MIL)* Luz en el tablero que advierte al conductor que se ha presentado una falla en uno de los circuitos monitoreados del tren de potencia.

**Manifold absolute pressure (MAP)** The MAP sensor provides the PCM with information about engine vacuum. Using this information, the PCM can calculate relative engine load.
*Sensor de presión absoluta del múltiple (MAP)* Sensor que proporciona información al PCM acerca del vacío en el motor. Con esta información, el PCM puede calcular la carga relativa del motor.

**Mass air flow (MAF) sensor** The MAF sensor is installed to inform the PCM how much air is being drawn into the engine. Using this information the PCM can calculate relative engine load.
*Sensor del flujo de masa de aire (MAF)* Se instala para informar al PCM de la cantidad de aire que consume el motor. Con esta información, el PCM puede calcular la carga relativa del motor.

**Mechanical energy** The ability to move objects. Mechanical energy is released any time an object is moved.
*Energía mecánica* Habilidad de mover objetos. Se libera energía mecánica cada vez que se mueve un objeto.

**Microprocessor** The component within the PCM that is charged with comparing data and making decisions.
*Microprocesador* Componente dentro del PCM que está a cargo de comparar datos y tomar decisiones.

**Minimum air rate** Establishes the minimum amount of air that should be allowed in the engine when the idle air control system is fully closed.
*Coeficiente mínimo de aire* Establece la cantidad mínima de aire que debe permitirse en el motor cuando el sistema de control de aire en marcha mínima está cerrado por completo.

**Misfire** A condition that keeps one or more of the engine cylinders from producing adequate power.
*Falla de chispa* Condición que evita que uno o más de los cilindros del motor produzca suficiente potencia.

**Misfire monitor** OBDII system test in which the PCM observes misfire indicators.
*Monitor de falla de chispa* Prueba de sistema OBD-II en la cual el PCM observa indicadores de falla de encendido.

**Mobile sources** Any emissions source that is not of fixed origin. Examples of mobile sources include motor vehicles, recreational products, locomotive, aircraft, and portable industrial equipment.
*Fuentes móviles* Cualquier fuente de emisiones que no es de origen fijo. Los ejemplos de fuentes móviles incluyen vehículos automotores, productos recreativos, locomotoras, aeronaves y equipo industrial portátil.

**Modules** Electronic components similar to computers that are designed to handle specific tasks.
*Módulos* Componentes electrónicos similares a computadoras que están diseñados para manejar tareas específicas.

**Monitors**   A diagnostic program that is designed to monitor the operation of specific components and systems.
*Monitores*   *Programa de diagnóstico que está diseñado para monitorear el funcionamiento de componentes y sistemas específicos.*

**Multiport fuel injection system (MPFI)**   See Port fuel injection (PFI).
*Sistema de inyección de combustible multipuerto (MPFI)*   *Véase Inyección de combustible para puerto (PFI).*

**Naturally aspirated**   Depends on the natural flow of the atmospheric pressure of air to fill the vacuum created within the cylinders.
*Aspirado de forma natural*   *Depende del flujo natural de la presión atmosférica del aire para llenar el vacío creado dentro de los cilindros.*

**Negative temperature coefficient**   The internal resistance of a negative temperature coefficient sensor decreases as its temperature rises.
*Coeficiente de temperatura negativa*   *Resistencia interna de un sensor de coeficiente de temperatura negativa disminuye al elevarse la temperatura.*

**Noble metal**   A metal, such as silver, gold, iridium, or platinum, that is resistant to oxidation and corrosion.
*Metal noble*   *Metal, como la plata, oro, iridio o platino, resistente a la oxidación y la corrosión.*

**Noid light**   A small test lamp that plugs directly into the injector harness to test the operation of the injector control circuit.
*Luz Noid*   *Lámpara pequeña de prueba que se enchufa directamente en el armazón del inyector para probar la operación del circuito de control del inyector.*

**Nonvolatile**   A type of memory storage within a computer in which the information stored there cannot be erased.
*No volátil*   *Tipo de almacenamiento de memoria en una computadora en el cual la información almacenada no se puede borrar.*

**Office of Transportation Air Quality (OTAQ)**   An office within the Environmental Protection Agency that is concerned with emissions from mobile sources.
*Oficina de Calidad del Aire en el Transporte (OTAQ)*   *Oficina dentro de la Agencia de Protección del Medio Ambiente que se ocupa con las emisiones de fuentes móviles.*

**On-board diagnostics (OBD)**   A standardized diagnostic software and hardware system used to detect performance problems that adversely affect emissions and engine performance.
*Diagnóstico a bordo (OBD)*   *Programa de diagnóstico estandarizado y equipo que se usan para detectar problemas de desempeño que afectan adversamente las emisiones y el rendimiento del motor.*

**Open loop**   The state of engine operation when the engine fuel and timing controls are based on a limited number of sensor inputs that are strictly compared to a control map.
*Ciclo abierto*   *Estado de operación del motor cuando el combustible del motor y los controles de sincronización se basan en un número limitado de entradas de sensor que se comparan estrictamente con un mapa de control.*

**Output shaft speed (OSS)**   The OSS provides the PCM information about the speed of the transmission output speed.
*Sensor de velocidad de la flecha de salida (OSS)*   *Sensor que proporciona información al PCM acerca de la velocidad de salida de la transmisión.*

**Overhead valve (OHV)**   An engine design in which the valves are located above the piston and are actuated by a camshaft in the cylinder block using rocker arms and pushrods.
*Válvula encima de la culata (OHV)*   *Diseño de motor en el cual las válvulas están colocadas sobre el pistón y se accionan con un árbol de levas en el bloque de cilindros mediante balancines y varillas de empuje.*

**Overlap**   The amount of time expressed in degrees of crankshaft rotation that the intake and exhaust valves are opened at the same time.
*Traslape*   *Cantidad de tiempo expresada en grados de giro del cigüeñal en los cuales se abren al mismo tiempo las válvulas de admisión y escape.*

**Oxides of nitrogen (NOx)**   Harmful gaseous emissions of the engine, comprised of compounds of nitrogen and varying amounts of oxygen, that are formed at the high combustion temperatures.
*Óxidos de nitrógeno (NOx)*   *Emisiones dañinas y gaseosas del motor que constan de compuestos de nitrógeno y cantidades variables de oxígeno que se forman a altas temperaturas de combustión.*

**Oxidizing catalyst**   A two-way catalytic converter that promotes the oxidation of HC and CO in an engine's exhaust stream, as distinguished from a three-way or reduction catalyst.
*Catalizador oxidable*   *Convertidor catalítico de dos vías que promueve la oxidación de HC y CO en el caudal de escape del motor, comparado con un catalizador de tres vías o de reducción.*

**Oxygen sensor (O$_2$S)**   The O$_2$S provides the PCM with information about the relative air fuel ratio that the engine is using. Some sensors have built in heating elements and are called heated oxygen sensors (HO$_2$S).
*Sensores de oxígeno (O$_2$S)*   *Sensores que proporcionan información al PCM acerca de la proporción de aire y combustible que usa el motor. Algunos sensores tienen elementos de calefacción integrados y se conocen como sensores de oxígeno calentado (HO$_2$S).*

**Palladium (Pd)**   A rare, valuable metallic element that is highly resistant to corrosion and is used as a catalytic agent in automotive catalytic converters of the oxidizing type.
*Paladio (Pd)*   *Elemento metálico valioso y raro que es altamente resistente a la corrosión y que se usa como agente catalítico en convertidores catalíticos automotores de tipo oxidante.*

**Parameter identification (PID)**   Used to identify parameters supplied by the PCM to the scan tool.
*Identificación de parámetros (PID)*   *Se usa para identificar los parámetros que proporciona el PCM a la herramienta examinadora de diagnóstico.*

**Parasitic loss**   The amount of power required to operate an accessory.
*Pérdida parasítica*   *Cantidad de potencia necesaria para operar un accesorio.*

**Photocell**   A component that produces a voltage when it is exposed to light.
*Fotocelda*   *Componente que produce un voltaje cuando se expone a la luz.*

**Pickup screen**   Removes large particles in the oil before it is directed into the oil pump.
*Pantalla captadora*   *Elimina partículas grandes del aceite antes de que se dirija a la bomba de aceite.*

**Pintle positions sensor (PPS)**   A linear potentiometer that measures the position of a linear actuator such as an EGR valve.
*Sensor de posición de pivote central (PPS)*   *Potenciómetro lineal que mide la posición de un actuador lineal, como una válvula EGR.*

**Pistons**   Aluminum slugs that are fitted into each cylinder and move up and down in the cylinder.
*Pistones*   *Trozos metálicos de aluminio que se ajustan dentro de cada cilindro y que se mueven hacia arriba y hacia abajo en el cilindro.*

**Platinum (Pt)**   A rare, valuable metallic element that is highly resistant to corrosion and is used as a catalytic agent in automotive catalytic converters of the oxidizing type.
*Platino (Pt)*   *Elemento metálico valioso y raro que es altamente resistente a la corrosión y que se usa como agente catalítico en convertidores catalíticos automotores de tipo oxidante.*

**Plenum**   The upper part of the intake system in which the air enters the throttle body and is distributed into different tubes that feed individual cylinders.
*Cámara de admisión*   *Parte superior del sistema de admisión en el cual el aire entra en el cuerpo del acelerador y se distribuye en tubos diferentes que alimentan cilindros individuales.*

**Plug gap**   The spacing of the electrodes.
*Separación de bujías*   *Espaciado de los electrodos.*

**Poppet nozzle**   A valve based on fuel pressure that utilizes a spring-loaded ball and seat to control the atomization and distribution of fuel.
*Boquilla de muñeca*   *Válvula basada en la presión del combustible que utiliza una bola y asiento accionados por resorte para controlar la atomización y distribución del combustible.*

**Port fuel injection (PFI)**   A fuel injection system that has one injector for every cylinder.
*Inyección de combustible en puerto (PFI)*   *Sistema de inyección de combustible que tiene un inyector por cada cilindro.*

**Ports**   Passages located within the cylinder head that connect the cylinders to the fuel and exhaust systems.
*Puertos*   *Pasajes localizados dentro de la cabeza del cilindro que conectan los cilindros a los sistemas de combustible y escape.*

**Position indicator pulse (PIP)**   An alternative phrase to "profile ignition pickup." The position indicator pulse provides cylinder position to the PCM from either a crankshaft or distributor sensor.
*Pulso indicador de posición (PIP)*   *Frase alternativa para "captación de encendido de perfil." El pulso indicador de posición proporciona la posición del cilindro al PCM desde un sensor de cigüeñal o de distribuidor.*

**Position indicator pulse (PIP) sensor**   A sensor that measures camshaft position and speed.
*Sensor posición de encendido de perfil (PIP)*   *Sensor que mide la posición y velocidad del cigüeñal.*

**Positive crankcase ventilation (PCV)**   An engine emissions control system, operating on engine vacuum, that recycles crankcase vapors and meters them into the intake stream to be burned.
*Ventilación positiva del cárter (PCV)*   *Sistema de control de emisiones del motor que opera en un vacío de motor y que recicla los vapores del cárter del cigüeñal y los mide hacia el caudal de entrada para quemarlos.*

**Positive temperature coefficient**   The internal resistance of a positive temperature coefficient sensor increases as its temperature rises.
*Coeficiente de temperatura positiva*   *Resistencia interna de un sensor de coeficiente de temperatura positiva aumenta al elevarse la temperatura.*

**Power**   A calculation of the rate at which work is done (horsepower).
*Potencia*   *Cálculo del coeficiente al cual se realiza el trabajo (caballos de fuerza).*

**Power steering pressure switch (PSPS)**   The PSPS alerts the PCM when power steering pressures are high, indicating additional load on the engine.
*Interruptor de presión de la dirección hidráulica (PSPS)*   *Alerta al PCM cuando se elevan las presiones de la dirección hidráulica, lo cual indica carga adicional en el motor.*

**Powertrain control module (PCM)**   The powertrain control module (PCM) is a microcomputer that is installed on a vehicle to monitor and control the actions of the powertrain systems, specifically the fuel, ignition, emission and transmission systems.
*Módulo de control del tren de potencia (PCM)*   *Microcomputadora que se instala en un vehículo para monitorear y controlar las acciones de los sistemas del tren de potencia; específicamente los sistemas de combustible, encendido, emisión y transmisión.*

**Preignition**   The ignition caused by pressure of mechanical compression combined with the natural heat that builds in the engine that can be enough to ignite an air/fuel mixture without providing a spark from the spark plugs.
*Preencendido*   *Encendido causado por la presión de compresión mecánica que combinada con el calor natural que se acumula en el motor puede ser suficiente para encender una mezcla de combustible y aire sin la intervención de una chispa de las bujías.*

**Pressure feedback EGR (PFE) sensor** A sensor used to measure exhaust backpressure.
*Sensor de retroalimentación de presión (PFE)* *Sensor que se usa para medir la contrapresión del escape.*

**Protocol** The language computers use to communicate with one another and diagnostic equipment.
*Protocolo* *Lenguaje que usan las computadoras para comunicarse entre sí y con el equipo de diagnóstico.*

**Pulsating DC** A DC signal that is chopped into an on-off signal.
*DC pulsante* *Señal de DC que se descompone en una señal de encendido o apagado.*

**Pulse width** A measurement of the amount of time that an actuator has been energized.
*Ancho de pulso* *Medida de la cantidad de tiempo en la cual se ha energizado un actuador.*

**Pulse width modulation** A method of controlling an actuator by turning it off and on a predetermined amount of time within a cycle.
*Modulación de ancho de pulso* *Método de controlar un actuador al apagarlo o encenderlo durante una cantidad predeterminada de tiempo dentro de un ciclo.*

**Purge** To remove gasoline or air from a component such as the charcoal storage canister.
*Purgar* *Quitar gasolina o aire de un componente, como un recipiente de almacenamiento de carbón.*

**Pyrometer** An instrument used to measure surface temperatures.
*Pirómetro* *Instrumento que se usa para medir la temperatura de superficies.*

**Quench area** Any internal portion of a combustion chamber that causes combustion to decrease because of the temperature drop in the air/fuel charge where the charge meets this area.
*Área de extinción* *Cualquier porción interna de una cámara de combustión que hace que disminuya la combustión debido a la caída de temperatura en la carga de aire y combustible donde la carga se encuentra con esta área.*

**Radiant energy** Energy produced by light.
*Energía radiante* *Energía producida por luz.*

**Radio frequency interference (RFI)** Electromagnetic disturbances that interrupt, or otherwise interfere with equipment operating in the radio frequency spectrum.
*Interferencia de radiofrecuencia (RFI)* *Perturbaciones electromagnéticas que interrumpen o interfieren con el equipo que opera en el espectro de radiofrecuencia.*

**Reducing catalyst** The section of a three-way catalytic converter that breaks NOx down into harmless nitrogen and oxygen through a reduction reaction.
*Catalizador de reducción* *Sección de un convertidor catalítico de tres vías que descompone NOx en nitrógeno y oxígeno inocuos a través de una reacción de reducción.*

**Reference voltage** A constant voltage of a known quantity that is provided by the PCM to various sensors.
*Voltaje por referencia* *Voltaje constante de una cantidad conocida que el PCM proporciona a varios sensores.*

**Relay** An electromagnetic switch that uses a relatively small current to control a larger current.
*Relé* *Interruptor electromagnético que usa una cantidad relativamente pequeña de corriente para controlar una corriente mayor.*

**Resistance** The opposition to electrical flow; any situation that makes it difficult for electrons to flow from atom to atom.
*Resistencia* *Oposición al flujo eléctrico. Cualquier situación que dificulte que los electrones fluyan de un átomo a otro.*

**Retarded** Used to describe that the amount of timing has been reduced.
*Retardado* *Se usa para describir que la cantidad de sincronización se ha reducido.*

**Rhodium (Rh)** A precious element that is used as a catalyst to reduce harmful exhaust pollutants.
*Rodio (Rh)* *Elemento precioso que se usa como catalizador para reducir contaminantes dañinos del escape.*

**Rich** An air/fuel ratio in which there is more than 1 pound of fuel present for every 14.7 pounds of air.
*Rica* *Relación de aire y combustible en la cual no hay presente más de una libra de combustible por cada 14.7 libras de aire.*

**Road octane** (R + M)/2 formula used to determine the octane level of gasoline.
*Octano en carretera* *Fórmula (R + M)/2 que se usa para determinar el nivel de octano de la gasolina.*

**Root cause** The underlying cause of a customer concern.
*Causa fundamental* *Causa subyacente de una preocupación del cliente.*

**Rotor** In a distributor-based (DI) ignition system, the component that switches the secondary voltage from the center terminal to the outer terminals of the distributor cap as the shaft rotates.
*Rotor* *En un sistema de encendido basado en el distribuidor (DI), el componente que conmuta el voltaje secundario de la terminal central a las terminales exteriores de la tapa del distribuidor al girar la flecha.*

**Saturated** A condition that occurs when the electrodes of an electronic component reach the point of maximum current flow.
*Saturado* *Condición que ocurre cuando los electrodos de un componente electrónico alcanzan el punto de máximo flujo de corriente.*

**Scan tool** A hand-held computer that communicates and performs diagnostic testing on the PCM.
*Herramienta examinadora de diagnóstico* *Computadora de mano que se comunica con y realiza pruebas de diagnóstico en el PCM.*

**Schmidt trigger** An electronic circuit that converts noisy or random signal into precise, square waves.
*Disparador Schmidt* *Circuito electrónico que convierte señales ruidosas o aleatorias en ondas cuadradas y precisas.*

**Schrader valve** A threaded fitting that uses a spring-loaded valve to control the flow of air or liquid.
*Válvula Schrader* *Acoplador roscado que usa una válvula accionada por resorte para controlar el flujo de aire o líquido.*

**Secondary air** Air that is pumped into the exhaust or catalytic converter to promote chemical reactions that reduce exhaust gas pollutants.
*Aire secundario* *Aire que se bombea en el escape o en el convertidor catalítico para promover reacciones químicas que reducen los contaminantes del gas de escape.*

**Secondary high voltage** Voltage developed from the secondary winding of the ignition coil.
*Alto voltaje secundario* *Voltaje desarrollado a partir del devanado secundario de la bobina de encendido.*

**Sensors** Electrical devices that convert mechanical conditions into electrical signals.
*Sensores* *Dispositivos eléctricos que convierten las condiciones mecánicas en señales eléctricas.*

**Sequential electronic fuel injection (SEFI)** A type of multiport injection system where individual fuel injectors are pulsed sequentially, one after another in the same firing order as the spark plugs.
*Inyección electrónica de combustible secuencial (SEFI)* *Tipo de sistema de inyección multipuerto donde los inyectores de combustible individuales se pulsan de manera secuencial, uno después de otro, en el mismo orden de disparo que las bujías.*

**Shell** The outer spark plug casing having a threaded end and hexagonal flats for installation and removal.
*Casquillo* *Cubierta exterior de la bujía que tiene un extremo roscado y caras planas hexagonales para instalar y quitar.*

**Signal converter** A part of the PCM where signals are converted from an analog signal to a digital signal.
*Convertidor de señales* *Parte del PCM donde las señales se convierten de análogas a digitales.*

**Signals** Varying amounts of voltage provided to relay a command or message; can be provided by the PCM or to the PCM.
*Señales* *Cantidad variable de voltaje que se proporciona para retransmitir un comando o mensaje al relé; pueden proporcionarse por el PCM o hacia el PCM.*

**Single overhead cam (SOHC)** An overhead valve design in which one camshaft per cylinder head is supported above the valves. The valves are actuated through the use of followers, tappets, or rocker arms.
*Árbol de levas sencillo encima de la culata (SOHC)* *Diseño de válvula encima de la culata en el cual un árbol de levas por cabeza de cilindro se apoya sobre las válvulas. Las válvulas se accionan mediante el uso de rodillos de leva, varillas de empuje o balancines.*

**Smog** A type of air pollution formed when sunlight causes chemical reactions in air pollutants, resulting in the formation of ozone and other compounds.
*Humo y niebla* *Tipo de contaminación del aire que se forma cuando la luz solar causa reacciones químicas en los contaminantes del aire, lo cual resulta en la formación de ozono y otros componentes.*

**Spark advance** The number of crankshaft degrees (BTDC) at which the air/fuel mixture is ignited.
*Avance de la chispa* *Cantidad de grados del cigüeñal (BTDC) a la cual se enciende la mezcla de aire y combustible.*

**Spark ignition (SI)** An engine-operating system in which the air-fuel mixture is ignited by an electrical spark.
*Encendido por chispa (SI)* *Sistema operado por motor en el cual la mezcla de aire y combustible se enciende mediante una chispa eléctrica.*

**Spark knock** A term used for detonation or ping.
*Golpeo por encendido* *Término que se usa para detonaciones o ruidos metálicos.*

**Spark output (SPOUT)** The timing advance signal developed by the PCM based upon engine operating conditions.
*Salida de chispa (SPOUT)* *Señal de avance de sincronización desarrollada por el PCM y basada en las condiciones de funcionamiento del motor.*

**Spark plug** An ignition component threaded into the cylinder head that contains two electrodes that extend into the cylinder. A high-voltage electrical signal jumps across the electrodes to ignite the compressed air/fuel mixture.
*Bujía* *Componente de encendido roscado en la cabeza del cilindro que contiene dos electrodos que se extienden dentro del cilindro. Una señal eléctrica de alto voltaje salta entre los electrodos para encender la mezcla comprimida de aire y combustible.*

**Stepper motor** A small DC motor that moves in small increments by applying electrical current to various windings.
*Motor escalonado* *Motor pequeño de CC que se mueve en incrementos pequeños al aplicar corriente eléctrica a devanados diversos.*

**Stoichiometric** An ideal air/fuel mixture of 14.7:1 (by weight) in which all of the fuel and oxygen enters completely into the combustion reaction, leaving only $H_2O$ and $CO_2$.
*Estequiométrica* *Mezcla ideal de aire y combustible de 14.71 (por peso) en el cual todo el combustible y oxígeno entra por completo en la reacción de combustión, dejando sólo $H_2O$ y $CO_2$.*

**Stroke** The movement of a piston from top dead center to bottom dead center.
*Carrera* *Movimiento de un pistón desde el punto muerto superior hasta el punto muerto inferior.*

**Test procedure** A logical process that begins by verifying the complaint and ends with the identification of the faulty component or system.
*Procedimiento de prueba* *Proceso lógico que inicia al verificar la queja y termina con la identificación del componente o sistema averiado.*

**Thermal energy**   Energy produced by heat.
*Energía térmica*   *Energía producida por calor.*

**Three-way catalyst**   Catalytic converter that controls carbon monoxide, hydrocarbons and oxides of nitrogen.
*Catalizador de tres vías*   *Convertidor catalítico que controla el monóxido de carbono, hidrocarburos y óxidos de nitrógeno.*

**Throttle position sensor (TP sensor)**   A sensor that reports throttle position information to the PCM.
*Sensor de posición del acelerador (TP sensor)*   *Sensor que proporciona información de la posición del acelerador al PCM.*

**Throws**   The offset portion of the crankshaft in which the connecting rods are attached.
*Vanales*   *Porción de compensación del cigüeñal a la cual se fijan las bielas.*

**Timing light**   A stroboscopic xenon lamp used to set ignition timing.
*Luz de sincronización*   *Lámpara estroboscópica de xenón que se usa para establecer la sincronización de encendido.*

**Top dead center (TDC)**   The point of a piston's stroke in which it has traveled to its highest point within the cylinder and can only move downward.
*Punto muerto superior (TDC)*   *Punto de la carrera de un pistón en el cual ha viajado a su punto más alto dentro del cilindro y sólo puede moverse hacia abajo.* ·

**Torque**   Force that is applied in a twisting motion.
*Torsión*   *Fuerza que se aplica en un movimiento giratorio.*

**Torque converter clutch (TCC)**   A computer-controlled lockup mechanism that activates a hydraulic clutch within the automatic transmission torque converter.
*Embrague de convertidor de torsión (TCC)*   *Mecanismo de bloqueo controlado por computadora que activa un embrague hidráulico dentro del convertidor de par motor de la transmisión automática.*

**Transaxle**   A special transmission combining the differential and transmission into one compact unit.
*Transeje*   *Transmisión especial que combina el diferencial y la transmisión en una unidad compacta.*

**Transformer**   An electrical component that can be used to step up or step down a time-varying signal.
*Transformador*   *Componente eléctrico que se puede usar para adelantar o atrasar una señal de tiempo variable.*

**Transient emissions testing**   Vehicle emissions testing using a chassis dynamometer that simulates driving situations.
*Pruebas de emisión transitoria*   *Prueba de emisiones de un vehículo mediante un dinamómetro de chasis que simula situaciones de manejo.*

**Transmission fluid temperature (TFT) sensor**   A thermistor located in the pan of the transmission that sends the PCM information about the temperature of the transmission fluid.
*Sensor de temperatura del líquido de la transmisión (TFT)*   *Termistor localizado en el cárter de la transmisión que envía información al PCM acerca de la temperatura del líquido de la transmisión.*

**Transmission range (TR) sensor**   The TR sensor provides the PCM with information regarding the specific transmission range that the operator has selected.
*Sensor de rango de transmisión (TR)*   *Sensor que proporciona información al PCM acerca del rango de transmisión específica que ha seleccionado el operador.*

**Trouble trees**   A trouble codes diagnosis chart written in a logical manner from the easiest most likely tests to the more difficult and less likely.
*Árboles de problemas*   *Tabla de diagnóstico de códigos de problema escrita de una manera lógica a partir de las pruebas más fáciles y más probables hasta las más difíciles y menos probables.*

**Turbo lag**   The delay until the engine produces a significant amount of exhaust flow to speed up the turbine.
*Retraso del turbo*   *Retraso que ocurre hasta que el motor produce una cantidad significativa de flujo de escape para acelerar la turbina.*

**Vacuum**   Any pressure less than that of the atmosphere.
*Vacío*   *Cualquier presión menor que la de la atmósfera.*

**Valve timing**   By timing the cam to the crank, the valves can open and close at precisely the right time for optimum engine operation.
*Sincronización de la válvula*   *Al regular la leva con el acodamiento del cigüeñal, las válvulas se pueden abrir y cerrar precisamente en el momento adecuado para el funcionamiento óptimo del motor.*

**Valve train**   Consists of all the components used to open the valves.
*Tren de válvulas*   *Consiste en todos los componentes que se usan para abrir las válvulas.*

**Variable dwell**   A process whereby the amount of ignition coil primary current can be varied, depending upon the amount of energy required to fire the spark plugs.
*Reposo variable*   *Proceso mediante el cual puede variar la cantidad de corriente primaria de la bobina de encendido, dependiendo de la cantidad de energía requerida para disparar las bujías.*

**Variable fuel vehicles (VFV)**   See Flexible fuel vehicles (FFV).
*Vehículos de combustión flexible (VFV)*   *Véase Vehículos de combustible flexible (FFV).*

**Variable reluctance sensor**   A position or speed sensor that creates a signal when the magnetic flux varies around a pickup coil.
*Sensor de reluctancia variable*   *Sensor de posición o velocidad que crea una señal cuando el flujo magnético varía alrededor de la bobina captadora.*

**Variable valve timing (VVT)**   Variable valve timing is a method of changing the relationship between intake valve opening/closing events and exhaust valve opening/closing events while the engine is running.
*Sincronización de válvula variable (VVT)*   *Sincronización de la válvula variable es un método de cambiar la relación*

*entre los eventos de apertura/cierre de la válvula de admisión y los eventos de apertura/cierre de la válvula de escape mientras el motor está en funcionamiento.*

**Vehicle emissions control information (VECI)**   An underhood decal that provides emissions information specifically related to the vehicle it is attached to.
***Información de control de emisiones del vehículo (VECI)*** *Calcomanía ubicada bajo el cofre que proporciona información sobre las emisiones relacionadas específicamente con el vehículo al cual está adherida.*

**Vehicle identification number (VIN)**   17-Digit number assigned to all vehicles that contains specific information about that particular vehicle; found in lower left-hand corner of the windshield.
***Número de identificación del vehículo (VIN)*** *Número de 17 dígitos asignado a todos los vehículos que contiene información específica sobre ese vehículo en particular y que se encuentra en la esquina inferior izquierda del parabrisas.*

**Vehicle speed sensor (VSS)**   The VSS provides the PCM with information about the speed at which the vehicle is traveling.
***Sensor de velocidad del vehículo (VSS)*** *El VSS le proporciona información al PCM acerca de la velocidad a la que viaja el vehículo.*

**Voltage**   The difference in potential between the positive side of a circuit and the negative side.
***Voltaje*** *Diferencia en potencial entre el lado positivo de un circuito y el lado negativo.*

**Voltage drop**   Amount of potential that is lost when passing through a resistance.
***Caída de voltaje*** *Cantidad de potencial que se pierde al pasar por la resistencia.*

**Volume**   The amount of space that is occupied by an object. The amount of air and fuel an engine's cylinder will hold.
***Volumen*** *Cantidad de espacio que ocupa un objeto. Cantidad de aire y combustible que puede contener un cilindro.*

**Volumetric efficiency (VE)**   A comparison of how much air the cylinders can hold when completely filled compared with how much is being drawn in.
***Eficacia volumétrica (VE)*** *Comparación de cuánto aire pueden contener los cilindros cuando están llenos por completo comparado con cuánto aire se atrae.*

**Warmup catalyst**   A small catalytic converter used to help increase the temperature of the main catalyst more rapidly.
***Catalizador de calentamiento*** *Pequeño convertidor catalítico que se usa para ayudar a aumentar más rápidamente la temperatura del catalizador principal.*

**Warmup–three-way catalyst (WU-TWC)**   A small catalytic converter that is designed to rapidly increase the temperature of the main converter. The WU-TWC catalyst also reduces NOx, HC and CO during warm-up.
***Calentamiento–Catalizador de tres vías (WU-TWC)*** *Convertidor catalítico pequeño diseñado para aumentar rápidamente la temperatura del convertidor principal. El catalizador WU-TWC también reduce NOx, HC y CO durante el calentamiento.*

**Wastegate**   A butterfly valve that controls the amount of exhaust gas allowed into a turbocharger.
***Compuerta de residuos*** *Válvula de mariposa que controla la cantidad de gas del escape que se permite en un turbocargador.*

**Waste spark**   An ignition spark that occurs on the exhaust stroke and that serves no purpose.
***Chispa residual*** *Chispa de encendido que ocurre en la carrera de escape y que no sirve propósito alguno.*

**Work**   The result of applying force to move an object.
***Trabajo*** *Resultado de aplicar fuerza para mover un objeto.*

**Zener diode**   A semiconductor device that allows current to flow in only one direction; however, if voltage applied to the circuit reaches a predetermined level, voltage can be allowed to flow backward.
***Diodo Zener*** *Dispositivo semiconductor que permite que la corriente fluya directamente solo en una dirección; sin embargo, si el voltaje que se aplica al circuito alcanza un nivel predeterminado, esto podría permitir que el voltaje fluya en la dirección opuesta.*

# Index